Synthetic Methods in Organic Electronic and Photonic Materials

A Practical Guide

Synthetic Methods in Organic Electronic and Photonic Materials

A Practical Guide

Timothy C. Parker
Georgia Institute of Technology, Atlanta, GA, USA
Email: *timothy.parker@chemistry.gatech.edu*

and

Seth R. Marder
Georgia Institute of Technology, Atlanta, GA, USA
Email: *seth.marder@chemistry.gatech.edu*

THE QUEEN'S AWARDS
FOR ENTERPRISE:
INTERNATIONAL TRADE
2013

Print ISBN: 978-1-84973-986-3

A catalogue record for this book is available from the British Library

Published by The Royal Society of Chemistry,
Thomas Graham House, Science Park, Milton Road,
Cambridge CB4 0WF, UK

Registered Charity Number 207890

Visit our website at www.rsc.org/books

Acknowledgements

This book grew out of courses we have taught over the years and the first acknowledgement must be given to the students that gave us useful feedback and took our pedagogical missteps in stride. We must also thank our research colleagues over the years—coworkers in companies, graduate students, post doctoral researchers, research scientists—for both introducing us to new subjects and constructively refining our knowledge of chemistry and physics. Certainly, thanks must go out to all the hard working colleagues across the globe that have contributed interesting and detailed work in the synthesis of new materials for a range of organic electronic and photonics applications; without that body of work this book, and indeed the vibrant field that exists today, would not exist.

Finally, we must thank individual colleagues that helped specifically with preparation of the manuscript through various contributions. In particular, Dr Stephen Barlow's (Georgia Tech) and Professor Natalie Stingelin-Stutzmann's (Imperial College) contributions to Chapter 2 were incisive, insightful, and all-round invaluable. In addition, Mr Fadi Jradi and Professor Bilal Kaafarani (American University of Beirut) provided critical help in proofreading many of the chapters and suggesting valuable additions, identifying areas for clarification, and spotting errors that escaped our tired eyes.

Synthetic Methods in Organic Electronic and Photonic Materials: A Practical Guide
By Timothy C. Parker and Seth R. Marder
© Timothy C. Parker and Seth R. Marder, 2015
Published by the Royal Society of Chemistry, www.rsc.org

Abbreviations

Ac₂O	Acetic anhydride	**DME**	1,2-Dimethoxyethane
ACN	Acetonitrile	**DMF**	Dimethylformamide
b.p.	Boiling point	**dppe**	1,3-Bis(diphenyl-phosphino)ethane, also DIPHOS
BINAP	2,2′-Bis(diphenyl-phosphino)-1,1′-binaphthyl		
Bpin	Pinacol boronate	**dppf**	Bis(diphenylphos-phino)ferrocene
Dabco	1,4-Diazabicy-clo[2.2.2]octane	**dppp**	1,3-Bis(diphenyl-phosphino)propane
DACH	1,2-Diaminocyclo-hexane	**dtbpy**	4,4′-Di(t-butyl) bipyridine
DBU	1,8-Diazabi-cyclo[5.4.0] undec-7-ene	**EA**	Electron affinity
		EDG	Electron donating group
DCE	1,2-Dichloroethane	**EDOT**	3,4-Ethylenedioxyth-iophene
DCM	Dichloromethane		
DIBAL-H	Diisobutylalumin-ium hydride	**EWG**	Electron withdrawing group
DIPA	Di(isopropyl)amine	**GC/MS**	Gas chromatography/ mass spectrometry
DIPHOS	1,3-Bis(diphenyl-phosphino)ethane, also dppe	**HOAc**	Acetic acid
		HPLC	High perfor-mance liquid chromatography
DMAc	N,N-Dimethylacet-amide		

Synthetic Methods in Organic Electronic and Photonic Materials: A Practical Guide
By Timothy C. Parker and Seth R. Marder
© Timothy C. Parker and Seth R. Marder, 2015
Published by the Royal Society of Chemistry, www.rsc.org

Hünig's base	Di(isopropyl)ethyl amine
HWE	Horner–Wadsworth–Emmons
IE	Ionization energy
LDA	Lithium diisoproylamide
***n*-Bu or ^{n}Bu**	Normal butyl
NBS	*N*-Bromosuccinimide
NCS	*N*-Chlorosuccinimide
Neocup	Neocuproine, 2,9-dimethyl-1,10-phenanthroline
NHS	*N*-Hydroxysuccinimide
NIS	*N*-Iodosuccinimide
***o*-tol**	*o*-Toluene, *ortho*-toluene
P3HT	Poly(3-hexylthiophene)
PCC	Pyridinium chlorochromate
PEG	Poly(ethylene glycol)
phen	1,10-Pphenanthroline
RT	Room temperature
***t*-Amyl**	2-Mmethyl-2-Bbbutanol
***t*-Bu or ^{t}Bu**	Tertiary butyl
TBAF	Tetrabutylammonium fluoride
TBDMS	*t*-Butyldimethylsilyl
TEA	Triethylamine
TES	Triethylsilyl
thexyl	*t*-Hexyl, tertiary hexyl
THF	Tetrahydrofuran
TLC	Thin layer chromatography
TMEDA	*N*,*N*,*N*,*N*-Tetramethylethylenediamine
TMP	2,2,6,6-Tetramethylpiperidine
TMS	Trimethylsilyl
TPD	Tetra-*N*-phenylbenzidine
Xphos	2-Dicyclohexylphosphino-2′,4′,6′-triisopropylbiphenyl

Contents

Synthetic Methods in Organic Electronic and Photonic Materials: A Practical Guide
By Timothy C. Parker and Seth R. Marder
© Timothy C. Parker and Seth R. Marder, 2015
Published by the Royal Society of Chemistry, www.rsc.org

CHAPTER 1

Introduction

1.1 HISTORICAL PERSPECTIVE

Organic electronics and photonics as a research field encompasses small molecule and polymer theory, materials synthesis, processing, and device design and fabrication. How does one define "organic electronics and photonics"? In the broadest sense, it can be defined as the study of devices and processes that operate through phenomena arising from electrons and/or photons interacting with materials that contain small molecule organics or organic polymers. For the practical purposes of this book, we will define organic electronics and photonics as a collection of technologies that include: (1) organic light emitting diodes (OLED); (2) organic photovoltaics (OPV); (3) dye sensitized solar cells (DSSC); (4) organic field effect transistors (OFET); (5) 2nd order nonlinear optical (NLO) chromophores and electro-optic (EO) polymers; (6) photo-refractive polymers; (7) electrochromic molecules and polymers; (8) two-photon absorption (TPA) molecules and polymers; (9) optical limiting molecules and polymers; and (10) 3rd order NLO polymers. Although this may seem like a long list of technologies to cover in one book, from a *synthetic* standpoint, this can be done because, fundamentally, the vast majority of organic electronic and photonic materials are *π-conjugated* molecules and polymers that have properties modified and tuned through organic synthesis. Throughout this book, many of the synthetic methods discussed

Synthetic Methods in Organic Electronic and Photonic Materials: A Practical Guide
By Timothy C. Parker and Seth R. Marder
© Timothy C. Parker and Seth R. Marder, 2015
Published by the Royal Society of Chemistry, www.rsc.org

for conjugated small molecules and polymers will have multiple practical examples cited from a variety organic electronic and photonic literature.

The field of organic electronics and photonics has grown extensively during the past twenty years. The driving force for this growth has been mostly the increasing importance of microelectronics, computers and mobile phones, and high bandwidth telecommunications in our daily lives. Research on the semiconducting properties of organic molecules was conducted as early as the 1910s, and the potential to modify such properties through organic synthesis,[1] as well as applications in "micromodule electronic components (printed and evaporated circuits)",[2] have been recognized since at least the 1960s. Perhaps the best way to illustrate the growing research interest and intensity in the field is by examining the number of scholarly articles that are cataloged as either "organic electronics" or "polymer photonics." In Figure 1.1, the number of articles found on Web of Science covering these topics shows some activity in the 1980s followed by growth during the 1990s and annual growth on average of greater than 20% until 2014. The foundation for much of this growth was largely laid down in the mid-to-late 1980s and 1990s by seminal work in the field including: a small molecule, double layer OLED,[3] a conjugated polymer light emitting diode[4] (PLED), and a phosphorescence-based light emitting diode[5] (PhOLED); an OPV with a small molecule donor layer and a small molecule acceptor layer[6] and then a polymer donor–fullerene acceptor bulk heterojunction[7] (BHJ) OPV; an efficient DSSC fabricated with low-cost processes and medium purity materials;[8] a polymer OFET,[9] and an all-polymer OFET fabricated through printing techniques;[10] a poled EO polymer,[11] a high EO activity polymer composite,[12] and a low drive voltage (V_π) optical modulator;[13] the observation of the photorefractive effect in polymer films;[14] a strongly nonlinear optical limiting material;[15] chromophores with large TPA cross section;[16] and more recently molecules with large third order NLO susceptibility (χ^3) and low nonlinear losses for all optical switching.[17] Thus, the seeds were sown for the impressive increase in research in organic electronics and photonics during the early part of the 21st century, which itself lays the foundation for further hypotheses and questions. It is strongly recommended that students and researchers read the foundational publications cited above both to appreciate how far the field has advanced and to understand the scientific motivations for new lines of inquiry.

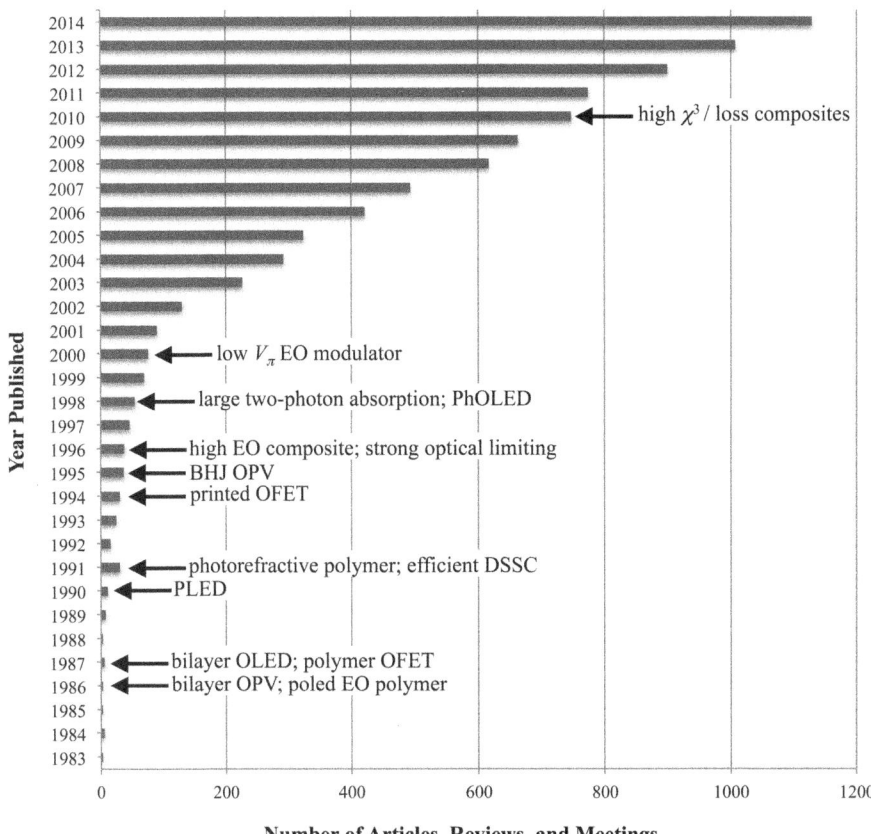

Figure 1.1   Growth of research in the field of organic electronics and photonics over the past 30 years. Seminal work is labeled briefly next to the bar for the year its journal publication appeared (references given in text).

1.2   THE IMPORTANCE OF ORGANIC SYNTHESIS

Organic electronic and photonic materials are typically highly polarizable π-conjugated molecules and polymers. As such, they often have design parameters that may be similar in certain aspects but different in others. For example, for NLO applications, in most cases intramolecular delocalization is highly desirable and in many cases intermolecular interactions are *undesirable*, as they tend to introduce excited state due to electronic coupling (interactions) between molecules, which can lead to unwanted linear or nonlinear absorptive properties, such as optical loss. In contrast, for organic semiconductors, in many cases intramolecular delocalization is essential and strong intermolecular interactions are *desirable* so that charges can move efficiently from molecule to molecule, or polymer chain to polymer chain. However, the factors that affect the π-conjugation within organic semiconductors and photonic materials are often fundamentally similar. For example, increased length of the polymer or molecular π-system typically leads to lower energy absorption of light; but the extent of the shift to lower energy depends not only on length but also on the degree to which different building blocks lead to delocalization—or disrupt delocalization—along the π chain axis. As such, one would the not expect the same absorption for similar-length materials that are differently comprised of triple bonds, double bonds, benzene, or thiophene rings. Furthermore, substitution of either the end of the π-system, or one of the backbone units of the π-system, with electron donors and/or acceptors can also further modulate the extent of delocalization along the chain in a dramatic fashion. Concomitant with expected changes in absorption properties induced by the aforementioned structural changes will come changes to the electron affinity and ionization potential, which are closely related to electrochemical reduction and oxidation potentials. Thus, by starting with simple π-conjugated groups one can build up complex and highly polarizable molecules with tunable properties related to electrical applications (such as ionization potential, electron affinity, charge mobility) and tunable optical properties such as the strength and position of the absorption band.

It is difficult to quantify exactly how much a role organic synthesis has played in both small molecule and polymer advances in the field. One way to at least illustrate this is by comparing "then and now" structures of materials in some of the seminal publications mentioned above to those in more recent noteworthy publications (Figure 1.2). The p-channel transport material dihexyl oligo-6-thiophene (**1**) used for printed OFET in 1994[10] has a thiophene as the only functional unit and straight hexyl chains as solubilizing groups; however, a more recent p-channel polymer[18] (**4**) has two more structurally complex functional units in the polymer chain (diketopyrrolopyrrole and thienothiophene) while also having two branched solubilizing groups (2-octyldodecyl) for each subunit. Another example is a recent small molecule n-channel transport material (**5**) used in ink-jet printed OFETs,[19] which also demonstrates increased structural complexity compared to **1** in that there are three functional units in the molecule (the peripheral naphthylene diimides, the core tetrazine, and the bridging thiophenes) instead of one. Another example is the 1995 BHJ OPV[7] polymer **2** having a single phenyl ring and alkene in the repeating unit compared to a more recent BHJ OPV example (**6**) that has a bifunctional benzodithiophene-fluorinated thienothiophene[20] subunit with multiple branched

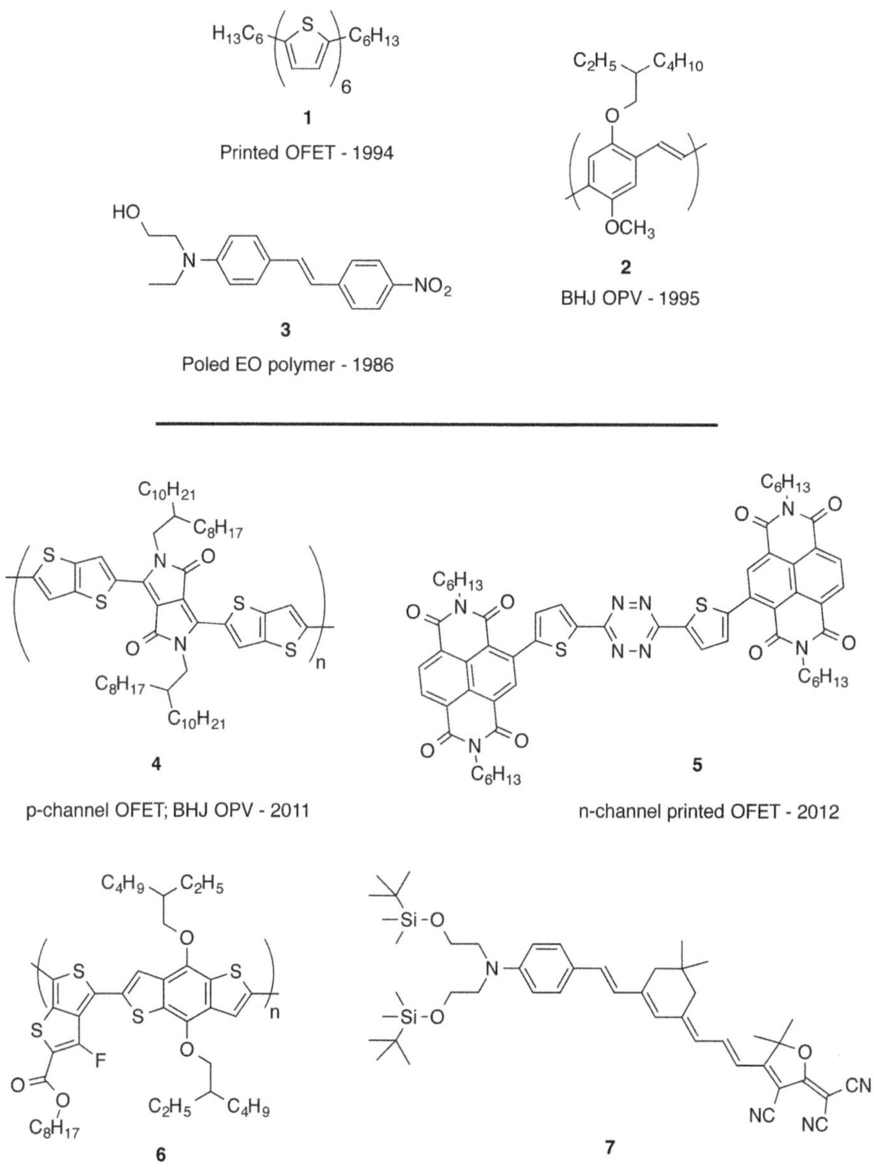

Figure 1.2 Chemical structures of materials appearing in early organic photonic and electronic publications (**1–3**) and structures appearing in more recent notable publications (**4–7**).

chain 2-ethylhexyl solubilizing groups and an additional *n*-octyl ester solubilizing group. A final example would be the 1986 poled EO polymer chromophore[11] (**3**) compared to a chromophore reported in 2000[13] (**7**): (1) the "donor" 2-hydroxyethyl aniline moiety in **3** is a protected bis(2-hydroxyethyl) aniline in **7** with *t*-butyldimethylsilyl (TBDMS) groups used as solubilizing substituents; (2) the "acceptor" 4-nitrophenyl moiety in **3** is a dicyanomethylene cyanofuran in **7**; and (3) the alkene "π-bridge" in **3** is a tetraene with a dimethylcyclohexene unit in **7**. Although these are just three examples, there are, of course, many other structural variations in the various aspects of organic electronics and photonics that have been explored. Some of these variations have been used to test hypotheses and discover important fundamental principles, and others have been used to map out structure–property relationships in a more Edisonian manner. Regardless of the research approach, organic synthesis has played a *central* role in moving these materials from promising demonstrations in 1980s and 1990s to the current state of some commercial applications (OLEDs, EO polymers) and continued intense development. Indeed, without organic synthesis, the structural variations required to validate hypotheses or discover new phenomenon would not be possible.

1.3   INTENT OF THIS BOOK AND RECOMMENDED READING

The materials chemist attempting to design materials for organic electronics and photonics needs to develop knowledge of how chemical and structural properties relate to the electrical and optical properties at both the molecular and materials level. There are numerous books and research papers that can serve as useful starting points to begin to understand the detailed interplay of structure and properties (many of which are still not well understood and so provide a still fertile ground for research), and many researchers involved in the development of organic materials often have the skills to explore structure/property relationships, or the ability to assemble complex molecules and materials; but only a small fraction of the community can do *both* effectively as the field is strongly interdisciplinary. As such, while we fully acknowledge that everyone cannot be expert in all things, years of experience suggest to us that even non-synthetic chemists involved in studying materials can benefit from having a rudimentary knowledge of how molecules and polymers are assembled, and certainly students and researchers entering the field can profit from having an understanding of the common reactions and strategies employed by organic synthetic chemists to synthesize π-conjugated molecules and polymers with tunable properties. Herein, we have chosen a variety of common building blocks whose structures are found in a diverse set of organic electronic and photonic materials and present to the reader most of the routes to synthesize these groups and then stitch them together into even more complex molecules and polymers. Since there are common reactions and strategies used to create functional parts of molecules and polymers, this book has been organized around synthesis of donor moieties (typically a collection of atoms with low ionization potential) and acceptor moieties (typically a collection of atoms with high electron affinities), and combination and extension of intermediates into larger π-systems with groups that may have double or triple bonds, aromatic rings, and/or hetero-aromatic rings. Throughout this book, some discussion of the reaction chemistry and principles is provided; but, mainly, many examples of synthetic methodologies are provided, all of which have well-described experimental write-ups in the primary literature, in order to provide the student or researcher with information about the predominant synthetic methodologies currently applied, in an efficient and organized manner.

One thing this book is not, and is not intended to be, is an introduction to synthetic laboratory techniques. In fact, we assume that the reader has had at least a successful course in college sophomore organic chemistry lecture and laboratory, and preferably more. Many of the reactions that are part of the scope of this book require advanced laboratory techniques and knowledge. There are a number of good books available that form a solid knowledge base of advanced synthetic laboratory techniques, some of which are listed below:

(a) B. S. Furniss, A. J. Hannaford, P. W. G. Smith and A. R. Tatchell, *Vogel's Textbook of Practical Organic Chemistry*, Long Group, UK, 5th edn, 1989.[21] Vogel's is a tour de force of practical organic chemistry techniques and reactions that have been developed in organic chemistry. It has lengthy sections on glassware, apparatuses, techniques, and perhaps most importantly, hazards that are presented in the synthesis of organic compounds and associated safety practices. The bulk of the book deals with synthetic preparations of the various organic functional groups including many well-described actual compound preparations. Although every synthesis lab would benefit greatly from having a copy of "Vogel", synthetic chemistry developed after about 1990, and also organometallic preparative methods, are not adequately represented;

(b) J. Leonard, B. Lygo and G. Procter, *Advanced Practical Organic Chemistry*, CRC Press, USA, 3rd edn, 2013.[22] This book has a number of strong points and forms a good complement to

Vogel's in that many modern synthetic techniques, apparatuses, and glassware are covered. One of the important aspects of this book is that is has a chapter devoted entirely to searching the chemical literature, which is particularly important since the number and quality of online searching tools has increased dramatically over the past twenty years. Any researcher should know this chapter well, since the most important thing a researcher can do is find out if someone else has done a task previously. Generally, there has been a saying in the field of chemical and materials science research: "A day in the library saves you two weeks in the lab," which actually may be an *under*estimate the time saved by properly searching the literature. Other highlights of this book include: a (effectively) first chapter that is dedicated to safety in the lab; separate chapters on running small scale and large scale reactions; and a table of common oxidizing and reducing agents;

(c) M. C. Pirrung, *The Synthetic Organic Chemist's Companion*, John Wiley and Sons, USA, 2007.[23] This book is also a very good compliment to the books listed above. It contains many explanations of modern synthetic methods as well as illustrations and pictures of specific products that are useful. It is helpfully organized around the synthetic process itself: planning the reaction, running the reaction, and following the reaction. In particular, following the reaction, *i.e.*, monitoring the starting material's disappearance and (hopefully!) the appearance of products, has a dedicated chapter to underscore the importance of these techniques in the synthetic process. Other highlights include "Safety Notes" throughout the book when glassware or techniques present certain hazards, a good section on keeping the laboratory notebook, and useful appendices that include a solvent selection grid.

Although the books listed above form a comprehensive knowledge base of practical organic chemistry, two other books we list below are very useful to have in the library of any organic synthetic group:

(d) W. L. F. Armarego and C. Chai, *Purification of Laboratory Chemicals*, Butterworth-Heinemann, UK, 7th edn, 2013.[24] This is the treatise on the purification of chemicals. It has good descriptions of purification techniques in general, a good description of purification of organic functional groups in general, and then a large collection of concise descriptions and some references used to purify specific compounds; and

(e) P. G. M. Wuts and T. W. Greene, *Greene's Protective Groups in Organic Synthesis*, John Wiley and Sons, USA, 4th edn, 2007.[25] Just about every protecting group for every functional group in organic chemistry is contained in these pages. The protection and deprotection methods are concisely described—many with illustrations—and all have references to the primary literature. There are reactivity charts for many of the functional groups that specify the stability to common reagents, pH, and reaction conditions to help guide the decision process on protective group selection. Hopefully you will never need a protective group in your intermediate of final target synthesis, but when you do, this book is invaluable.

Finally, in addition to the knowledge base contained in the books above, organic electronic and photonic materials often are polymers. Although examples of polymer syntheses under certain reaction categories are provided in this book, particularly in the organometallic coupling section, there are other methods of making π-conjugated polymers, such as electropolymerization, that are outside the scope of this book. Fortunately, the book *Conjugated Polymers: A Practical Guide to Synthesis*, ed. K. Müllen, J. R. Reynolds and T. Masuda, RSC Publishing, UK, 2014,[26] is a thorough guide to polymer synthesis, purification, and characterization. The book is organized around chapters dedicated to major π-conjugated polymer classes, often with several different chemistries used in polymerization reactions in the class as well as analyses of polymerization strategies. Each chapter is loaded with polymerization procedures for specific compounds including many intermediates. Given the specialization of polymer synthesis, and the intent of this current book to be a broad introduction to synthetic methods used across the spectrum of organic electronic and photonic materials, the authors believe that *Conjugated Polymers: A Practical Guide to Synthesis* and this current book are very complementary sources of modern organic π-conjugated materials synthesis.

1.4 SOME NOTES ABOUT SAFETY

The books recommended above on synthetic lab techniques all contain sections or chapters devoted to safety, which has been pointed out for each text. Although this is not a lab technique manual, we would like to particularly point out the importance of lab safety once again: *maintaining safe lab practices and demanding the same from labmates is the most important aspect of your chemical research.* It is the most important aspect to the researcher, his/her labmates, and the Institution at which the

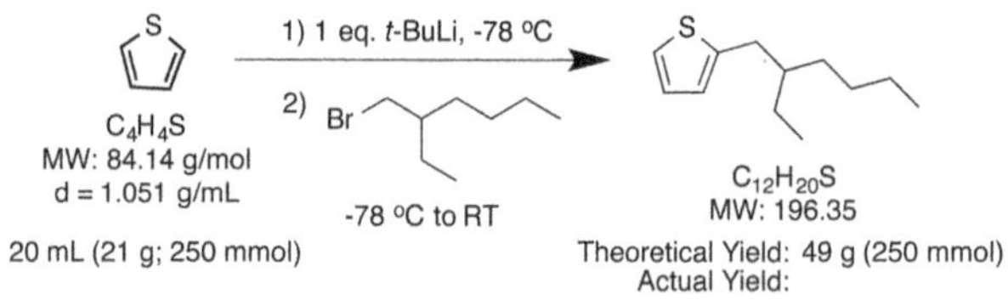

Chemical	MW or M	Amount	mmol	Equiv.
t-BuLi, 1.5 M in pentane, Alfa Aesar	1.5 M	170 mL	255	1.02
2-Ethylhexylbromide, d = 1.086 g/mL, Aldrich	193.12 g/mol	58.0 g	300	1.20
THF, DriSolv – Anhydrous, EMD-Millipore	-	1 L	-	-

Hazard Assessment			
Chemical	Hazards	Source	Actions
Thiophene	Highly flammable liquid and vapour; Harmful if swallowed; Causes skin irritation; Causes serious eye damage; Toxic if inhaled; May cause respiratory irritation	MSDS	Avoid vapor. Open and dispense ONLY in fume hood, keep all glassware in hood until dry.
t-BuLi, 1.5 M in pentane	Highly flammable liquid and vapour; Catches fire spontaneously if exposed to air; In contact with water releases flammable gases which may ignite spontaneously; May be fatal if swallowed and enters airways; Causes severe skin burns and eye damage.	MSDS	Pyrophoric. Transfer to an addition funnel under dry argon by cannula. Labmate standing by in case of fire.
2-Ethylhexylbromide	Combustible liquid; Causes skin irritation; Causes serious eye irritation; May cause respiratory irritation.	MSDS	Handle in fume hood as much as possible
THF	Highly flammable liquid and vapour; May be harmful if swallowed; Causes mild skin irritation; Causes serious eye irritation; May cause respiratory irritation; Suspected of causing cancer.	MSDS	Handle in fume hood as much as possible.
Product	Unknown, treat as toxic and flammable.		Avoid skin contact, breathing, ingestion. Keep away from flame
By-product: LiBr	Harmful if swallowed; Causes skin irritation; May cause an allergic skin reaction; Causes serious eye irritation.	MSDS	Dispose of in hazardous waste
By-product: isobutane	Extremely flammable gas, bp = -12 °C	MSDS	Reaction vented away from ignition sources into hood

Figure 1.3   An example of a hazard assessment demonstrating documentation of the hazards and, more importantly, actions considered to address those hazards for starting materials, products, and known by-products of an organic reaction.

research is being performed. The practical lab technique books recommended above cover the basics of personal protection, fire hazards, toxic hazards, and other hazards very well; however, one of the most important and fundamental parts of lab safety is situational awareness. Questions the researcher should ask herself are: what specific chemicals are being used by the researcher to start, what are the by-products and expected significant side-products (if any), and what are the hazards associated with each of them. One way to address this is to do a hazard assessment as part of planning for the reaction. Most of the hazards associated with a commercially available material should be available in multiple formats such as hard copies in shipments, in binders in a safety office, on the labels, or on Materials Safety Data Sheets (MSDS) that are required for shipping in the US and are available from chemical vendors online.

In Figure 1.3, we have included an example of how a hazard assessment might appear in a lab notebook *before* any process takes place. In this example, note that thiophene and especially *t*-butyl lithium present very significant hazards and should be treated with an abundance of caution. Also note that there is a hazard assessment for the solvent (THF) and the two by-products (lithium bromide and isobutane). As the product you are making might be an unknown material, the hazards may not be known, so you should treat them as toxic and respect them accordingly. Performing this type of hazard assessment is a strong first step in planning precautionary measures for both the handling of the materials *and* the response to any accident that may occur; and being properly cautious and prepared for specific accident threats greatly reduces the risks associated with organic synthesis.

REFERENCES

1. H. Inokuchi and H. Akamatu, *Solid State Phys.*, 1961, **12C**, 93.
2. A. Bradley and J. P. Hammes, *J. Electrochem. Soc.*, 1963, **110**, 543.
3. C. W. Tang and S. A. VanSlyke, *Appl. Phys. Lett.*, 1987, **51**, 913.
4. J. H. Burroughes, D. D. C. Bradley, A. R. Brown, R. N. Marks, K. Mackay, R. H. Friend, P. L. Burns and A. B. Holmes, *Nature*, 1990, **347**, 539.
5. S. R. Forrest, M. A. Baldo, D. F. O'Brien, Y. You, A. Shoustikov, S. Sibley and M. E. Thompson, *Nature*, 1998, **395**, 151.
6. C. W. Tang, *Appl. Phys. Lett.*, 1986, **48**, 183.
7. G. Yu, J. Gao, J. C. Hummelen, F. Wudl and A. J. Heeger, *Science*, 1995, **270**, 1789.
8. B. O'Regan and M. Grätzel, *Nature*, 1991, **353**, 737.
9. H. Koezuka, A. Tsumura and T. Ando, *Synth. Met.*, 1987, **18**, 699.

10. F. Garnier, R. Hajlaoui, A. Yassar and P. Srivastava, *Science*, 1994, **265**, 1684.
11. K. D. Singer, J. E. Sohn and S. J. Lalama, *Appl. Phys. Lett.*, 1986, **49**, 248.
12. M. Ahlheim, M. Barzoukas, P. V. Bedworth, M. Blanchard-Desce, A. Fort, Z.-Y. Hu, S. R. Marder, J. W. Perry, C. Runser, M. Staehelin and B. Zysset, *Science*, 1996, **271**, 335.
13. Y. Shi, C. Zhang, H. Zhang, J. H. Bechtel, L. R. Dalton, B. H. Robinson and W. H. Steier, *Science*, 2000, **288**, 119.
14. S. Ducharme, J. Scott, R. Twieg and W. Moerner, *Phys. Rev. Lett.*, 1991, **66**, 1846.
15. J. W. Perry, K. Mansour, I. Y. S. Lee, X. L. Wu, P. V. Bedworth, C. T. Chen, D. Ng, S. R. Marder, P. Miles, T. Wada, M. Tian and H. Sasabe, *Science*, 1996, **273**, 1533.
16. M. Albota, D. Beljonne, J. L. Bredas, J. E. Ehrlich, J. Y. Fu, A. A. Heikal, S. E. Hess, T. Kogej, M. D. Levin, S. R. Marder, D. McCord-Maughon, J. W. Perry, H. Rockel, M. Rumi, C. Subramaniam, W. W. Webb, X. L. Wu and C. Xu, *Science*, 1998, **281**, 1653.
17. J. M. Hales, J. Matichak, S. Barlow, S. Ohira, K. Yesudas, J. L. Bredas, J. W. Perry and S. R. Marder, *Science*, 2010, **327**, 1485.
18. H. Bronstein, Z. Chen, R. S. Ashraf, W. Zhang, J. Du, J. R. Durrant, P. S. Tuladhar, K. Song, S. E. Watkins, Y. Geerts, M. M. Wienk, R. A. Janssen, T. Anthopoulos, H. Sirringhaus, M. Heeney and I. McCulloch, *J. Am. Chem. Soc.*, 2011, **133**, 3272.
19. K. Hwang do, R. R. Dasari, M. Fenoll, V. Alain-Rizzo, A. Dindar, J. W. Shim, N. Deb, C. Fuentes-Hernandez, S. Barlow, D. G. Bucknall, P. Audebert, S. R. Marder and B. Kippelen, *Adv. Mater.*, 2012, **24**, 4445.
20. H.-Y. Chen, J. Hou, S. Zhang, Y. Liang, G. Yang, Y. Yang, L. Yu, Y. Wu and G. Li, *Nat. Photonics*, 2009, **3**, 649.
21. B. Furniss, A. Hannaford, P. Smith and A. Tatchell, *Vogel's Textbook of Practical Organic Chemistry*, Long Group, UK, 5th edn, 1989.
22. J. Leonard, B. Lygo and G. Procter, *Advanced Practical Organic Chemistry*, CRC Press, USA, 3rd edn, 2013.
23. M. Pirrung, *The Synthetic Organic Chemist's Companion*, John Wiley and Sons, USA, 2007.
24. W. Armarego and C. Chai, *Purification of Laboratory Chemicals*, Butterworth-Heinemann, UK, 7th edn, 2013.
25. P. Wuts and T. Greene, *Greene's Protective Groups in Organic Synthesis*, John Wiley and Sons, USA, 4th edn, 2007.
26. *Conjugated Polymers: A Practical Guide to Synthesis*, ed. K. Mullen, J. Reynolds and T. Masuda, The Royal Society of Chemistry, Cambridge, UK, 2014.

CHAPTER 2

Changing Material Properties

2.1 INTRODUCTION

The suitability of organic molecules and polymers for electronic and photonic applications depends primarily on their electronic structure, although, as discussed below, other characteristics such as process-ability and microstructure can also play a crucial role. The electronic structure determines the ease of removal of an electron from the material (oxidation), the ease with which the molecule can capture an electron (reduction), the wavelengths at which the molecule absorbs and emits light, and the strength of absorption and emission. In many cases, trends in the ease of oxidation and reduction and in the energy of the lowest lying electronic transitions can be rationalized by consideration of the frontier molecular orbitals (FMOs) in organic materials, *i.e.*, the highest occupied molecular orbital (HOMO) and the lowest unoccupied molecular orbital (LUMO). Thus, the design of organic electronic and photonic materials relies upon being able to rationally change the energy of the HOMO (E_{HOMO}) and/or LUMO (E_{LUMO}), and, in some cases, to control the spatial extent of these orbitals.

Other important material properties are associated with the *inter-molecular arrangement* of small molecules and/or polymers from the molecular to the micrometre scale. Electronic properties can be affected by the intermolecular arrangement and/or the molecular conformation for long-chain molecules such as π-conjugated polymers. For example, aggregates of molecules or aggregated polymers often

Synthetic Methods in Organic Electronic and Photonic Materials: A Practical Guide
By Timothy C. Parker and Seth R. Marder
© Timothy C. Parker and Seth R. Marder, 2015
Published by the Royal Society of Chemistry, www.rsc.org

show very different absorption spectra depending on the specific intermolecular arrangement within these aggregates. In other cases, the intermolecular arrangement determines overlap between FMOs, which is important for properties such as charge transport, charge generation, and photoluminescence.[1,2] The intermolecular arrangement may be repeated on longer length scales of tens of nanometres to microns to give ordered *molecular packing*. The molecular arrangement and packing in a material over nanometre to micrometre distances often leads to micron-sized domains that may detected by, for example, optical microscopy and light scattering. The shape, form, and size of these domains may be referred to as the *morphology* of the material, whereas more quantitative details of molecular conformations, intermolecular arrangement, and packing *within* domains (*e.g.*, degree of crystallinity, crystalline quality, dimensions and orientation of crystals/crystalline moieties, liquid crystallinity, *etc.*) may be referred to as *nanostructure* to *microstructure* depending on what scale is being addressed. Both morphology and, within domains, molecular arrangement and nano-to-microstructure, is important in determining charge transport in π-conjugated materials.[3] In photonic materials, phase domains can lead to undesirable light scattering centers and higher optical loss than in an amorphous polymer.[4,5]

Material qualities that are often *practically* important are solubility and processability. To a first order, solubility is important to aid purification and to characterize the molecular or polymeric structure by spectroscopy; however, solubility is also important for the formation of thin films by solution deposition techniques such as spin-deposition, drop-casting, and screen printing or other printing/coating techniques, such as gravure printing and wire or blade coating, to name a few. For these solution deposition techniques, in addition to solubility, obtaining thin films having desired thickness and uniformity typically requires control of viscosity, solvent evaporation rate, solidification mechanism (solute undergoing liquid/liquid *vs.* solid/liquid phase separation), and wettability. Moreover, the structure and morphology of small molecule and polymer thin films—and hence the overall performance of materials—are influenced, and in some cases may be controlled by, various deposition techniques.[6–8] In printed organic electronics, solubility is important to the formulation of ink compositions for device fabrication that, for example, do not block the nozzle when deposited by ink-jet printing, dry at the desired rate, and wet the substrate.[9] The combined characteristics of solubility and how readily films can be prepared from solution (or melt) is typically referred to as *processability*.

In addition to electronic properties, solubility, and processability, interfacial interactions between material layers in a device or between materials and electrodes also significantly impact device performance or may change the *in situ* material properties relative to those measured in individual thin films or solutions.[10–12] It is often the case that trade-offs are necessary, or that changing one material property will critically impact another property and reduce performance of a device, or in the case of reduced processability and stability, may even hinder the ability to characterize molecular structure or fabricate devices the first place. Thus, taken as a whole, optimizing the performance of materials and devices at a minimum involves tuning the characteristics of the FMOs, imparting processability *and* having desirable molecular order from the nanometre to the micrometre scale *and* favorably controlling the overall morphology. *Managing all these tasks simultaneously is a significant challenge, and it is also the heart of the science and the art of developing useful materials.* In this chapter, general strategies to tune the electronic properties of materials, factors that influence processability, and the critical challenge of morphology control are introduced.

2.2 KEY ELECTRONIC PROPERTIES

Key material properties that are measured and can be most straightforwardly modified by synthesis are summarized in Figure 2.1, which has been adapted from a recent publication.[13] The optical gap (E_g^{opt}) and the lowest energy absorption maximum (E_{max} or λ_{max} in wavelength) measure fundamental aspects of absorption of light. Ionization energy (IE) measures the energy associated with oxidation of a material (M $\rightarrow$ M$^+$) and electron affinity (EA) measures energy associated with reduction of a material (M $\rightarrow$ M$^-$). Note that E_g^{opt} and E_{max} measure the *energy difference* between the ground state (S_0) and the first excited state (S_1) and that IE and EA measure energy changes associated with *forming a cation or anion*, respectively, and that none of these properties actually measures E_{HOMO}, E_{LUMO}, or the difference between them. Although Koopmans' theorem equates the IE to $-E_{HOMO}$, this is only an approximation in polyatomic molecules since the remaining electrons may rearrange when ionization occurs and the relevant orbital in the neutral molecule (N) will have a different energy to the corresponding orbital in the ionized molecule ($N - 1$).[14] E_{HOMO} and E_{LUMO} are theoretical quantities and are not the properties *measured* in organic materials.[13] As such, throughout this book, discussion of E_{HOMO} and E_{LUMO} are typically used only the in the context

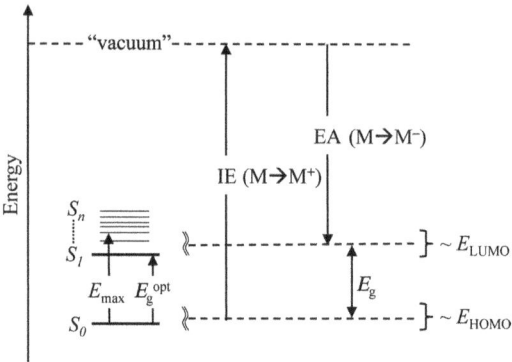

Figure 2.1 Summary of key electronic properties and of organic electronic and photonic materials. S_0: ground state; S_1: first electronic excited state; S_n: higher electronic excited states; E_{max}: lowest energy absorption maximum; E_g^{opt}: the optical gap; IE: ionization energy for removing an electron from a neutral material to the vacuum level (M → M⁺); EA: electron affinity for attaching an electron from the vacuum level to a neutral molecule (M → M⁻); E_{HOMO}: the energy of the highest occupied molecular orbital; E_{LUMO}: the energy of the lowest occupied molecular orbital. The brackets adjacent to E_{HOMO} and E_{LUMO} indicate that IE, EA, and E_{max} and E_g^{opt} are only respective approximations of E_{HOMO}, E_{LUMO}, and E_{HOMO}-E_{LUMO}.

of quantum chemical calculations or molecular orbital theory and not as direct measures of material properties.

The properties summarized in Figure 2.1 are briefly discussed below to provide additional definition and context as well as to point out certain aspects that might often be overlooked in the organic electron and photonics literature. Additionally, leading references are provided for phenomena associated with each property and main techniques typically used to measure them.

2.2.1 Optical Properties

There are many good books and reviews that cover basic to advanced principles of light absorption and photophysics of molecules. Two such books in particular are *Modern Molecular Photochemistry* by Turro, Scaiano, and Ramamurthy[15] and *Symmetry and Spectroscopy* by Harris and Bertolucci.[16] From the perspective of a book on materials synthesis, only an understanding of what quantities E_g^{opt}, E_{max}, and ε_{max} measure will be discussed here. The "optical gap" (E_g^{opt}) for an isolated molecule is a measure of the energy of the *adiabatic* excited-state, *i.e.*, the *relaxed* excited state in question above that of the ground, which for most organic molecules is a singlet (S_1). Indeed, it is often referred to as $E_{0,0}$ (where the subscripts indicate the vibrational level of interest in the ground- and excited electronic state respectively) to indicate that it is the estimated energy for the lowest vibrational states of the two electronic states in question. It is often experimentally estimated by the "onset" of the lowest energy absorption, or by the intersection point of the normalized absorption tail and fluorescence spectra. The absorption onset in the solid state may be seen at significantly different energies than in solution due to aggregation effects. Accordingly care must be taken to confirm the exact conditions of the measurement and which measure is most relevant to the application at hand. Assuming that the excited states are well separated and so there is no significant overlap between the lowest energy transitions, the energy, E_{max}, of the lowest-energy absorption maximum (corresponding to longest wavelength maximum, λ_{max}) represents (assuming only one state is involved) the *vertical* transition between the ground state (S_0) and the lowest energy optically allowed excited state *with the ground state geometry* and thus, for molecules where ground and excited state have significantly different geometries, the maximum corresponds to formation the electronic excited state in a vibrationally excited state.

Finally, ε_{max} is the molar absorptivity at λ_{max} and is a measure of the strength of the optical absorption at that particular wavelength. Related quantities that quantify the strength of absorption by a particular transition are the *oscillator strength* (f), which is proportional to the area of under a plot of absorptivity *vs.* energy for a particular transition, and the transition dipole moment, which is in many cases well approximated as being proportional to the square root of f divided by that of E_{max}.

In some, but by no means all, classes of π-conjugated organic molecules, the transition from S_0 to the lowest singlet excited state, S_1, is responsible for the absorption onset and lowest energy absorption maximum, and is fairly well described as being derived from promotion of an electron from the HOMO to the LUMO. In such cases, *trends* in E_g^{opt} or E_{max} seen for a series of similar compounds often correlate well with trends in $E_{LUMO}-E_{HOMO}$ as obtained from quantum-chemical calculations and shifts in the electronic absorption bands can be engineered by consideration of substituent effects on these orbital energies. However, it is important to note that a simple one electron HOMO → LUMO promotion is an approximation that ignores important electron–electron repulsion interactions; and there are many cases where this approximation breaks down, in particular where significant configuration interactions result in multiple different orbitals being involved in the S_0–S_1 transition or in describing the S_1 state, or where the S_0–S_1 transition is extremely weak or quantum mechanically forbidden for symmetry reasons (*e.g.*, as in extended polyenes such as β-carotene).[17]

2.2.2 Ionization Energy

The ionization energy (E_i or IE, also historically known as ionization potential) of a material is defined as the "minimum energy required to eject an electron out of a neutral atom or molecule in its ground state" according to the International Union of Pure and Applied Chemists (IUPAC).[18] An equivalent definition is the energy required to remove an electron from the ground state of a neutral species to yield the ground state of the corresponding cation and an electron at rest beyond the Coulombic influence of the cation (often referred to as "at the vacuum level"). For π-conjugated materials, IE can be measured by a variety of methods,[19] but ultraviolet photoelectron spectroscopy[20] (UPS) is currently most common and has been extensively reviewed.[21,22] Two specific IEs to note are: (1) the *vertical* IE (E_{vi} or IE$_v$), which refers to an ionization that occurs from the ground state M to M$^+$ without a change in geometry; and (2) the *adiabatic* IE (E_{ai} or IE$_{ad}$), which refers to an ionization to the lowest energy vibrational state of M$^+$. IE$_{vi}$ and IE$_{ad}$ can be estimated from the maximum and the low-energy onset, respectively, of the lowest-IE feature in the UPS spectrum. UPS can be used to measure the IE of both gas-phase molecules and of solids. For molecular materials where both quantities can be measured, the solid-state IE is considerably lower than that in the gas-phase, often by up to ~1 eV, due to the stabilization of the ion state by polarization of surrounding molecules in the solid.

2.2.3 Electron Affinity

The electron affinity (E_{ea} or EA) of a material is defined by IUPAC as the "energy released when an additional electron is attached to a neutral atom or molecule".[18] According to this definition, EA is positive if the negative ion (M$^-$) is more stable than the neutral species (M) plus an isolated electron at the vacuum level;[23] solid-state EA values defined this way are indeed positive for most but not all π-conjugated molecules. It is worth noting, however, that an alternative definition of EA as the energy *absorbed* on attachment of an electron has also been widely used in the literature; according to this definition, values are *opposite in sign but equal in magnitude* to those according to the IUPAC definition. As with IE, the *vertical* EA (EA$_v$) is the energy released when the anion is formed at the ground-state geometry and the adiabatic EA (EA_{ad}) is the overall energy when the relaxed anion M^- is formed. Several different methods have been used to determine the overall EA as well as EA_v and EA_{ad}, each with their own strengths and weaknesses, and generally EA has been more challenging to measure than IE.[19,23] Inverse photoemission spectroscopy (IPES) has recently been used with success in determining the solid state EA of organic electronic materials; however, the sensitivity of the technique is much lower than for UPS.[24]

2.2.4 Electrochemistry

The ease with which molecules can be ionized or can capture an electron can also be assessed using electrochemistry, either using a solution of the analyte or a thin film of the analyte on the working electrode immersed in electrolyte solution, the latter approach being especially useful for polymers. Accordingly, and because of the specialized (and expensive) instrumentation required for UPS and IPES, IE and EA are often estimated from the electrochemical oxidation and reduction of the neutral species with varying degrees of accuracy.[19,25] It has been argued that this can be best accomplished by obtaining a value of the electrochemical oxidation and/or reduction potential for a particular compound relative to a reference electrode (*e.g.*, standard hydrogen electrode (SHE), saturated calomel electrode (SCE), or silver chloride/silver electrode) and then adding the estimated difference between the reference electrode and the vacuum level (*i.e.*, the reference for IE and EA measurements) to adjust the values to the estimated IE and EA. However, even assuming that the absolute value of the reference electrode *vs.* vacuum is reliably known, this approach will in fact give estimates of the *solution-phase* IE and EA, which will in general differ from those in the solid state due to differences in solvation and solid-state polarization energies. Nonetheless, *trends* in electrochemical potentials for a set of related molecules will generally follow those in IE and EA and in frontier orbital energies and, thus, correlations between IE and oxidation potentials and between EA and reduction potentials are used to estimate IE/EA values from electrochemical data; although the reader is advised that in many cases this is without a clear statement of the assumptions made regarding solvation effects, and consequently caution should be exercised in reporting and assessing these correlations.

Another complication is that the thermodynamic standard redox potential, E^0, well approximated by the halfwave potential $E_{1/2}$, is not readily determined for all molecules, such as in cases where the redox process is not reversible, or especially, for polymers. Also, in some cases either the reduction or the oxidation does not fall within the usable potential window for the solvent used in the experiment, which results in some authors reporting: (1) when the reduction falls outside the useable potential window, the EA as the IE estimated from electrochemistry plus E_g^{opt} or (2) when the oxidation falls outside the useable potential window, the IE as the EA estimated by electrochemistry minus the E_g^{opt} (see Figure 2.1 for the basis of this estimation). Thus, given the large number of assumptions and variations that are encountered in estimating IE from EA from electrochemical data, *due caution* should be exercised in comparing redox-estimated IE and EA values between materials *and in particular between studies*.

2.3 DONORS AND ACCEPTORS

As pointed out in Section 2.1 above, although IE and EA values are not directly equivalent to E_{HOMO} and E_{LUMO}, and E_{max} and E_g^{opt} are not necessarily equivalent to the $E_{HOMO}-E_{LUMO}$, assuming changes in these electronic properties to be roughly parallel to changes in E_{HOMO} and E_{LUMO} can be advantageous for considering molecular factors that influence electronic properties. Using this consideration, relatively straightforward perturbation theory on frontier molecular orbitals (FMOs) can be used to predict how changes in structure will influence electronic properties. Thus, modifying the energy of the FMOs of π-conjugated materials may be accomplished in molecules and polymers by, alone or in combination: (1) adding/removing a pendant group that can donate electrons into the π-system (an electron "donor", "D"); (2) adding/removing a pendant group that can withdraw/accept electrons from the π-system (an electron "acceptor", "A"); (3) substituting an atom with a more/less electronegative atom while maintaining π-conjugation; and (4) adding/removing groups on the molecule or polymer that modify torsional strain and decrease/increase overall π-conjugation. The first two strategies explicitly involve using "donors" and "acceptors" to modify the properties of π-conjugated materials and the third does so implicitly since increasing or decreasing electronegativity in a π-system potentially provides a new π-system that is a donor or acceptor itself relative to the unperturbed system. Thus, *donors and acceptors are two central concepts in organic electronic and photonic materials*; however, what is meant by a "donor" and an "acceptor" should be taken in strict context of the property and the application under discussion. Fundamentally, something is a donor or an acceptor only *relative* to something else, as is discussed below.

A donor or an acceptor is classified by its effect on the energy of the FMOs and the electron distribution in the materials *relative* to another part of the molecule and/or relative to another molecular moiety to which it is being compared. These effects can often be analyzed using perturbational molecular orbital (PMO) theory.[26,27] Herein, PMO theory is used only in a *qualitative* manner, which assumes that (1) only relatively small changes are being compared; (2) that bonding between molecular moieties being compared are nearly the same, and that there is no large torsional changes that disrupt π-conjugation; and (3) only orbitals on each moiety that are nearest in energy mix to form molecular orbitals of the final system, and mixing of other orbitals does not significantly impact FMO energies. Although these are admittedly significant restrictions, especially compared to the quantitative information obtainable by quantum mechanical calculations, such restrictions do allow qualitative discussion on a simple level that informs definitions of donors and acceptors and their expected fundamental effects on critical material properties.

In the simplest case, a perturbation could be applied to a system treating the groups or molecular fragments as *monocentric*,[28] *i.e.*, the two moieties in each case are treated as "atomic fragments" and are only considered to donate one electron and one orbital to the new molecule. Such an analysis is shown in Figure 2.2 for three molecules: X–X; X–Y; and X–Z, where X, Y, and Z individually have different energies. The dimer X–X is formed from two energetically equivalent fragments (F2.2a) to form one molecular orbital (MO) lower in energy (F2.2b) and another MO higher in energy (F2.2c). If both X fragments contribute one electron, the lower energy MO would be the HOMO and the higher energy MO would be the LUMO, which has a node (F2.2d) between

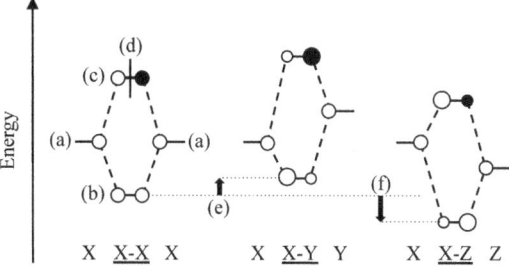

Figure 2.2 Monocentric perturbational analysis of molecular fragments X, Y, and Z where each fragment is treated effectively as an atom, contributing one orbital each to the new molecule.

atom X centers. Note that the orbital coefficients, denoted by circle size, are the same since the individual X fragments are equivalent in energy. In the case of X–Y, Y has a higher energy than X, so the HOMO (and LUMO) of X–Y would be higher in energy (F2.2e) than those of X–X and the MO coefficients would be changed such that there is a higher coefficient on X in the HOMO of X–Y, which would mean X is an electron *acceptor* compared to Y within X–Y (or conversely, Y is a donor compared to X within X–Y). However, in the case of X–Z, Z is lower in energy than X, so the HOMO (and LUMO) would be lower in energy (F2.2f) than those of X–Y and X–X, and the MO coefficients would be higher on Z in the HOMO of X–Z, *i.e.*, X is an electron *donor* compared to Z within X–Z. Thus, within the series presented in Figure 2.2, and with respect to changing the energy of the FMOs, it can be said that the donor *strength* is Y > X > Z and conversely that the acceptor *strength* is Z > X > Y. Also note that, assuming parallels in electronic properties with E_{HOMO} and E_{LUMO}, the trend in IE would be X–Z > X–X > X–Y, *i.e.*, X–Y has the lowest energy required to remove the electron in the HOMO to the vacuum level, is the easiest to oxidize, and has the smallest IE. The EA trend would be X–Z > X–X > X–Y. Generally, a stronger donor makes the IE (and often the EA) of the system decrease and a stronger acceptor makes the EA of the system (and often the IE) increase.

The situation is more complex when one considers bonding not as monocentric but as an *intermolecular union*[28] between two molecules, each with HOMOs and LUMOs of their own that are used to construct the HOMO and LUMO of the new molecule. In this case, as shown in Figure 2.3, the HOMOs from the two molecules X (F2.3a) form the HOMO of X–X (F2.3b), and the LUMOs from the two molecules of X (F2.3c) form the LUMO of X–X (F2.3d). The HOMO–LUMO gap

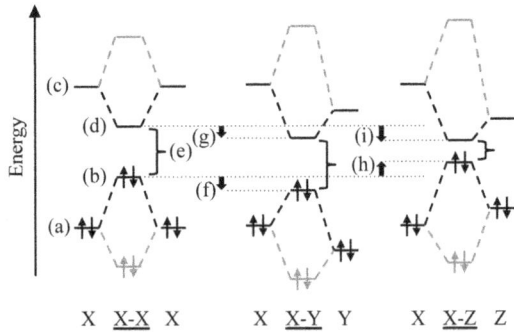

Figure 2.3 Intermolecular union perturbational analysis of molecular fragments X, Y, and Z, where each fragment is treated effectively as a molecule, contributing two molecular orbitals each to the new molecule.

of each new molecule is highlighted as a bracket in Figure 2.3 (*e.g.*, F2.3e). With a molecule such as Y that has lower E_{HOMO} and E_{LUMO} than X, a corresponding decrease in E_{HOMO} (F2.3f) and E_{LUMO} (F2.3g) can be expected, but in the case shown for X–Y, there is not much change in the HOMO–LUMO gap. However, when a molecule such as Z is considered, where E_{HOMO} of Z is higher than E_{HOMO} of X, and the E_{LUMO} of Z is lower than E_{LUMO} of X, the new molecule X–Z has higher E_{HOMO} (F2.3h) and lower E_{LUMO} (F2.3i). As a result, the HOMO–LUMO gap would be significantly narrower in X–Z than in X–X and X–Y. Assuming E_{HOMO}, E_{LUMO}, and the HOMO–LUMO gap as reasonable estimates for IE, EA, and E_g^{opt}, respectively, then this hypothetical case demonstrates the critical importance of context in discussing donors and acceptors. One could say that overall Y is the best acceptor since it both increases IE and increases EA; but Z is the best acceptor for increasing EA. At the same time, Z is actually a *donor* relative to X and Y in the context of IE since the IE of X–Y is lowest. At the same time, X and Y have nearly the same effect on E_g^{opt} while Z significantly narrows E_g^{opt}, and in many photonic applications, based on the smaller E_g^{opt} alone, Z would be considered as a relatively strong acceptor. Thus, in both reading and contributing to the literature of organic photonics and electronics, exercising due caution and seeking context in discussing "donors" and "acceptors" is the best approach.

Finally, it is often the case that donating or accepting of electrons by a group will be governed by the specifics of bonding rather than simple electronegativity. An example of this is nitrogen, which is more electronegative than carbon; but, as shown in Figure 2.4, it can be a

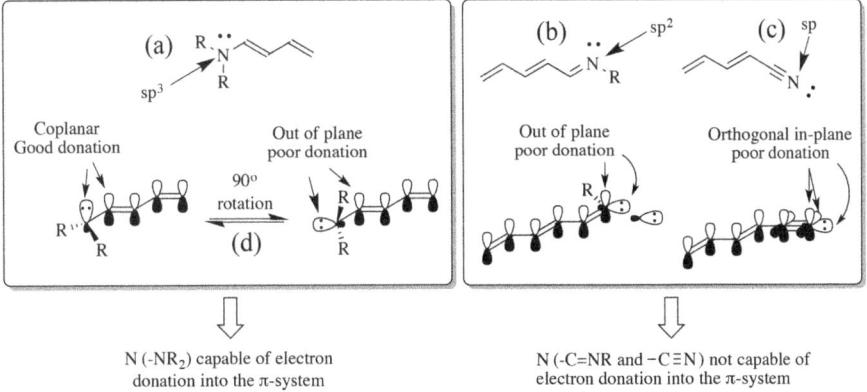

Figure 2.4 Bonding of a nitrogen atom where the lone pair donates effectively into a π-system (a), and where the lone pair of electrons donates poorly into the π-system (b) and (c).

π-donor when sp³-hybridized and pendant to a π-system such as in an amine (F2.4a, where the lone pair of electrons can overlap with and donate into a π-system) or a π-acceptor when sp²-hybridized or sp-hybridized such as in an imine or a nitrile, respectively (F2.4b and F2.4c, where the lone pair cannot overlap with/donate into a π-system efficiently due to geometrical bonding restraints). Any molecular modification that would increase the torsional angle of the amine–alkene (F2.4d) bond would tend to decrease electron donation until a maximum disruption is reached at 90°. Although such torsional strain is typically not used to modify frontier orbital energies, it may often help explain unanticipated results if geometrical restraints are not or cannot be fully considered beforehand.

2.3.1   Pendant Donor and Acceptor Groups

Adding pendant electron donor or acceptor groups to part of the π-conjugated system can often modify the energy of the FMOs in π-conjugated materials. The strength of many pendant donor and acceptor groups have been measured and reported as substituent parameters.[29] The most widely reported of these are the Hammett substituent parameters (σ_o, σ_m, and σ_p) based on the ionization of benzoic acids, which measure a combination of σ-releasing/withdrawing ability and π-releasing/withdrawing ability. However, one limitation of Hammett parameters is that substituents only *indirectly* affect the acidic oxygen through resonance. To address this limitation, other substituent parameters based on *direct* resonance delocalization of either a positive charge (σ_p^+) or negative charge (σ_p^-) have been developed. All these substituent parameters may be divided into "inductive parameters", or "field parameters" (F), that measure σ-releasing/withdrawing ability and "resonance parameters" (R) that measure π-releasing/withdrawing ability. Although σ-releasing/withdrawing groups may change the FMO energies in a predictable manner (especially for highly electronegative atoms such as F), they tend to have less impact on the FMOs of π-conjugated materials than π-releasing/withdrawing groups that formally may donate or accept electrons into π-orbitals since π-orbitals are typically the HOMO and LUMO of π-conjugated materials (although, lone pairs in σ orbitals can be the HOMO in some heterocycles). However, one important caveat is that for the resonance parameter to have a significant effect, the atom to which the pendant group is attached must have sufficient π-orbital coefficient. In cases where there is little to no π-orbital coefficient on the atom to which the pendant group is attached, then inductive effects may dominate.

Thus, assuming sufficient π-orbital coefficient on the attachment atom, perhaps the most relevant substituent parameters for π-conjugated molecules are the resonance parameters, which measure the strength of π-donors and π-acceptors *relative to hydrogen*. Resonance parameters have been determined for a number of functional groups and atoms, and the parameter R^+ measures the strength of π-donors (stabilizing a positive center by resonance through a π-system) and R^- measures the strength of π-acceptors (stabilizing a negative center by resonance through a π-system).

Selected R^+ values for pendant π-donor groups are presented in Table 2.1, along with F values to compare σ-releasing/withdrawing ability. Increasingly negative values of R^+ indicate stronger π-donation, and increasingly positive F values indicate increasing σ-withdrawing ability. Although trends will always be subject to the specifics of bonding in particular systems, there are some general trends that are useful as working knowledge in both synthesis and eventual properties of materials: (1) nitrogen is a stronger donor than oxygen (compare Entry 2 *vs.* 5 and Entry 3 *vs.* 6); (2) substitution of a hydrogen on a heteroatom with an alkyl group increases donor strength (compare Entry 4 *vs.* 6 *vs.* 7); (3) a phenyl substituent on a heteroatom decreases donor strength compared to an alkyl group (compare Entry 2 *vs.* 3 and Entry 5 *vs.* 6); and (4) increased electronegativity within a group of the periodic table does not necessarily decrease donor strength (compare Entry 1 *vs.* 3, where the increased electronegativity of oxygen correlates with a larger F parameter). Also note that the relatively electronegative heteroatoms in the pendant groups in Table 2.1 will tend to be inductively σ-withdrawing (positive F values) or not withdrawing when attached to a π-system. Again, it is important to note that these trends are for pendant donors and that specifics of bonding may also play a dominant role, especially with respect to their *positions* in the π-system and not merely the donor or acceptor strength; in particular, donors and acceptors attached to atoms in a π-system having a node

Table 2.1 Resonance (R^+) and Field (F) parameters of some pendant electron π-donors.

Entry	1	2	3	4	5	6	7	8
D	$\pi-S-CH_3$	$\pi-O-Ph$	$\pi-O-CH_3$	$\pi-N\begin{smallmatrix}H\\H\end{smallmatrix}$	$\pi-N\begin{smallmatrix}H\\Ph\end{smallmatrix}$	$\pi-N\begin{smallmatrix}H\\CH_3\end{smallmatrix}$	$\pi-N\begin{smallmatrix}CH_3\\CH_3\end{smallmatrix}$	$\pi-N\begin{smallmatrix}CH_2CH_3\\CH_2CH_3\end{smallmatrix}$
R^+	−0.83	−0.87	−1.07	−1.38	−1.43	−1.78	−1.85	−2.08
F	0.23	0.37	0.29	0.08	0.03	−0.03	0.15	0.01

in a HOMO and/or LUMO will often have negligible impact on the respective E_{HOMO} or E_{LUMO} compared to the unperturbed system. An illustrative example of this is substitution effects at various positions on the π-conjugated bridge in cyanine dyes and their correlation with orbital coefficients in the HOMO and LUMO.[30]

Selected R^- and F values for σ- and π-acceptor pendant groups are presented in Table 2.2, with increasingly positive R^- values indicating a stronger π-acceptor. For reference, the R^- and F for the prototypical π-acceptor group $-NO_2$ are 0.62 and 0.65, respectively. Again, trends are subject to specifics of bonding, but in general: (1) groups that may have a resonance form/s where a negative charge is delocalized onto an electronegative atom are the strongest π-acceptors (compare R^- for Entry 1 *vs.* 2–8); (2) electron donating/releasing groups *in conjugation* with a π-acceptor moieties tend to decrease π-acceptor strength (compare R^- for Entry 2 *vs.* 6); (3) substitution of the π-acceptor with an σ-withdrawing group increases acceptor strength (compare Entry 4 *vs.* 7); (4) multiple substitutions of a simple acceptor often increases overall acceptor strength (compare Entry 3 *vs.* 5 *vs.* 8). Note that for acceptors, the inductive effect (F) is often as strong as the resonance effect (R^-), in contrast to most donor groups (Table 2.1). As will be presented below, especially with regards to organic photonics, some of the best acceptor groups are composed of multiple strong acceptors such as the cyano group ($-CN$).

2.3.2 Heterocyclic Donor and Acceptors

Another way to modify the FMO energy of π-conjugated materials is by modifying the aromatic moieties by substituting heteroatoms for carbon. There are two basic ways this can occur, as demonstrated in Figure 2.5. Compared to benzene, one is to substitute a carbon atom with a heteroatom such that the lone pair of electrons cannot participate in the aromatic π-system, such as in pyridine (F2.5a), and the

Table 2.2 Resonance (R^-) and Field (F) parameters of some pendant electron π-acceptors.

Entry	1	2	3	4	5	6	7	8
A	$\pi-CF_3$	$\pi-C(=O)OMe$	$\pi-CN$	$\pi-S(=O)_2CH_3$	$\pi-C(NC)=CH-CN$	$\pi-C(=O)H$	$\pi-S(=O)_2CF_3$	$\pi-C(NC)=CH-C(CN)_2$
R^-	0.27	0.41	0.49	0.60	0.63	0.70	0.89	1.05
F	0.38	0.34	0.51	0.53	0.57	0.33	0.74	0.65

other is to substitute a $-CH=CH-$ fragment such that a lone pair of electrons can donate into the π-system, such as in pyrrole (F2.5b). This is analogous to the bonding situation for nitrogen discussed above in Figure 2.4, and the consequences are the same: when the lone pair of a heteroatom is part of the aromatic π-system it tends to act as a donor and decrease IE (and maybe EA as well) whereas when the lone pair does not participate in the aromatic π-system it acts as an acceptor and tends to increase IE due to the effect of its electronegativity on the HOMO. Additionally, both substituting with a heteroatom that participates in the π-system and one that does not is also possible, such as in imidazole (F2.5c). The effect of heterocyclic substitutions on the gas-phase IEs of several heterocyclic compounds is shown in Table 2.3.[22] Atoms that are better donors according to comparable R^- values and have a lone pair that can participate in aromatic delocalization tend to decrease IE relative to benzene (compare Entry 1–3 to 4). Heteroatoms with lone pairs that do not participate in π-delocalization tend to increase IE (compare Entry 1 to 3; Entry 4 to 6; and Entry 2 to 5). Additional heteroatoms also tend to increase IE (compare Entry 5 to 8 and Entry 6 to 7). EAs are not given in Table 2.3 due to a lack of comparative data.

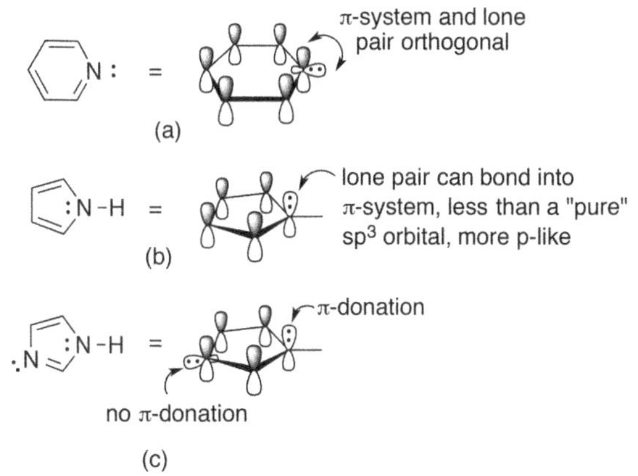

Figure 2.5   Bonding of heteroatoms and lone pair participation in heteroaromatics: (a) pyridine; (b) pyrrole; and (c) imidazole.

Table 2.3   Gas phase ionization energy (IE) of some heteroaromatics.

Entry	1	2	3	4	5	6	7	8
Structure	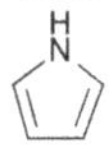		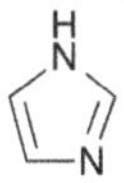		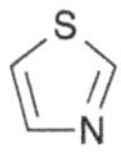	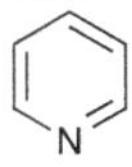	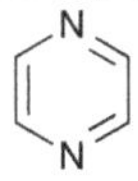	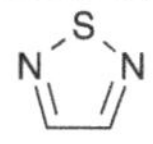
IE/eV	8.22	8.87	8.96	9.24	9.43	9.80	10.18	10.15

Another important capability of heterocycles is that aromaticity differences between different heterocyclic structures can be used to change material properties upon incorporation into π-conjugated materials. The aromaticity of heterocyclic compounds has been determined by a number of different methods, which include methods based on the determination/estimation of resonance stabilization energy (RSE) and methods based on structural comparison of bond order, electron releasing/withdrawing ability of heteroatoms in the ring, and/or internal bond angles in the ring. Some of these methods give contradictory rankings of heteroaromaticity; however, Katritzky and coworkers have performed a statistical analysis to determine the most relevant models,[31] and based on this analysis, the Bird indices of some common heterocycles are shown in Table 2.4 along with their RSE.[32] The Bird index is a structural analysis based on the uniformity of the bond orders of peripheral bonds in a π-system, where a π-system having perfect uniformity amongst the peripheral bonds, such as benzene, has a Bird index of 100.[33,34] What can be said is that, compared to benzene, substitution of an sp^2 carbon with a heteroatom that does not donate a lone pair of electrons into the π-system, such as in pyridine and pyrazine, does not decrease aromaticity as much (compare Entry 1 to 2 and 3 and Entry 4 to 5) as substitution of an ethylene fragment with heteroatoms that delocalize a lone pair of electrons into the π-system (compare Entry 1 to 4–7). Also, furan has significantly less aromatic character than thiophene or pyrrole (compare Entry 4 and 6–7).

A consequence of the differing aromaticity is that conjugation through the π-system can be changed itself and used to stabilize/destabilize certain resonance forms and hence material properties (Figure 2.6). An example of this is in the area of dipolar donor–acceptor compounds where the lower aromatic stabilization of thiophene compared to benzene (compare Table 2.4, Entry 1 *vs.* 4) makes intramolecular donor–acceptor charge transfer through a π-system where

Table 2.4 Bird indices and selected resonance stabilization energy (RSE) as a measure of aromaticity of some heteroaromatics.

Entry	1	2	3	4	5	6	7
Structure							
Bird	100.0	85.7	88.8	66.0	64.0	59.0	43.0
RSE/kJ mol^{-1}	150	117		122		90	68

benzene aromaticity is disrupted (F2.6a) less efficient than through a π-system when thiophene aromaticity is broken (F2.6b) and, thus, leads to a larger degree of ground-state charge separation in the thiophene structure. Likewise, the thiophene structure will have less ground-state charge separation than in a polyene (F2.6c) where there is no aromaticity change from one resonance form to another. In organic electronics, the interplay of aromaticity has been used to stabilize "quinoidal" structures along the polymer backbone to decrease the E_g^{opt} (F2.6d–f).[35] This has been used with an interplay between the aromaticity of thiophene and benzene to stabilize a quinoidal form (F2.6d′) since the stabilization of benzene is greater. A more subtle approach between a main-chain thiophene (F2.6e) and a

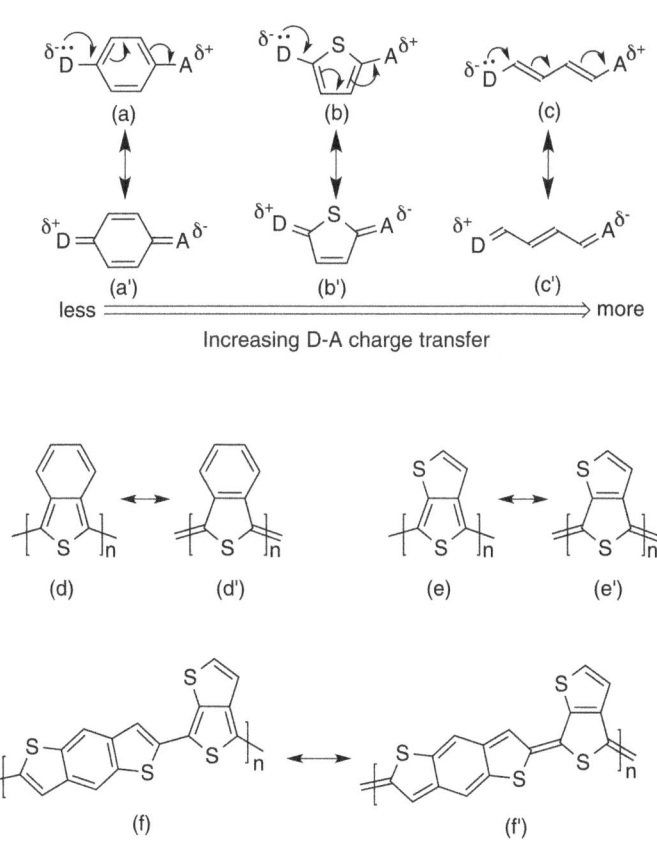

Figure 2.6 The effect of aromatic stabilization on donor and acceptor charge separation and quinoidal resonance forms in some π-conjugated structures.

fused pendant thiophene (F2.6e′) has also been used to affect the resonance structure balance. Another example is using both main-chain benzene aromatic stabilization (F2.6f) and pendant thiophene stabilization (F2.6f′) to provide an additional way to fine-tune the effect.

As will be seen throughout this book, heterocycles have been used widely in organic electronics and photonics and form the basis of most state-of-the-art organic electronic and optical materials. This is because heterocyclic chemistry offers a tremendous variety of structures and preparative chemistry,[32] which allows for the synthesis of a great diversity of donors, acceptors, and π-conjugated bridges. Additionally, heterocycles such as thiophene are often more thermally and chemically stable than equivalent polyene fragments. Figure 2.7 attempts to summarize the strategies molecule and polymer designers can use to change donor, acceptor, and/or π-conjugation strength in a material and hence the electronic and photonic properties. The full scope of synthetic methodologies used to manipulate the material properties is powerful: pendant donor and acceptor groups can be used on aromatic or heteroaromatic structures to achieve a variety of FMO energies across a broad range while at the same time being used to balance resonance forms through an interplay of aromatic and heteroaromatic stabilization. The downside is that changing one property through synthesis often changes one or more properties in an interlinked manner that is often still unpredictable. Thus, being

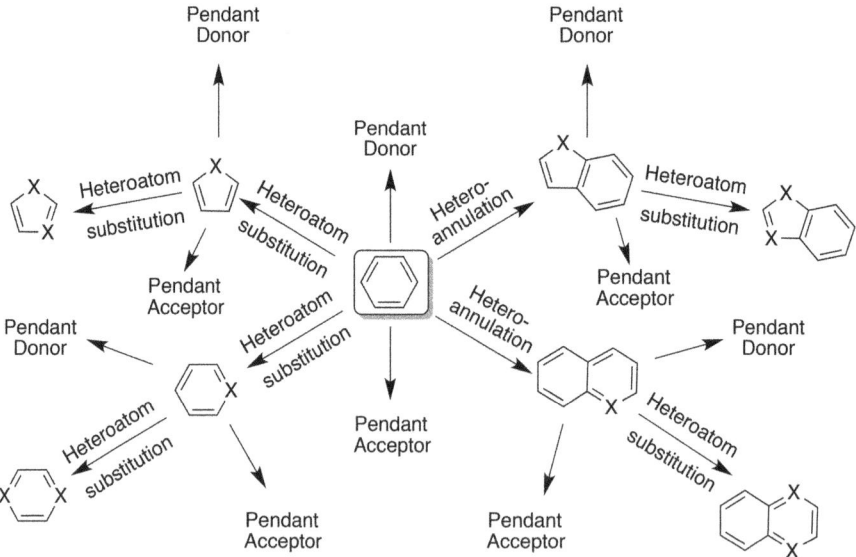

Figure 2.7   Heteroaromatic transformation and diversification of benzene.

able to discover the "perfect" material in a targeted or combinatorial manner is often fraught with unintended consequences; however, being able to probe and develop structure–property relationships by rational, hypothesis-driven research, which can then be extended into development of better materials, is perhaps the biggest advantage bestowed by the many different synthetic chemical methods currently available to the material scientist.

2.3.3 Donors and Acceptors in the Context of Organic Electronics

Two critical processes for all organic electronic devices are moving electrons into and out of devices. Injecting an electron into the device typically requires forming an anion from the neutral π-conjugated material. Conversely, moving an electron out of the device (*i.e.*, hole injection) requires forming a cation from the neutral π-conjugated material. For example, if the IE of a hole-transport material in an OLED is high, there can be large energetic barrier to hole injection from an electrode, requiring a large driving voltage to overcome. On the other hand, a large electron injection barrier will be encountered for a material with a low EA, unless very low work function materials (defined below) are used as electrodes. Section 2.2.2 introduced the energy required to form a cation from the neutral species as the IE, and Section 2.2.3 the energy released when an electron is attached to a material as the EA. The related quantity for an electrode is the work function (ϕ), which is the minimum work needed to extract electrons from the Fermi level. The property ϕ is a *thermodynamic* quantity and can be used along with EA and IE of the π-conjugated material to give an idea of the energy change upon electron or hole injection; however, it should be noted that interface effects between the organic and electrode as well as between organics can affect the overall energetics as well and is an active and important area of research.[10,36]

Device energetics is often illustrated in energy-device diagrams as shown in Figure 2.8. For the simplest case, the electron injection (F2.8a) is slightly energetically uphill ($\phi_c >$ EA, *i.e.*, the energy to extract an electron from the metal is greater than the energy released on attaching the electron to the material) so that injection will not occur spontaneously but can be driven by application of a field between the anode and the cathode. Once the electron is in the material, it can move through the device and to the anode (F2.8b). Since IE is so much larger than the ϕ of both electrodes, hole injection cannot occur, and the device is "electron only." A similar situation is operative for hole injection in a "hole only" device (F2.8c), where the barrier to hole

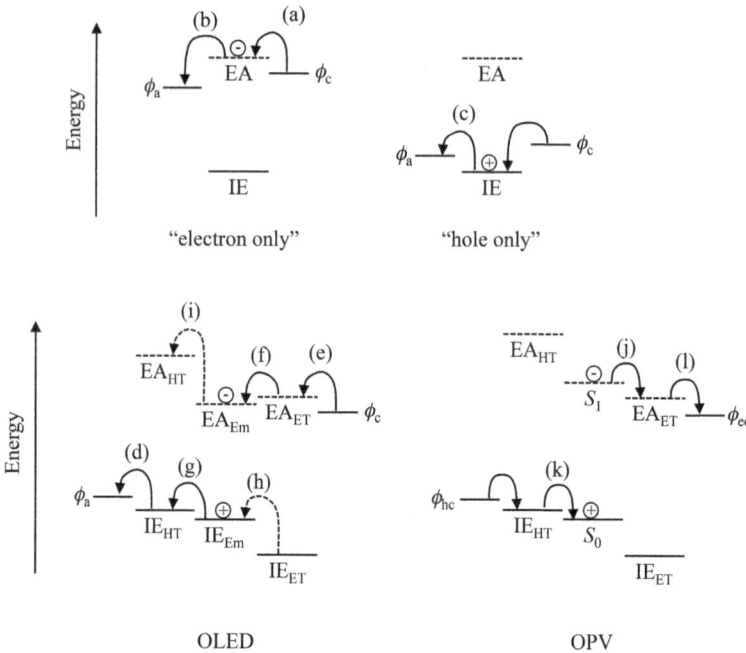

Figure 2.8 Relationship between ionization energy (IE) and electron affinity (EA) of organic materials and work function (ϕ) of electrode materials. ϕ_a: anode work function; ϕ_c: cathode work function; ϕ_{hc}: hole-collecting electrode work function; ϕ_{ec}: electron-collecting electrode work function; S_0: electronic ground state; S_1: first electronic excited state; subscript "HT" indicates a hole transport material; subscript "ET" indicates an electron transport material; subscript "Em" indicates an emissive material; arrows indicate the direction of electron transport with hole transport occurring in the opposite direction. Dashed arrows are processes that are too energetically unfavorable to occur appreciably under normal device operation.

injection can be seen as "pushing" the holes down from the electrode to the material. For an OLED, which tends to be multilayered, the situation is more complex since hole injection (F2.8d) occurs in a "hole transporting" material (HT) and electron injection (F2.8e) occurs into an "electron transporting" material (ET). The charges migrate through their respective materials to an emissive layer, where both get injected (F2.8f and F2.8g) and ideally are confined in the device by a "hole blocking" energy barrier (F2.8h) and an "electron blocking" energy barrier (F2.8i). At this point, within a simple one electron picture, the single electron in the LUMO and the single electron in the HOMO (the hole) are equivalent to an excited state that can emit as light. The situation is conceptually reversed in an OPV, but the reference properties may be different. Here, the "excited state oxidation

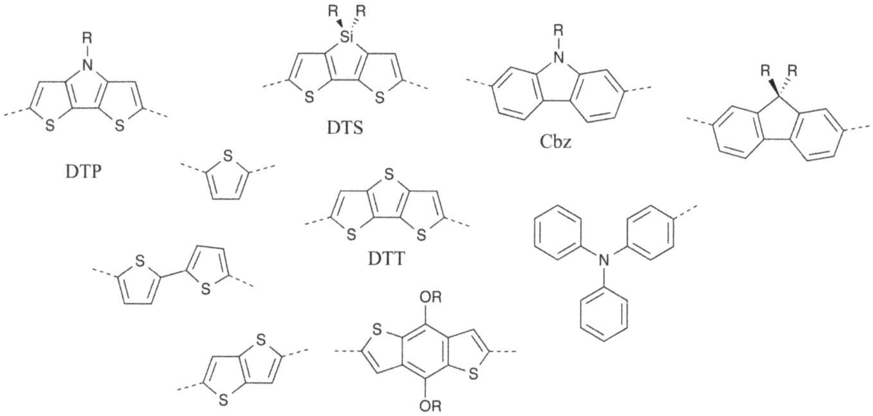

Figure 2.9 Some heterocyclic donors incorporated into organic electronic and photonic materials.

energy" (S_1^*) is an important concept where an excitation by the absorption or light creates an excited state S_1 that then injects (F2.8j) an electron into an ET material and leaves behind a hole (F2.8k). Electron transfer often takes place from a photoexcited donor to the acceptor and so the excited-state ionization energy of the donor (IE^*_{donor}) and the EA of the acceptor ($EA_{acceptor}$) (as determined in the material system) are key quantities. IE^*_{donor} may be estimated by $IE_{donor} - E_g^{opt}{}_{donor}$ and the driving force for charge-separation may be estimated as $IE^*_{donor} - EA_{acceptor}$, which will be exothermic if $EA_{acceptor} > IE^*_{donor}$, although other factors may also influence estimation of the driving force.[37,38] Once charge separation has occurred, the hole migrates to a hole-collecting electrode (ϕ_{hc}) while the electron migrates (F2.8l) to an electron-collecting electrode (ϕ_{ec}). The exact processes that govern the splitting of the electron/hole pair after excitation in OPVs are still under research.[39] Thus, donors and acceptors are critical for organic electronics in order to control the IE and EA of the materials to match the ϕ of conventional electrodes and to match the IEs and EAs of other materials in devices.

There are a number of good reviews that compile donor and acceptor moieties that have been used to construct π-conjugated materials in the context of organic electronics.[35,40,41] Some of the more widely employed donors are summarized in Figure 2.9. These include dithienopyrrole (DTP), dithienosilole (DTS), and dithienothiophene (DTT), carbazole (Cbz), and triarylamines, with fluorene-based donors typically having the weakest donating ability. Note that, there are number of sites on these donors for pendant solubilizing or functional groups (R).

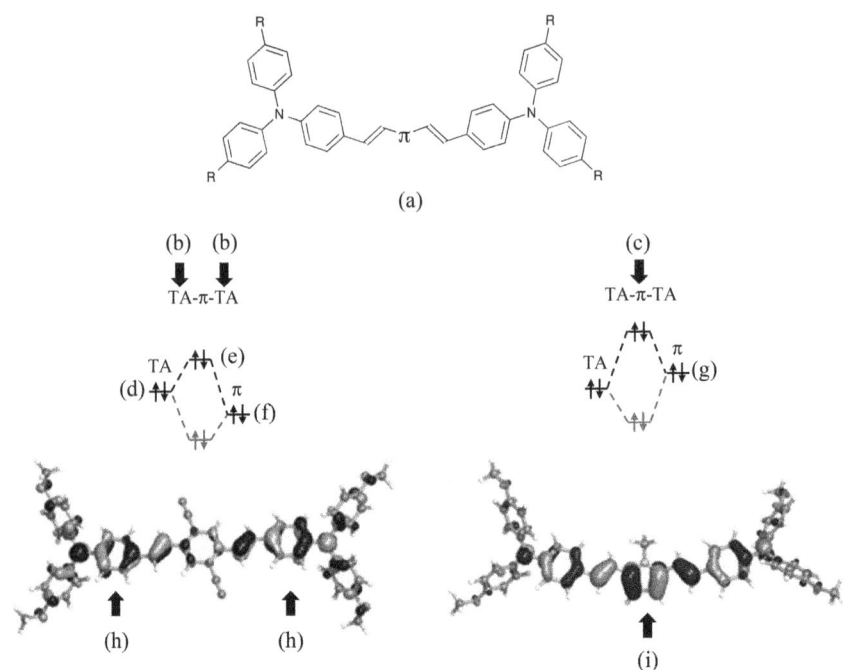

(a)

(b) (b)

TA-π-TA

(c)

TA-π-TA

TA (e)

(d)   π

(f)

TA (g)

π

(h)   (h)

(i)

Figure 2.10   Effect of π-bridge energy on the HOMO of triarylamine (TA) substituted π-bridges including perturbational analysis and B3LYP/3-21G*
calculated HOMO orbitals. When the π-bridge (π) is relatively low
in energy (f) compared to the TA end groups (d), HOMO coefficients
are greater on the TA moieties (b) and (h). When the π is relatively high
in energy (g) compared to the TA end groups, HOMO coefficients are
greater on the π moiety (c) and (i). Adapted with permission from
J. Am. Chem. Soc., 2012, **134**, 10146. Copyright 2012 American Chemical
Society.

These donor moieties have been incorporated into a variety of different polymers and small-molecule materials, with a number of different
properties and ranges for both IE and EA; however, since IE and EA are
determined and estimated in many different ways with many different
assumptions, straightforward quantitative comparison of the donor
strength of these donors as a series is problematic.

There are studies that are informative with respect to general strategies for manipulating the IE and the electron distribution in the
HOMO. One such study is summarized in Figure 2.10 and Table 2.5
on bis(triarylamine) compounds (F2.10a) with differing π-bridges and
IEs estimated from electrochemical measurements.[42] When π is benzene with electron *accepting* CN groups, the IE of the material is highest, and the IE is decreased when donating alkoxy groups are added
(compare Entry 1 *vs.* 2). When the CN groups are removed (Entry 3),

Table 2.5 Electrochemically estimated ionization energies (IE) of bis(triarylamine) π-bridge molecules with the general formula illustrated in Figure 2.10. IEs estimated from reported $E_{1/2}^{ox}$ (V) + 4.4 eV.

Entry	1	2	3	4	5	6
π						
R	H	$-OC_4H_9$	$-OMe$	$-OMe$	$-OC_4H_9$	$-t$-Bu
IE	4.9	4.7	4.6	4.5	4.2	4.1

the IE is decreased further. When the lower IE thiophene is substituted for the benzene, the IE decreases again (compare Entry 3 *vs.* 4). The lowest IE materials are with the lowest-IE bridged pyrrole where substitution with donating alkoxy groups decreases IE to the lowest value in the series (Entry 6). One interesting aspect of this study is that the location of the electron density in the HOMO changes from the triarylamine end groups (TA, F2.10b) to the low IE pyrrole bridge (F2.10c). This is in accordance with a perturbational analysis that puts the energy of the triarylamine (TA) groups (F2.10d) closer to the HOMO (F2.10e) when the π-bridge itself has a higher IE (F2.10f) than TA. When the IE of the π-bridge decreases past that of TA, such as with pyrrole (F2.10g), the HOMO is now closer in energy to the π-bridge, and consequently the orbital coefficients are larger on the π-bridge. The HOMOs calculated with B3LYP/6-31G* of models of the compounds in Entry 2 (R = OMe, F2.10h) and Entry 5 (R = OMe, F2.10i) are shown in Figure 2.10 to emphasize the difference in orbital coefficients. With this example, it can be seen that addition and subtraction of donor and acceptor groups, as well as substitution of aryl groups with different IEs, is an effective way to both change IE and the HOMO energy, and ultimately change the HOMO electron distribution itself.

A number of acceptors that have been used extensively in organic electronics are shown in Figure 2.11. These are arranged in *approximate* order of decreasing acceptor strength from left to right. In particular, 2,1,3-benzothiadiazole (BT) has been incorporated into a number of materials for many applications.[35,41] However, stronger acceptors have been developed in particular for OPV applications, and to a lesser extent for n-channel OFET applications, and include iso-indigo (I), pyrrolo[3,4-*c*]pyrrole-1,4-dione (DPP), thieno[3,4-*c*]pyrrole-4,6-dione (TPD), and thieno[3,4-*b*]thiophene esters (TTE). Comparing

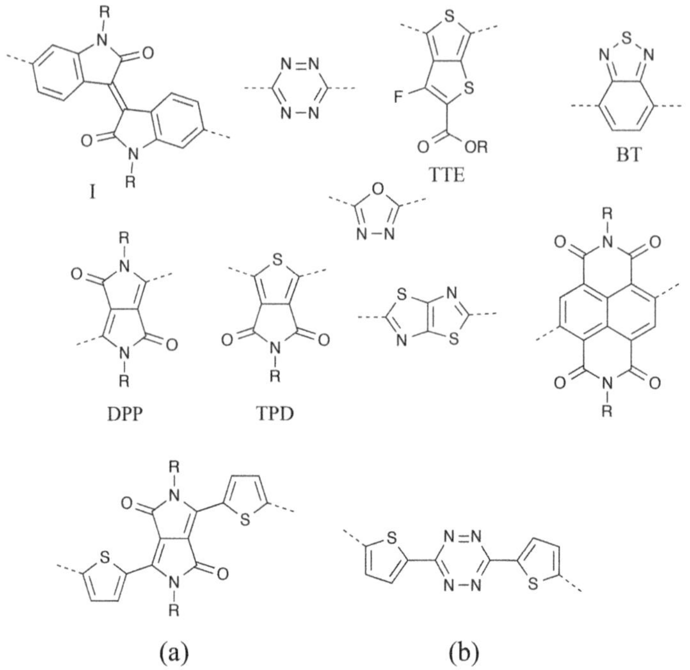

Figure 2.11 Some heterocyclic acceptors incorporated into organic electronic and photonic materials. (a) DPP acceptor with pendant thiophene groups; (b) tetrazine acceptor with pendant thiophene groups.

between acceptors is particularly difficult since some acceptors are incorporated into materials with integrated electron donor groups for synthetic efficacy, such as DPPs (F2.11a) and tetrazines (F2.11b), and therefore are difficult to quantitatively compare with acceptors such as isoindigo and TPD that can be incorporated "cleanly". As illustrated in later chapters, this is typically because the Stille, Suzuki, and halogenated reagents required for the cross couplings used to synthesize many acceptor-containing materials are often difficult to obtain or have low reactivity on strongly electron-accepting substrates.[43] These synthetic challenges are perhaps the main reason why electron transport (n-channel) materials have been less developed than p-channel materials.[43] Note that, as with the donors in Figure 2.9, there are number of sites on the acceptors for pendant solubilizing or functional groups (R).

As with the donors, there are studies that have demonstrated illustrated some general principles of acceptor design. One such study is on the Cbz-containing polymers **I** in Table 2.6, which incorporate several different acceptors groups in a directly comparable manner.[44] For example, the effect of substituting a heteroatom for a carbon can be

Table 2.6 Estimated ionization energy (IE) and electron affinity (EA) of polymers incorporating heterocyclic acceptors. For the details of IE and EA estimation: see ref. 44 for polymer **I** and ref. 45 for polymer **II**.

Entry	1	2	3	4	5	6	7
A							
EA	3.42	3.67	3.60	3.80	3.2	3.7	4.0
IE	5.46	5.52	5.45	5.53	4.9	4.9	4.7

seen in reduction of both the EA and IE of the materials (compare Entry 1 *vs.* 2 and Entry 3 *vs.* 4). Also, heteroannulating the benzene core with a heterocycle having higher IE, such as from pyrazine to thiadiazole (Table 2.3, Entry 7 *vs.* 8), gives an increase in EA but not necessarily in IE (compare Entry 1 *vs.* 3 and Entry 2 *vs.* 4). Another study on DTP polymers, **II** in Table 2.6, illustrates an unpredictable affect of heteroannulation.[45] Going from BT (Entry 5) to the pyrazine annulated acceptor (Entry 6) to the thiadiazole annulated acceptor (Entry 7) results in a marked increase in EA, but also a *decrease* in IE. This is a manifestation of the perturbational analysis example shown in Figure 2.3 for the "X–Z" compound.[46] In other words, the benzobisthiadiazole (BBT, Entry 7) is a strong acceptor relative to BT in the context of EA, but is actually a *weaker* acceptor (or a stronger donor) relative to BT in the context of IE. This is one of many examples of organic electronic and photonic materials where close attention must be paid to the *context and relative nature* of the properties and the applications of the materials being considered.

2.3.4   Donors, π-Bridges, and Acceptors in the Context of Organic Photonics

Donor, acceptors, and the π-conjugated bridges between them are also fundamentally important in organic photonics as the means by which critical properties are changed. In organic photonics, the major concern is material properties that interact with light or arise from the interaction with light, such as the refractive index (n), absorptivity, and fluorescence. In some cases, it is *changes* in these properties that drive device operation, such as with electric-field-driven changes in refractive index for χ^2 (second-order nonlinear, electro-optic) materials and devices[47,48] or changes in light intensity for optical limiting and χ^3 (third-order nonlinear)[49] materials and devices. In multi-photon absorption and fluorescent materials, *nonlinear* absorption of two or more photons enables excitation effectively only at the focal point of the light beam and creates excited states that can then be used in fluorescence imaging or microfabrication.[50]

The polarization of, and electronic transition between, ground and excited states of organic chromophores are often key chromophore characteristics that substantially determine the important properties of the materials. A very simple model for ground state polarization can be thought about in terms of the bond length alternation (BLA) in a polymethine π-bridge between a donor (D) and an acceptor (A). This concept is illustrated with prototypical D–π–A resonance structures in Figure 2.12, where BLA is defined as the average difference in lengths between the single bonds (σ) and the double bonds (π) in the in the polyene π-bridge. Normal σ-bond lengths are ~1.48 Å and normal π-bond lengths are ~1.32 Å. At the "polyene limit" (F2.12a), there is a certain amount of ground-state polarization (represented by the size of the hemisphere around D) and a maximum (negative) amount of BLA. When the electrons from D are formally transferred to A, the "charge-transfer limit" (F2.12b) is reached and the polarization for the chromophore is at or near a maximum. In between the polyene limit and the charge-transfer limit, charge transfer is balanced at the "cyanine limit" (F2.12c) and the electronic structure is well-described in terms of equal contributions from the neutral and charge-transferred resonance structures. The implication of the resonance structures F2.12a–c is that for a given π-bridge, changing the BLA can be achieved by changing the donor strength of D and/or the acceptor strength or A. For constant D and/or A, changing BLA may

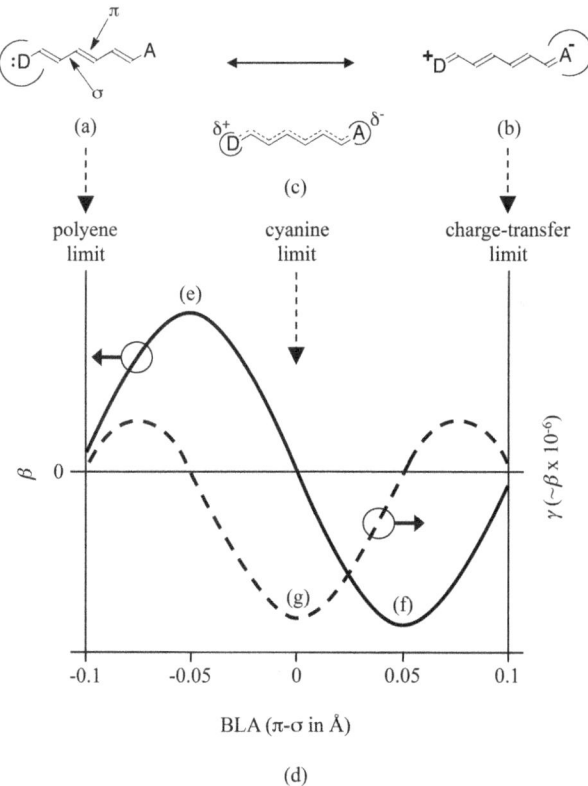

Figure 2.12 Bond length alternation (BLA) and qualitative relationship to first hyperpolarizability (β) and second hyperpolarizability (γ).

be accomplished by changing the π-bridge to be less/more aromatic in either neutral or charge-separated resonance forms (as illustrated in Figure 2.6). The importance of BLA (and hence the ground state and excited state/s polarization) has been demonstrated by correlating it to the molecular 2nd-order optical nonlinearity (β) and the molecular 3rd-order optical nonlinearity (γ).[51-53] Qualitative β and γ vs. BLA graphs are shown in Figure 2.12d. One important point is that β is maximum in a positive sense between the polyene limit and the cyanine limit (F2.12e) and maximum in a negative sense between the cyanine limit and the charge-transfer limit (F2.12f). Also, γ is maximum in a negative sense at the cyanine limit (F2.12g). Thus, given that molecular nonlinearities can be optimized by "moving along" the BLA curve, relative donor and acceptor strengths and degree of π-conjugation through the π-bridge are as critical in organic photonics as they are in organic electronics.

As with the donor and acceptor discussion for organic electronics, the discussion is limited to examples of donors, acceptors, and π-bridges that help illustrate some more general concepts, since there are a number of good reviews that explore donors, acceptors, and π-bridges in the context of photonics.[4,47] What is typically defined structurally as a "donor," "acceptor," and "π-bridge" in photonics is illustrated in Figure 2.13, and as is illustrated throughout the text, these definitions are based on the synthetic intermediates used to make the chromophores. Generally, a "donor" is a substituted aromatic or heteroaromatic low IE fragment from which alkene- and/or heterocycle-containing π-bridges are attached/extended to an acceptor group that is generally a high EA heterocycle. Certain D, π, and A groups area also shown in Figure 2.13.

Table 2.7 illustrates a number of chromophores with varying D, π, and A.[54-59] As discussed in Section 2.3 with perturbational analysis, decreasing the ionization energy of a donor and/or increasing the electron affinity of an acceptor generally narrows the HOMO–LUMO gap of the D–A system. Since the HOMO–LUMO gap is a rough estimate of optical absorption, one useful method for comparing the effects of varying D and A strength, as well as the efficiency of D–A coupling through π, is by examining the λ_{max} of chromophores. Using λ_{max} for such comparisons of charge-transfer in chromophores is instructive if: (1) other parts of the chromophores are constant; (2) the solvents used for the measurement are the same and concentrations are similar; and (3) the chromophores are on the polyene side of the cyanine limit such that their solvatochromatic properties are both positive. With these conditions, increasing donor strength, acceptor strength,

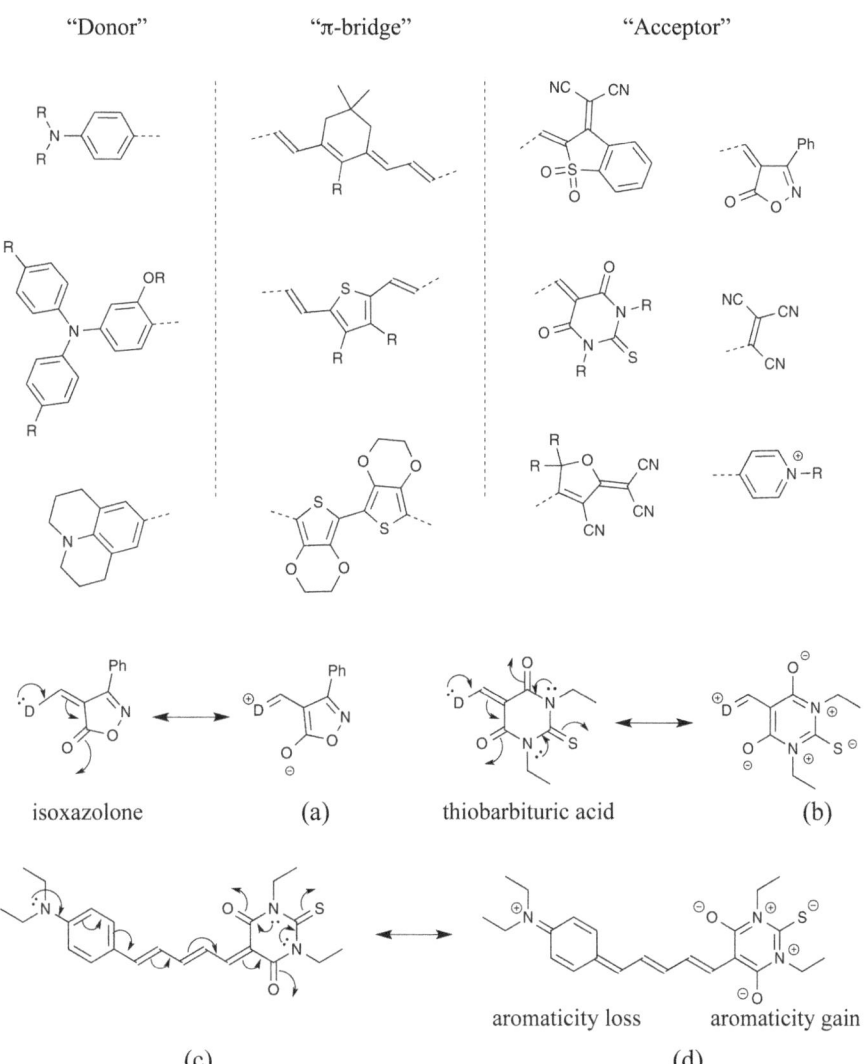

Figure 2.13 Some donor, π-bridges, and acceptors in the context of organic photonics. Aromatic stabilization in isoxazolone accetpor (a) and thiobarbituric acid acceptor (b) and the interplay of aromaticity loss and gain in
a "push–pull" chromophore (c) and (d).

or π conjugation red shifts λ_{max} and decreasing donor strength, accep-
tor strength, or π conjugation blue shifts λ_{max}. As mentioned above,
Table 2.7 shows only *relatively few* of the more widely used D, π, and
A groups.

Some general trends for donor, π-bridge, and acceptor strength
illustrated in Table 2.7 are: (1) heterocyclic acceptors typically are

Table 2.7 Comparison of intramolecular charge-transfer donor and acceptor strengths in chromophores with varying π-bridges. Numbers in brackets refer to references.

1

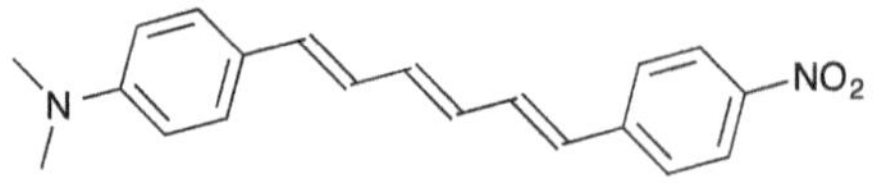

$\lambda_{max} = 458$ nm (54)

2

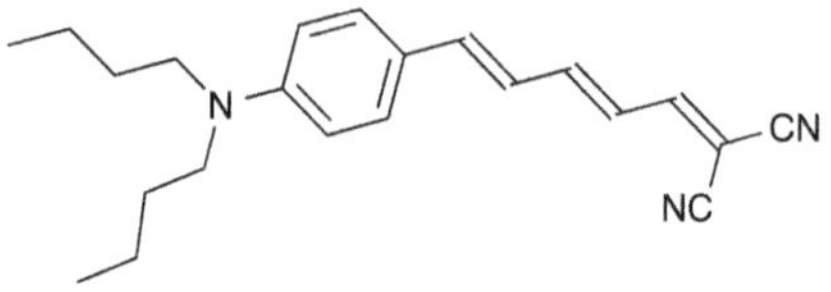

507 (55)

3

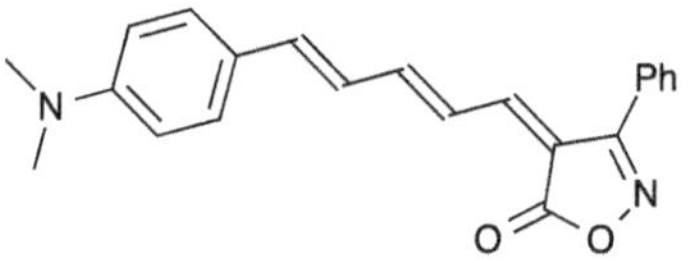

562 (54)

4

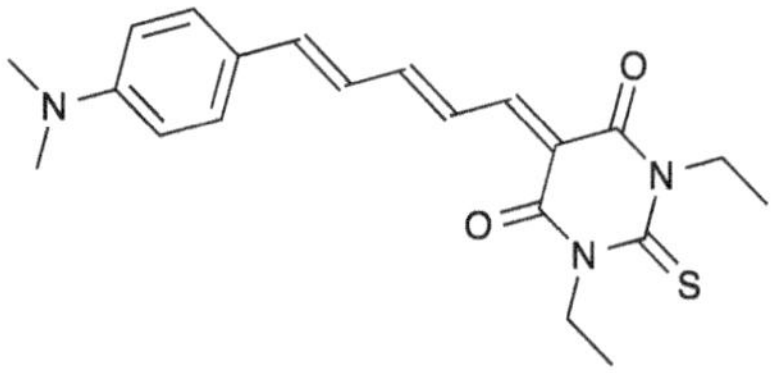

604 (56)

5

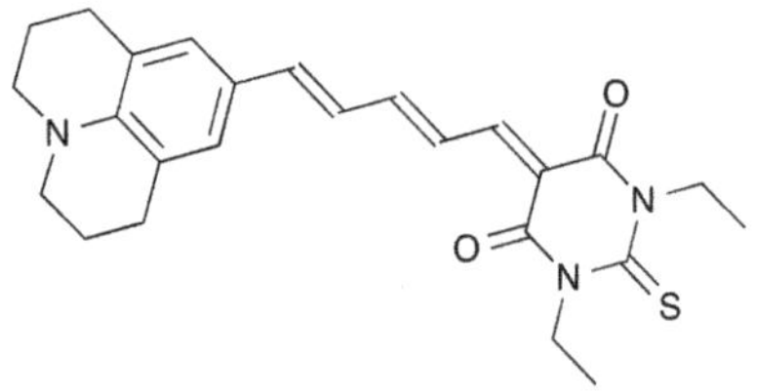

680 (54)

6

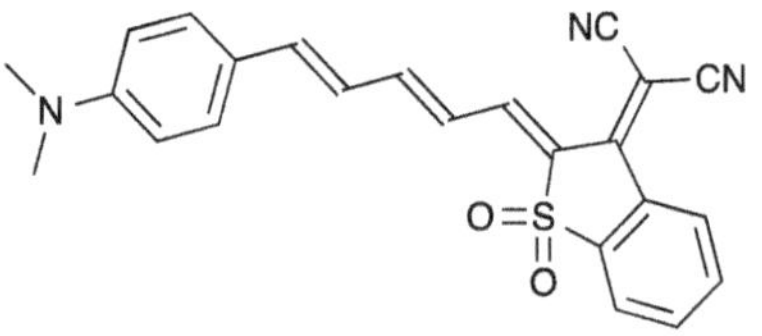

744 (56)

7

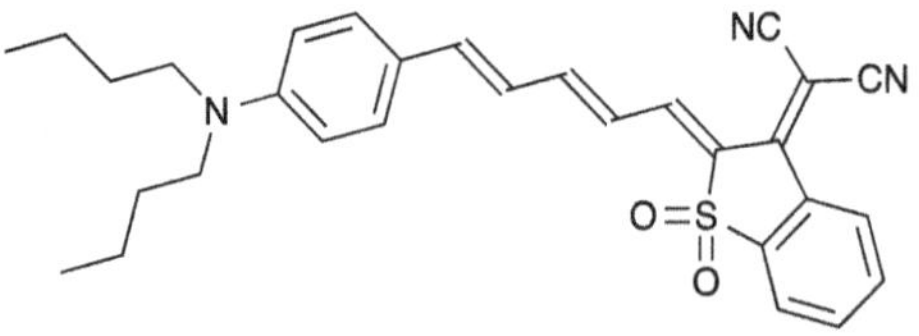

770 (56)

8   744 (56)

9   685 (57)

10   719 (58)

11   749 (58)

12   774 (57)

13   781 (57)

14   694 (59)

stronger than a combination of "simple" acceptors (compare Entry 2 *vs.* 3), especially when the heterocycle can have aromatic character in the charge separated form (Figure 2.13, isoxazolone and thiobarbituric acid); (2) increasing the number and/or strength of simple acceptor groups in the heterocycle can increase acceptor strength (compare Entry 3 *vs.* 4 *vs.* 6); (3) locking a dialkylamine group from rotation, such as in julolidine, can increase donor strength (Entry 4 *vs.* 5), although some of the increase is likely due to replacing a hydrogen on each *N*-methyl group and the phenyl ring with alkyl groups, which also increases donor strength consistent with the R^+ resonance parameter in Table 2.1 (compare Entry 6 *vs.* 7 in Table 2.1); (4) replacing an alkene with a thiophene decreases the efficiency of a π-bridge even though the number of double bonds increases by 1 (compare Entry 7 *vs.* 8), which is because formal electron donation through the bridge breaks the heteroaromaticity of the thiophene; (5) increasing the number of simple donors (*e.g.*, –OMe) on a donor group can increase donor strength (compare Entry 9 *vs.* 10 *vs.* 11); (6) adding a simple inductive withdrawing group can increase acceptor group strength (compare Entry 9 *vs.* 12); (7) replacing alkyl groups on an amine with phenyl groups decreases donor strength (compare Entry 13 *vs.* 14).

It should be noted that while Table 2.7 illustrates changes in donor, π-bridge, and acceptor strength, it is not necessarily the case the strongest acceptor or strongest donor is most appropriate since the performance of a material may not require the strongest donor and acceptor. As illustrated in Figure 2.8, acceptor or donor strength can be too strong for certain applications by, for example, lowering EA or raising IE outside of the range of other materials in the system. Likewise, as shown by the *β vs.* BLA curve in Figure 2.12, *β* peaks at a certain BLA and adding a stronger donor and/or acceptor will overpolarize the system and actually decrease *β*. Another reason that the strongest donors and acceptors might not be the best material choice is that stronger donors and acceptors are often too unstable in the device operating conditions. This is particularly true of thermal and photochemical stability of chromophores for electro-optic polymer applications, which must be poled at temperatures up to 250 °C for several minutes to induce electro-optic activity and then devices must operate for years at high optical power.[48] For organic electronic applications, materials that are vacuum deposited must be sufficiently stable to be sublimed at elevated temperatures and must have photochemical stability for up to 30 000 hours for OLEDs.[8]

Generally, alkenes increase the effective conjugation of a π-system compared to a direct aryl–aryl linkage while alkynes, although increasing the physical length of the molecule or polymer, tend to decrease the conjugation of a π-system compared to a direct aryl–aryl linkage. This is largely because: (1) the energetic difference between the aryl sp²-hybridized carbons and the alkyne sp-hybridized carbons results in less overlap between the aryl–alkyne π-systems compared to the overlap between aryl–alkene π-systems formed from two energetically more similar aryl sp² and alkene sp² carbons; and (2) the higher s-character in the sp-hybridized carbons shortens the alkyne bond (~1.18 Å) compared to the alkene bond (~1.32 Å), leading to localization of π-electrons in the alkyne and thus poor π-delocalization through an alkyne compared to an alkene.[60–62] An illustrative example is the comparison of the D–π–A chromophores in Figure 2.14, which differ only in that F2.14a has an aryl–aryl bond in the π-bridge whereas F2.14b has an alkene and F2.14c has an alkyne.[63] Whereas F2.14a has a λ_{max} = 533 nm the alkene analogue F2.14b has a λ_{max} = 546 nm (a 13 nm red shift) indicating a greater conjugation for alkene analogue F2.14b compared to parent F2.14a. Conversely, alkyne analogue has a λ_{max} = 515 (an 18 nm blue shift) indicating a *shorter* conjugation length for alkyne analogue F2.14c compared to parent F2.14a. Also note in Figure 2.14 that the IE for alkene analog F2.14b (5.21 eV) is smaller than parent F2.14a (5.33 eV) while the IE for alkyne analog F2.14c (5.47 eV) is larger than for parent F2.14a. Although several factors may be responsible for the IE trend, including the degree of conjugation between the donor triarylamine and the rest of the molecule, a likely large contributing factor is that alkenes are generally electron releasing groups (R^+ = −0.29) while alkynes are generally electron withdrawing groups (R^- = 0.31) at least in part due to the higher s-character in the sp-hybridized carbons of the alkyne compared to the sp²-hybridized carbons in alkenes.[29]

One factor that is important to consider mainly in direct aryl–aryl linkages, and to a lesser extent in aryl–alkene linkages, is steric bulkiness around one or both aryl groups (or aryl–alkene groups) disrupting π-conjugation. This is illustrated in Figure 2.15 for an example of a phenyl–thiophene linkage (F2.15a) where proximal hydrogens on each aryl group have little steric repulsion (F2.15b) that allows through conjugation as represented by the "quinoidal" resonance form (F2.15c); however, when there is large steric interactions between the proximal substituents (F2.15d) a torsion (F2.15e) about the single bond can result in the two aryl rings being out of plane (F2.15f), which may decrease or effectively disrupt π-conjugation depending on the degree of twisting. The sterics-induced twisting may be particularly an issue in π-conjugated polymers since alkyl groups are often added

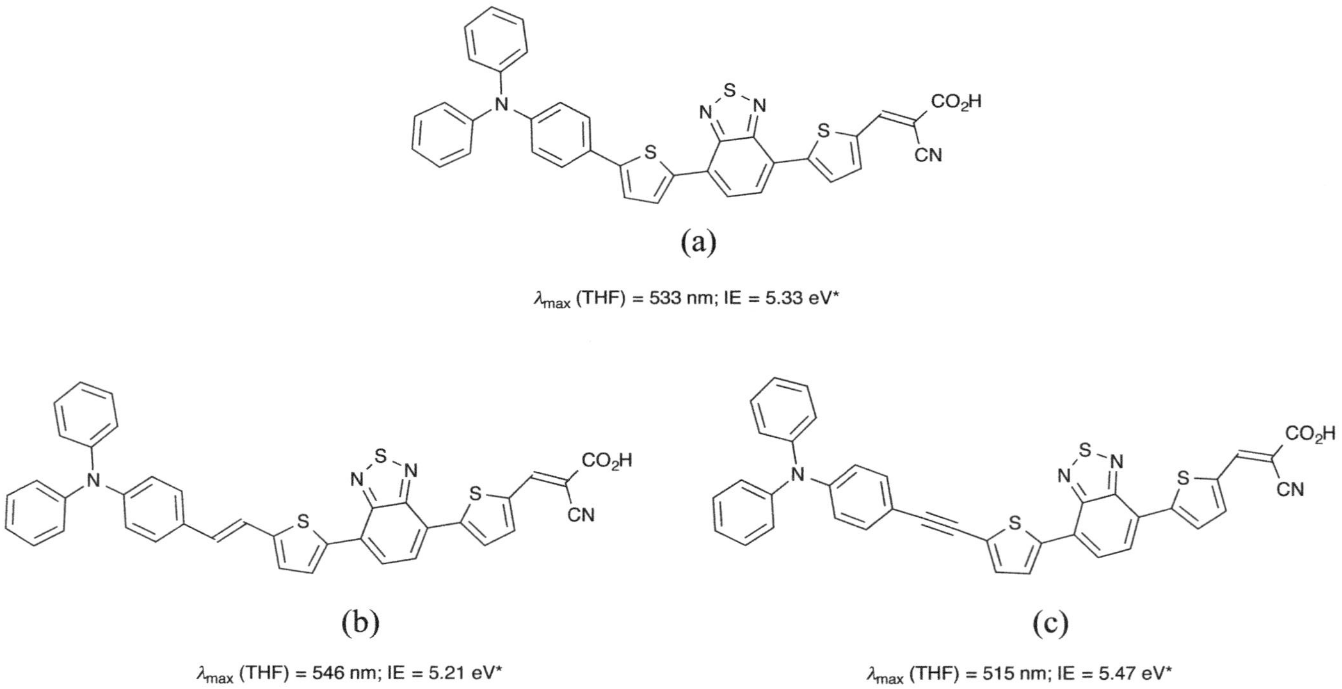

(a)

λ_{max} (THF) = 533 nm; IE = 5.33 eV*

(b)

λ_{max} (THF) = 546 nm; IE = 5.21 eV*

(c)

λ_{max} (THF) = 515 nm; IE = 5.47 eV*

*IE = reported eE_{ox} (vs. NHE) + 4.28 eV

Figure 2.14 Differences in conjugation and ionization energy between a triarylamine donor and cyanoacetic acid acceptor through a benzothiadiazole π-bridge linked by a single bond (a); an alkene (b), and an alkyne (c).

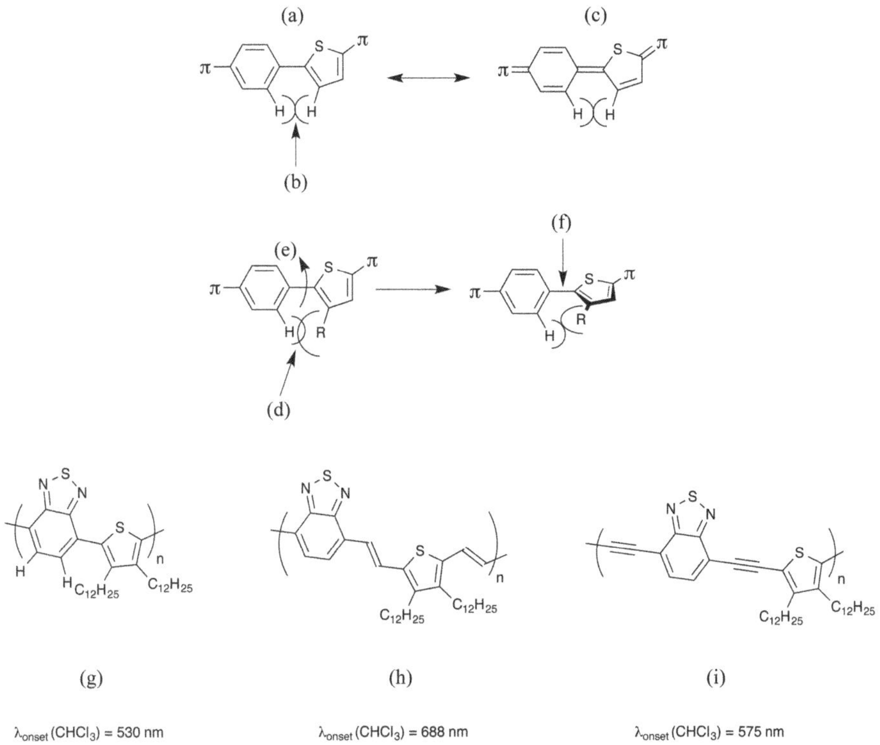

Figure 2.15 Effect of sterically induced torsions on π-conjugation.

to impart solubility and processability. An example of the effects of such sterics-induced twisting is shown in Figure 2.15 for a series of BT–thiophene D–A polymers.[64] BT–thiophene copolymer F2.15g is, as expected, blue shifted compare to alkene-containing copolymer F2.15h (530 nm → 688 nm), but is also blue shifted compared to the *alkyne*-containing polymer F2.15i (530 nm → 575 nm), which is an indication of reduced conjugation in F2.15g compared to F2.15i (the opposite of the trend seen for alkyne F2.14c *vs.* F2.14a in Figure 2.14). This break in conjugation is largely due to twisting induced by the steric repulsion between the dodecyl groups on the thiophene ring subunit and the hydrogens on the benzothiadiazole subunit in F2.15g compared to both F2.15h and F2.15i.

2.4 STABILITY

For vacuum-deposited organic electronic devices, initial thermal stability is critical for being able to form the thin film at elevated temperatures where evaporation may occur. For spin deposited and printed electronics and photonics, thermal stability is generally an issue, for example, if another part of the device fabrication process requires elevated temperatures, if a thin film of the material itself requires thermal annealing, if there is high resistive heating during operation, *etc.* For electronic devices that either emit or absorb light, and typically for all photonic devices, photochemical stability is a concern. For electro-optic polymers, both photostability and thermal stability are required for longer-term device operation and fabrication, respectively. As such, much work has been devoted to increasing photochemical and thermal stability of chromophores for 2nd-order NLO, and general principles that have been developed here are often application to other materials in organic electronics and photonics. Table 2.8 lists a relatively few examples that illustrate some general principles.[57,59,65] It is important to point out that the chromophore (and any material) will only be as stable as its weakest link (*e.g.*, having a very stable π-bridge may not be enough if the acceptor is very unstable). Often, some donor or acceptor strength is sacrificed for a gain in stability; but sometimes the performance can be regained by adding simple donors or acceptors to the more thermally stable group.

The chromophores in Table 2.8 illustrate some general principles: (1) "protecting" alkene π-bridges by locking them with rings generally increases thermal stability (compare Entry 1 *vs.* 2) and may increase photochemical stability by preventing *cis/trans* isomerization and sterically preventing approach of other species (such as singlet oxygen), but the efficiency of the π-bridge decreases somewhat; (2) replacing an alkene in a bridge, even a protected one, with a thiophene increases thermal stability (compare Entry 3 *vs.* 4); (3) replacing alkyl groups on an amine with phenyl groups tends to increase thermal stability (compare Entry 6 *vs.* 4 and 5) and photochemical stability. Note that although the π-donor strength decreases on substitution of the alkyl groups (compare Entry 6 *vs.* 4), donor strength can be at least partially made up by adding simple donor groups (compare Entry 4 *vs.* 5 *vs.* 6). Admittedly, this is an abbreviated discussion of this important topic; however, these are the main stability trends seen in organic materials, and the reader is directed towards thorough reviews for more in depth discussion and other examples.[40,47,48,66]

2.5 PROCESSABILITY AND MORPHOLOGY

Much of the ultimate goal of π-conjugated materials is solution processing through coating or printing techniques—slot-dye coating, gravure and reverse gravure printing, ink-jet printing—that allow

Table 2.8 Thermal stability of chromophores as a function of π-bridge and donor group.

	Chromophore	T_d	λ_{max}	Reference
1		174	644	57
2		252	619	57
3		220	722	59
4		259	694	59
5		260	745	59
6		209	781	65

deposition over large areas and in large scale.[9,67,68] Historically, processability of π-conjugated materials has been effected by derivatizing
the π-functional backbone with side chains that are flexible and therefore add solubility to both small molecules and polymers for both electronics and photonics applications. More recently, side-chains have
been used more functionally to modify and in some cases increase the
performance of materials in both organic electronic[69] and photonic[47]
applications. Ultimately, these side-chains and the manner in which
the thin films are processed affects the intermolecular arrangement
and can result in different phases and domains and thus affect the
both the *nano-to-microstructure* and *morphology* of the resulting film.[6]
According to IUPAC, morphology is defined as the "shape, optical
appearance, or form of phase domains in substances, such as high
polymers, polymer blends, composites, and crystals"[18] that occurs
typically on the tens of nanometres to microns length scale; the specific intermolecular arrangement and packing of the molecules (*i.e.*
crystal structure, the crystal quality, orientation and dimensions, as
well as the degree of crystallinity) may more usefully fall under the
term *nanostructure* to *microstructure* depending on the scale. As will be
discussed below, these nano-to-microstructural and morphological
features are critical for key material performance parameters in both
electronics and photonics, and may not be uniform throughout the
thickness of the thin film;[70] however, this brings about one possible
pitfall in π-conjugated materials research that is ever-present: in order
to change the processability of materials, the intermolecular order and
overall morphology is often changed as well, and then so too is the
material performance. Moreover, successfully predicting how molecules will arrange and pack due to changes in the chemical structure
is still a very challenging area of research. Thus, the interrelationship
of a material's nano-to-microstructure, processing, and performance
is an expanding and impactful area of research, which includes developing new methods and tools to study thin-film nano-microstruc-
ture[71] and advanced means to predict which solidification pathway
leads to which solid-state microstructure.[6]

Various terms encountered in discussion of molecular arrangement
and packing over various length scales are illustrated in Figure 2.16
using a prototypical π-conjugated material terphenyl, which here
is represented as being essentially flat. Such a molecule has a face
(F2.16a), an edge (F2.16b), and an end (F2.16c) that can interact with
other molecules, polymer chains, or parts thereof. These π-moieties
may be oriented in an arrangement that is face-to-face (F2.16d), face-
to-end (F2.16e), face-to-edge (F2.16f), and edge-to-edge (not shown).
On larger length scale, these molecules can orient themselves in a

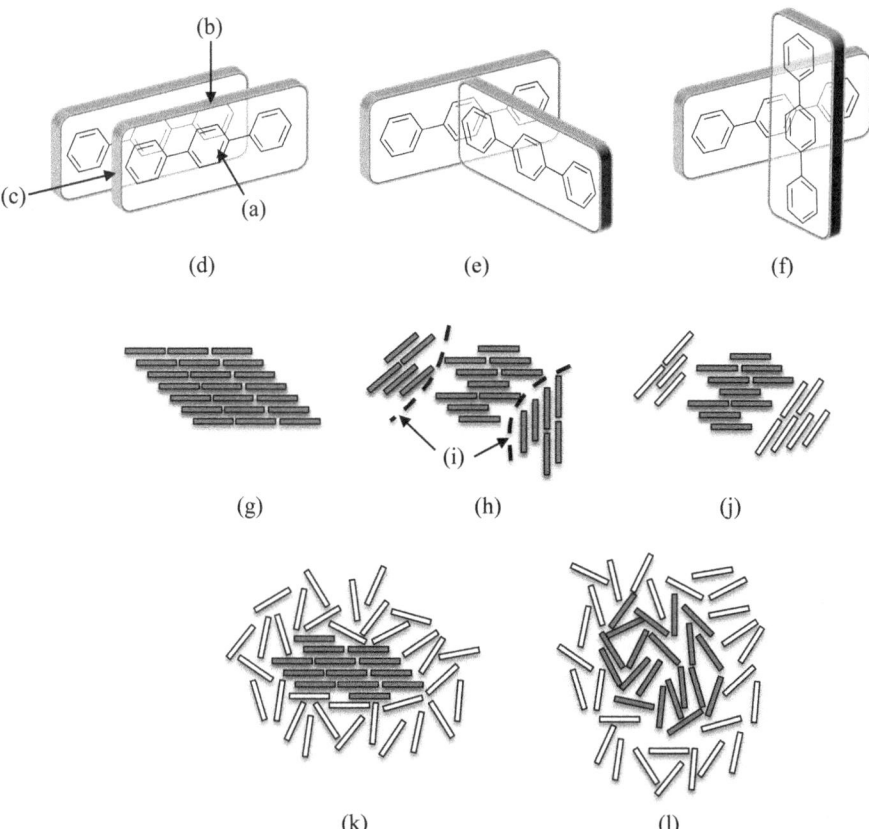

Figure 2.16 Intermolecular interactions and microstructure: (a) "face" of terphenyl; (b) "edge" of terphenyl; (c) "end" of terphenyl; (d) "face-to-face" interaction; (e) "end-to-face" interaction; (f) "edge-to-face" interaction; (g) packing in a crystal; (h) crystallite domains of the same material; (i) grain boundaries; (j) crystallite domains of different materials; (k) crystallite in an amorphous host; (l) amorphous domain in an amorphous host.

number of ways, as illustrated in Figure 2.16, including: (1) a crystalline arrangement that is highly regular throughout the material (F2.16g); (2) a polycrystalline material comprised of randomly oriented crystalline domains (F2.16h) where areas of highly ordered crystalline domains are separated from other domains by grain boundaries (F2.16i); and in the case of two component systems: (3) polycrystalline systems comprised of crystallite domains of each material (F2.16j) or (not shown) intermixed co-crystal structures; (4) systems where one component forms a crystallite while the other is a molecularly disordered amorphous phase (F2.16k); and (5) material phases separated into two different amorphous phases (F2.16l). It should be noted that the clean disjointed grain boundaries illustrated in Figure 2.16 are an oversimplification, and grain boundaries can have different regular structure of their own or may consist of narrow amorphous regions. Finally, only simple face-to-face and edge-to-face type interactions across grain boundaries have been illustrated whereas there are many cases of face-to-edge or edge-to-edge type interactions as well. Taken together, and adding other nano-to-microstructural and morphological possibilities not shown, it is clear there are a number of different ways in which structure can change, and it is often the case that change occurs in a manner that is deleterious to device performance.

In organic electronics, charge transport[3,72] is a key process and the impact of the nano-to-micro scale molecular order on charge transport in π-conjugated materials,[73] OFETs,[74] and on device performance in a bulk heterojunction-OPV (BHJ-OPV)[75] has been studied for some time. In the case of charge transport and OFETs specifically, certain nano–microstructural features that are both favorable and deleterious to charge mobility have been identified whereas in BHJ-OPV both the structure and phase morphology of the donor–acceptor blends have been used to manipulate polymer OPV efficiencies. Intrachain charge transport in polymers (*i.e.*, transport of the charge up and down the polymer chain) is often efficient whereas charge hopping between polymer chains depends on orientation of the π-systems that allow orbital overlap. For face-to-face interactions, the hopping process can be very effective if the π-faces are closely packed *and* if there is appropriate orbital overlap (*i.e.*, short interchain distances are a necessary but not sufficient criteria for effective overlap, in addition the orbitals must have the appropriate phase relationships relative to one another for "spatial" overlap to lead to significant productive electronic orbital overlap), but face-to-end and face-to-edge are often much less efficient since the π-orbitals of neighboring chains are largely orthogonal.[1,2] This is particularly true for polymers that have side chains for processability, such as poly-3-hexylthiophene (P3HT).[76] As illustrated in Figure 2.17 for a 6-monomer fragment, P3HT may form lamellar structures (F2.17a) that in a prototypical ordered arrangement have a π-stacking direction and an alkyl stacking direction. Orbital overlap along the face-to-face π-stacking direction allows relatively facile *inter*chain charge hopping while interchain charge hopping along the edge-on, alkyl stacking direction is comparatively much less favorable, particular since the π-conjugated backbones are separated by insulating alkyl chains. This has been manifested by charge mobility in different orientations of P3HT with respect to the substrate in OFETs (F2.17b and c), where the source and drain electrodes in the device are situated such that hole transport must be *parallel* to the substrate.

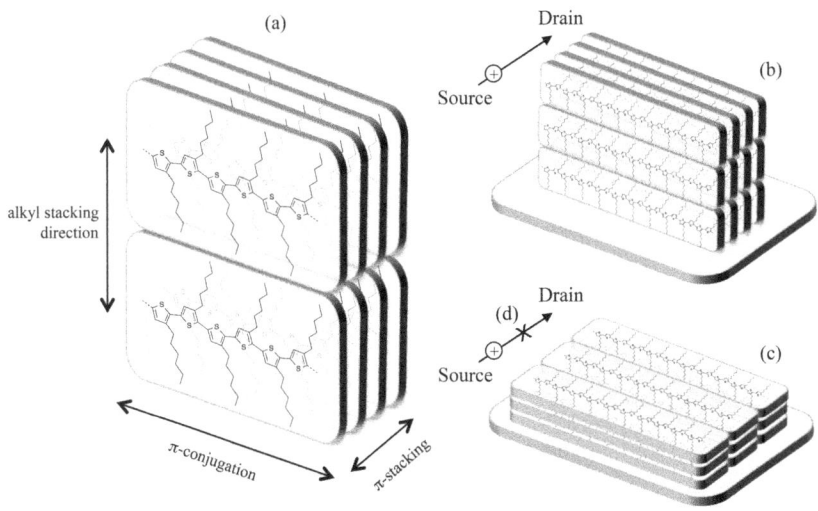

Figure 2.17 Packing arrangement illustration in poly(3-hexylthiophene) (P3HT): (a) anisotropy of hole transport in crystalline P3HT; (b) edge-on orientation of P3HT in reference to the substrate; (c) face-on orientation of P3HT in reference to the substrate; and (d) charge transport that cannot occur efficiently (ref. 77).

As a consequence, polymer chains oriented with the π-stacking direction parallel to the substrate (F2.17b) resulted in ~100 times greater charge mobility than when the π-stacking direction was perpendicular to the substrate (F2.17c). In this case, when the π-stacking direction is perpendicular to the substrate and the electrode configuration favors charge transport parallel to the substrate, charge transport in P3HT must occur in the alkyl stacking direction (F2.17d) where there is negligible orbital overlap between the π-conjugated chains.[77] In another study that demonstrates many points about chemical structure differences, resulting changes in molecular packing, and their effect on charge-transport,[78] P3HT was found to have a solid-state structure and hole mobility that depends on the chain length of the polymer, with high molecular weight polymers (MW) having relatively higher mobility than the low molecular weight polymers. Although both polymers comprised lamellar crystals (Figure 2.18), it was proposed that the longer polymer chains in the high MW polymers (F2.18a) have bridges (F2.18b) between fibrillar crystallites that allow hole transport (F2.18c) between the crystalline moieties, whereas in the low MW case (F2.18d) there is an abundance of grain boundaries (F2.18e) that prevent hole transport between crystallites (composed of chain-extended crystals) and reduce overall mobility.

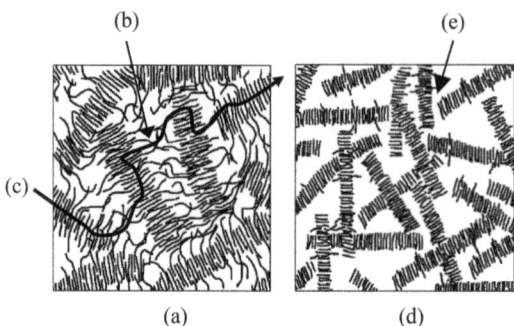

(b)   (e)

(c)

(a)   (d)

Figure 2.18   Illustration of packing in poly(3-hexylthiophene) (P3HT) having higher
MW and lower MW: (a) packing in higher MW P3HT; (b) bridges between
fibrillar crystallites; (c) charge transport through the material; (d) packing in lower MW P3HT; and (e) grain boundaries. A larger number of
grain boundaries exist in the lower MW P3HT. Adapted with permission
from *Synth. Met.*, 2006, **156**, 815. Copyright © 2006, Elsevier.

In photonics, optical loss is a critical material property that is often
negatively impacted by morphological features, and in particular the
phase morphology. In guided wave optical devices, two of the main
mechanisms for optical loss are absorption and scattering of the operating light, which ultimately causes a need for expensive amplifica-
tion of the optical signal or loss of the signal altogether.[4] In photonics,
a lack of phase separation is often necessary to have low optical loss.
As in conjugated polymers, the chromophores in photonic applications have π-surfaces that can stack face-to-face (Figure 2.19), which
is compounded in 2nd order NLO chromophores by an anti-parallel
arrangement of the dipoles that increases the interactions that occur
upon aggregation (F2.19a). When the chromophores aggregate, their
optical absorption tends to broaden and they can phase separate
into large crystallites or amorphous phases within a polymer host.
The broadening of the optical spectrum can cause direct absorptive
loss and the large particles typically cause scattering loss of operating
wavelengths used in telecommunications (1310–1550 nm). Microscopy of thin films for examples shown in Figure 2.19 illustrates the
importance of the phase morphology to optical loss in both 2nd order
and 3rd order NLO materials for chromophores in an "amorphous
polycarbonate" (APC). The structures of APC (F2.20a) and the chromophores are shown in Figure 2.20. The chromophore letters in Figure
2.20 correspond to the images of their composites in Figure 2.19. In
the first set of images (F2.19b–d), atomic force microscopy (AFM) was
used to resolve morphological features on the 5 μm scale. In F2.19b,
the chromophores were not very compatible with the polymer matrix

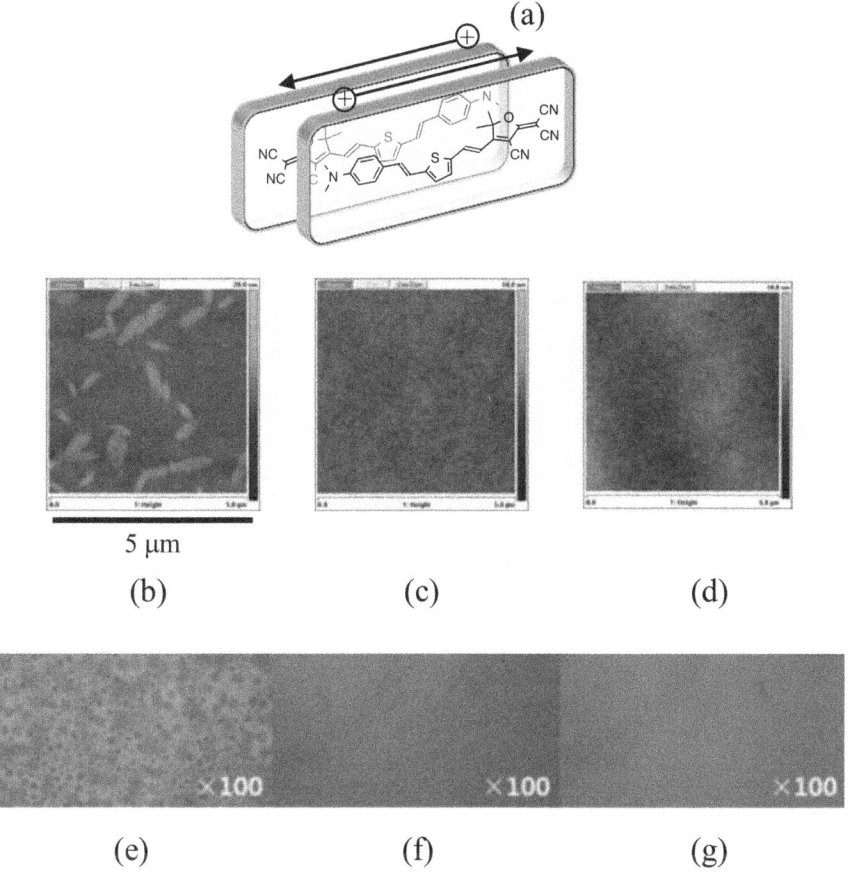

Figure 2.19 Anti-parallel interaction of dipolar chromophores (a) and morphology of thin films having variable optical loss. (b)–(d) Reprinted with permission from *Proc. SPIE*, 2006, **6243**, 62430G. Copyright © 2006 SPIE, Bellingham, WA USA. (e)–(g) Reprinted with permission from *Adv. Mater.*, 2012, **24**, OP326. Copyright © 2012 WILEY-VCH Verlag GmbH & Co. KGaA, Weinheim.

and formed crystallites on the order of 1 µm in size. However, the other two chromophore–polymer blends showed no such crystallite features, and the optical loss of F2.19b was twice that of F2.19c and F2.19d.[79] In a study on 3rd order NLO cyanine–polymer composites, phase separation in one of the materials was evident at 100× magnification in an optical microscope (F2.19e), and relatively less apparent in F2.19f and F2.19g. The composite shown in F2.19e had an optical loss that was too high to be reliably measured.[80]

Many other examples of the effects of the solid-state intermolecular order and morphology on both organic electronic[73–75] and photonic[4,47]

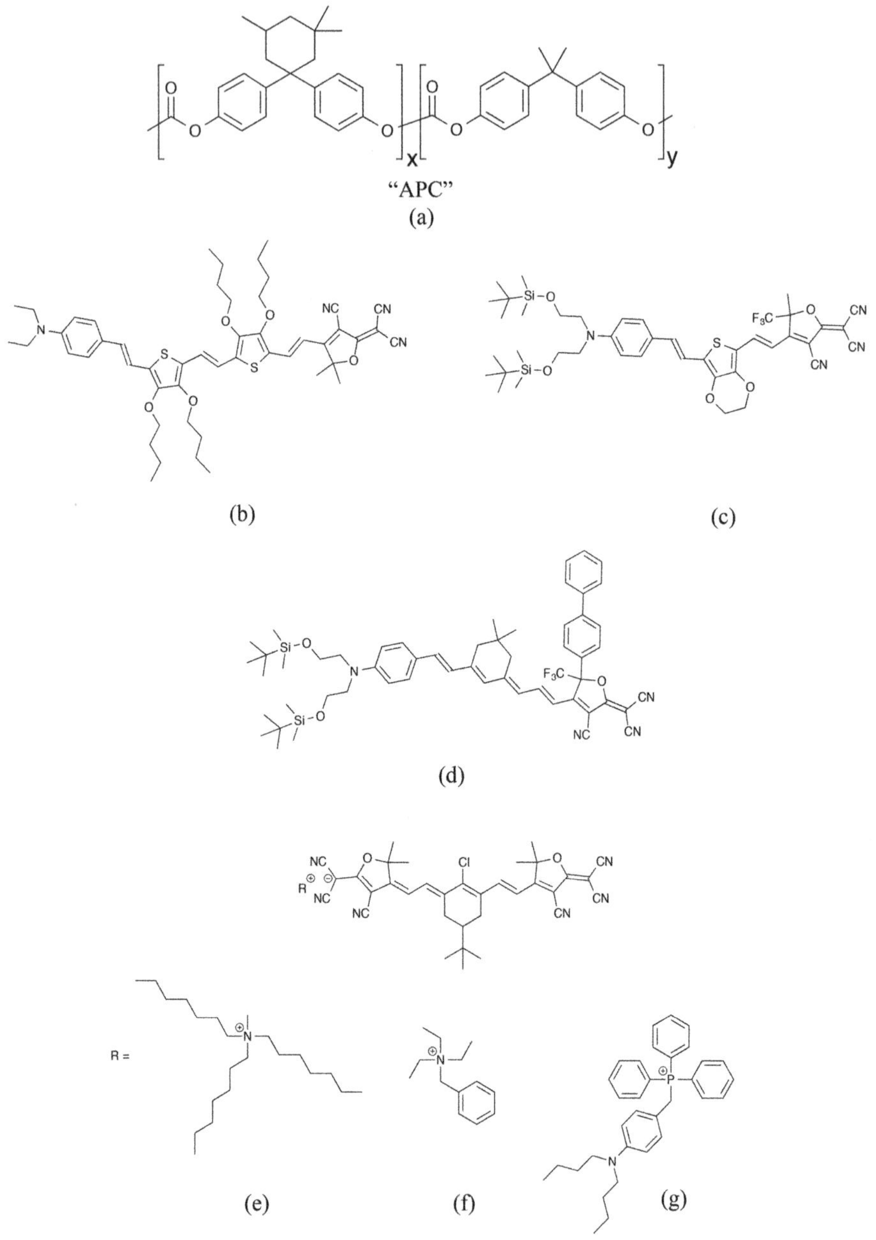

Figure 2.20   Structures of the various materials discussed in Figure 2.19.

materials are discussed in the thorough reviews listed above, and the reader is encouraged to explore the literature even further for additional recent examples. Addressing the complexity and unpredictability of the molecular order and arrangement over all length scales, leading to different nano–microstructures and morphologies, and their effect on material properties is a research area that will continue to expand theoretically, computationally, and experimentally.

2.6 SYNTHESIS OVERVIEW

In the above sections, a brief introduction was given as to *why* material properties would be changed and what properties are generally important. The balance of this book is devoted to the *how* of making molecules and polymers from different donors, π-bridges, and acceptors. The art of the science is defining an experimental plan to test a hypothesis that can provide meaningful insight about how molecular structure changes affect certain properties. The art of synthesis involves obtaining enough materials that fit the experimental design as safely as possible within a reasonable period of time and with the minimum number of synthetic and purification steps. Toward that end, each synthetic step is typically related to (1) adding side-chains that can effect solubility, morphology, and/or processability; (2) adjusting the IE or EA of the materials by adding simple pendant donor and acceptor groups, by heteroannulation of the π-core, and/or by coupling to other donor or acceptor aromatic moieties; (3) extending the π-system through annulation or by coupling to another π-system; and (4) functionalizing the π-system to allow any one or a combination of the (1)–(3). This book is organized around the synthesis and functionalization of donors; the synthesis, functionalization, and extension of π-bridges; the synthesis and functionalization of acceptors; and bringing all the constituent π-units together to form materials. As will be illustrated, the explosive growth in research on organic electronic and photonic materials in the last 30 years has resulted in the development of a rich synthetic toolbox for both small molecules and polymers.

REFERENCES

1. J. L. Bredas, D. Beljonne, V. Coropceanu and J. Cornil, *Chem. Rev.*, 2004, **104**, 4971.
2. V. Coropceanu, J. Cornil, D. A. da Silva Filho, Y. Olivier, R. Silbey and J. L. Bredas, *Chem. Rev.*, 2007, **107**, 926.

3. M. Jaiswal and R. Menon, *Polym. Int.*, 2006, **55**, 1371.

4. H. Ma, A. K. Y. Jen and L. R. Dalton, *Adv. Mater.*, 2002, **14**, 1339.

5. W. Caseri, *Macromol. Rapid Commun.*, 2000, **21**, 705.

6. A. A. Virkar, S. Mannsfeld, Z. Bao and N. Stingelin, *Adv. Mater.*, 2010, **22**, 3857.

7. Y. Yao, H. Dong and W. Hu, *Polym. Chem.*, 2013, **4**, 5197.

8. S. Biswas, O. Shalev and M. Shtein, *Annu. Rev. Chem. Biomol. Eng.*, 2013, **4**, 289.

9. A. C. Arias, J. D. MacKenzie, I. McCulloch, J. Rivnay and A. Salleo, *Chem. Rev.*, 2010, **110**, 3.

10. A. Kahn, N. Koch and W. Gao, *J. Polym. Sci., Part B: Polym. Phys.*, 2003, **41**, 2529.

11. V. Coropceanu, H. Li, P. Winget, L. Zhu and J.-L. Brédas, *Annu. Rev. Mater. Res.*, 2013, **43**, 63.

12. Y. Don Park, J. A. Lim, H. S. Lee and K. Cho, *Mater. Today*, 2007, **10**, 46.

13. J.-L. Bredas, *Mater. Horiz.*, 2014, **1**, 17.

14. P. Atkins and R. Friedman, *Molecular Quantum Mechanics*, Oxford University Press, New York, USA, 5th edn, 2011.

15. N. Turro, J. Scaiano and V. Ramamurthy, *Modern Molecular Photochemistry of Organic Molecules*, University Science Books, USA, 2010.

16. D. Harris and M. Bertolucci, *Symmetry and Spectroscopy: An Introduction to Vibrational and Electronic Spectroscopy*, Dover Publications, Inc., New York, 1989.

17. J.-L. Brédas, J. Cornil, D. Beljonne, D. A. dos Santos and Z. Shuai, *Acc. Chem. Res.*, 1999, **32**, 267.

18. A. McNaught and A. Wilkinson, *IUPAC. Compendium of Chemical Terminology*, Blackwell Scientific Publications, Oxford, 2nd edn, 1997.

19. K. Seki and K. Kanai, *Mol. Cryst. Liq. Cryst.*, 2006, **455**, 145.

20. D. Turner, C. Baker, A. Baker and C. Brundle, *Molecular Photoelectron Spectroscopy*, Wiley-Interscience, London, 1970.

21. E. Heilbronner, J. Maier and E. Haselbach, in *Phys. Methods Heterocycl. Chem.*, ed. A. R. Katritzky, Academic, New York, 1974, vol. 6, ch. 1, pp. 1–52.

22. C. Rao and P. Basu, in *Phys. Methods Heterocycl. Chem.*, ed. R. Gupta, Wiley, New York, 1984, ch. 3, pp. 231–279.

23. E. C. M. Chen and W. E. Wentworth, *Mol. Cryst. Liq. Cryst.*, 1989, **171**, 271.

24. H. Yoshida, *Anal. Bioanal. Chem.*, 2014, **406**, 2231.

25. P. I. Djurovich, E. I. Mayo, S. R. Forrest and M. E. Thompson, *Org. Electron.*, 2009, **10**, 515.

26. E. Anslyn and D. Dougherty, in *Modern Physical Organic Chemistry*, University Science Books, USA, 2006, ch. 14, pp. 807–875.
27. M. Dewar and R. Dougherty, *The PMO Theory of Organic Chemistry*, Plenum, New York, 1975.
28. M. Dewar and R. Dougherty, in *The PMO Theory of Organic Chemistry*, Plenum, New York, 1975, ch. 2, pp. 57–71.
29. C. Hansch, A. Leo and R. Taft, *Chem. Rev.*, 1991, **91**, 165.
30. J. Fabian and H. Hartmann, in *Light Absorption of Organic Colorants. Theoretical Treatment and Empirical Rules*, Springer-Verlag, Berlin, 1980, ch. 7, pp. 162–197.
31. A. R. Katritzky, M. Karelson and N. Malhotra, *Heterocycles*, 1991, **32**, 128.
32. J. Joule and K. Mills, *Heterocyclic Chemistry*, John Wiley and Sons, Chichester, UK, 5th edn, 2010.
33. C. W. Bird, *Tetrahedron*, 1985, **41**, 1409.
34. C. W. Bird, *Tetrahedron*, 1986, **42**, 89.
35. H. Zhou, L. Yang and W. You, *Macromolecules*, 2012, **45**, 607.
36. J. Hwang, A. Wan and A. Kahn, *Mater. Sci. Eng., R*, 2009, **64**, 1.
37. C. W. Schlenker and M. E. Thompson, *Top. Curr. Chem.*, 2012, **312**, 175.
38. G. Ren, C. W. Schlenker, E. Ahmed, S. Subramaniyan, S. Olthof, A. Kahn, D. S. Ginger and S. A. Jenekhe, *Adv. Funct. Mater.*, 2013, **23**, 1238.
39. J. L. Bredas, J. E. Norton, J. Cornil and V. Coropceanu, *Acc. Chem. Res.*, 2009, **42**, 1691.
40. X. Guo, M. Baumgarten and K. Müllen, *Prog. Polym. Sci.*, 2013, **38**, 1832.
41. P.-L. T. Boudreault, A. Najari and M. Leclerc, *Chem. Mater.*, 2011, **23**, 456.
42. S. Barlow, C. Risko, S. A. Odom, S. Zheng, V. Coropceanu, L. Beverina, J. L. Bredas and S. R. Marder, *J. Am. Chem. Soc.*, 2012, **134**, 10146.
43. J. E. Anthony, A. Facchetti, M. Heeney, S. R. Marder and X. Zhan, *Adv. Mater.*, 2010, **22**, 3876.
44. N. Blouin, A. Michaud, D. Gendron, S. Wakim, E. Blair, R. Neagu-Plesu, M. Belletete, G. Durocher, Y. Tao and M. Leclerc, *J. Am. Chem. Soc.*, 2008, **130**, 732.
45. X. Zhang, T. T. Steckler, R. R. Dasari, S. Ohira, W. J. Potscavage, S. P. Tiwari, S. Coppée, S. Ellinger, S. Barlow, J.-L. Brédas, B. Kippelen, J. R. Reynolds and S. R. Marder, *J. Mater. Chem.*, 2010, **20**, 123.
46. T. C. Parker, D. G. Patel, K. Moudgil, S. Barlow, C. Risko, J.-L. Brédas, J. R. Reynolds and S. R. Marder, *Mater. Horiz.*, 2014, **2**, 22.

47. L. R. Dalton, W. H. Steier, B. H. Robinson, C. Zhang, A. Ren, S. Garner, A. Chen, T. Londergan, L. Irwin, B. Carlson, L. Fifield, G. Phelan, C. Kincaid, J. Amend and A. Jen, *J. Mater. Chem.*, 1999, **9**, 1905.

48. L. R. Dalton, P. A. Sullivan and D. H. Bale, *Chem. Rev.*, 2010, **110**, 25.

49. J. M. Hales, S. Barlow, H. Kim, S. Mukhopadhyay, J.-L. Brédas, J. W. Perry and S. R. Marder, *Chem. Mater.*, 2014, **26**, 549.

50. M. Rumi, S. Barlow, J. Wang, J. W. Perry and S. R. Marder, *Adv. Polym. Sci.*, 2008, **213**, 1.

51. C. Gorman and S. Marder, *Proc. Natl. Acad. Sci. U. S. A.*, 1993, **93**, 11297.

52. S. R. Marder, C. B. Gorman, F. Meyers, J. W. Perry, G. Bourhill, J. L. Bredas and B. M. Pierce, *Science*, 1994, **265**, 632.

53. F. Meyers, S. R. Marder, B. M. Pierce and J. L. Bredas, *J. Am. Chem. Soc.*, 1994, **116**, 10703.

54. S. R. Marder, L. T. Cheng, B. G. Tiemann, A. C. Friedli, M. Blanchard-Desce, J. W. Perry and J. Skindhoj, *Science*, 1994, **263**, 511.

55. V. Alain, L. Thouin, M. Blanchard-Desce, U. Gubler, C. Bosshard, P. Günter, J. Muller, A. Fort and M. Barzoukas, *Adv. Mater.*, 1999, **11**, 1210.

56. M. Ahlheim, M. Barzoukas, P. V. Bedworth, M. Blanchard-Desce, A. Fort, Z.-Y. Hu, S. R. Marder, J. W. Perry, C. Runser, M. Staehelin and B. Zysset, *Science*, 1996, **271**, 335.

57. X. Piao, X. Zhang, S. Inoue, S. Yokoyama, I. Aoki, H. Miki, A. Otomo and H. Tazawa, *Org. Electron.*, 2011, **12**, 1093.

58. A. Otomo, I. Aoki, H. Miki, H. Tazawa and S. Yokoyama, *US Pat.*, 2012/0172599, 2012.

59. Y.-C. Shu, Z.-H. Gong, C.-F. Shu, E. M. Breitung, R. J. McMahon, G.-H. Lee and A. K. Y. Jen, *Chem. Mater.*, 1999, **11**, 1628.

60. L. T. Cheng, W. Tam, S. R. Marder, A. E. Stiegman, G. Rikken and C. W. Spangler, *J. Phys. Chem.*, 1991, **95**, 10643.

61. A. Bhaskar, G. Ramakrishna, Z. Lu, R. Twieg, J. M. Hales, D. J. Hagan, E. Van Stryland and T. Goodson, 3rd, *J. Am. Chem. Soc.*, 2006, **128**, 11840.

62. S. Barlow, C. Risko, V. Coropceanu, N. M. Tucker, S. C. Jones, Z. Levi, V. N. Khrustalev, M. Y. Antipin, T. L. Kinnibrugh, T. Timofeeva, S. R. Marder and J. L. Bredas, *Chem. Commun.*, 2005, 764.

63. D. H. Lee, M. J. Lee, H. M. Song, B. J. Song, K. D. Seo, M. Pastore, C. Anselmi, S. Fantacci, F. De Angelis, M. K. Nazeeruddin, M. Grätzel and H. K. Kim, *Dyes Pigm.*, 2011, **91**, 192.

64. J. Li, M. Yan, Y. Xie and Q. Qiao, *Energy Environ. Sci.*, 2011, **4**, 4276.
65. Y.-J. Cheng, J. Luo, S. Hau, D. H. Bale, T.-D. Kim, Z. Shi, D. B. Lao, N. M. Tucker, Y. Tian, L. R. Dalton, P. J. Reid and A. K. Y. Jen, *Chem. Mater.*, 2007, **19**, 1154.
66. M. Jørgensen, K. Norrman and F. C. Krebs, *Sol. Energy Mater. Sol. Cells*, 2008, **92**, 686.
67. F. C. Krebs, *Sol. Energy Mater. Sol. Cells*, 2009, **93**, 394.
68. M. Helgesen, R. Søndergaard and F. C. Krebs, *J. Mater. Chem.*, 2010, **20**, 36.
69. J. Mei and Z. Bao, *Chem. Mater.*, 2014, **26**, 604.
70. S. S. Lee and Y. L. Loo, *Annu. Rev. Chem. Biomol. Eng.*, 2010, **1**, 59.
71. S. S. Lee and Y. L. Loo, *Annu. Rev. Chem. Biomol. Eng.*, 2010, **1**, 59.
72. V. Coropceanu, J. Cornil, D. A. da Silva Filho, Y. Olivier, R. Silbey and J. L. Bredas, *Chem. Rev.*, 2007, **107**, 926.
73. Y. Olivier, D. Niedzialek, V. Lemaur, W. Pisula, K. Mullen, U. Koldemir, J. R. Reynolds, R. Lazzaroni, J. Cornil and D. Beljonne, *Adv. Mater.*, 2014, **26**, 2119.
74. H. N. Tsao and K. Mullen, *Chem. Soc. Rev.*, 2010, **39**, 2372.
75. M. He, M. Wang, C. Lin and Z. Lin, *Nanoscale*, 2014, **6**, 3984.
76. A. Salleo, *Mater. Today*, 2007, **10**, 38.
77. H. Sirringhaus, P. J. Brown, R. H. Friend, M. M. Nielsen, K. Bechgaard, B. M. W. Langeveld-Voss, A. J. H. Spiering, R. A. J. Janssen, E. W. Meijer, P. Herwig and D. M. de Leeuw, *Nature*, 1999, **401**, 685.
78. J.-M. Verilhac, G. LeBlevennec, D. Djurado, F. Rieutord, M. Chouiki, J.-P. Travers and A. Pron, *Synth. Met.*, 2006, **156**, 815.
79. R. Dinu, D. Jin, D. Huang, M. K. Koenig, A. M. Barklund, Y. Fang, T. C. Parker, Z. Shi, J. Luo and A. K. Y. Jen, Enabling Photonics Technologies for Defense, Security, and Aerospace Applications II. *Proc. SPIE*, 2006, **6243**, 62430G.
80. Z. Li, Y. Liu, H. Kim, J. M. Hales, S. H. Jang, J. Luo, T. Baehr-Jones, M. Hochberg, S. R. Marder, J. W. Perry and A. K. Jen, *Adv. Mater.*, 2012, **24**, OP326.

CHAPTER 3

Donors

3.1 SYNTHESIS OF DONORS

The synthesis of donors used in materials for organic electronics and photonics can be in large part grouped into chemistry involving either thiophenes or aryl amines. This can be seen by examining some of the donors shown in Figure 3.1 that have been used widely themselves or which appear as building blocks in small molecules and polymers. Often, the synthesis of new donors involves benzannulating or heteroannulating parent thiophene or arylamine compounds to increase π-surface area, stability, or electron donating strength. Additionally, donors are often derivatized with straight- and branched-chain alkyl groups to increase processability (solubility) and/or may include functional groups for crosslinking to render the resulting materials insoluble. This chapter covers some of the basic reactivity and reactions of thiophene and aryl amine compounds that can serve as the basis for synthesis of functionalized donors that in turn can be reacted with other reagents or π-aryl moieties for incorporation into various π-conjugated materials, which will be covered in later chapters on functionalization, π-extension, and C−C coupling reactions.

3.2 THIOPHENE CHEMISTRY

Much of the thiophene chemistry encountered in materials science involves reactivity of the π-system of thiophene or nucleophilic reactivity of the σ-anions of thiophene obtained from metallation of the

Synthetic Methods in Organic Electronic and Photonic Materials: A Practical Guide
By Timothy C. Parker and Seth R. Marder
© Timothy C. Parker and Seth R. Marder, 2015
Published by the Royal Society of Chemistry, www.rsc.org

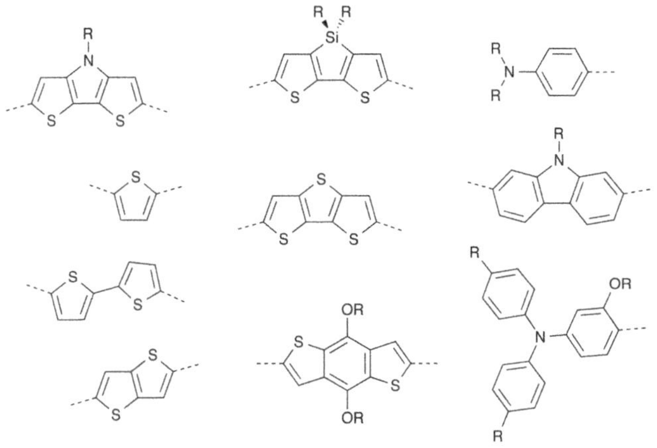

Figure 3.1 Some common donors used in organic electronics and photonic materials.

thiophene C−H or C−halogen bonds. It is often the case that the differing π and σ-reactivity can be used to control the regioselectivity of certain reactions. In this section, the fundamental π *vs.* σ reactivity in thiophenes is presented, and more in-depth discussion of donor functionalization reactions that utilize thiophene π *vs.* σ reactivity is presented in later chapters. Note that, as with all heterocycles, the chemistry of thiophene encountered in materials science represents a useful subset of the more rich and varied chemistry of thiophene known to the synthetic community at large, and review of thiophene chemistry in a broader context can be inspiring to the materials scientist.[1]

As an aromatic compound, thiophene has aromatic reactivity towards electrophiles and nucleophiles (when a π-withdrawing group is present) that is analogous to the reactivity of benzene. Figure 3.2 shows the general structure of thiophene with the numbering scheme illustrated around the periphery (F3.2a). The 2 and 5 positions are equivalent by symmetry and are also referred to as "α" positions. The 3 and 4 positions are also equivalent by symmetry and are referred to as "β" positions. The π-reactivity towards electrophilic substitution can be rationalized by examining resonance structures F3.2b−e shown in Figure 3.2. All the resonance structures are zwitterionic with a positive charge on the sulfur atom with negative charge on both the α carbons (F3.2c,d) and the β carbons (F3.2b,e). Thus, a hybrid resonance structure of thiophene would have partial negative charges on all the carbons, which is consistent with thiophene being an electron donating groups that releases electron density into π-systems. The *regioselectivity* of thiophene electrophilic aromatic substitution is that the α positions are the most active followed by the β positions, suggesting that resonance structures F3.2c and F3.2d contribute more character to a hybrid representation. This may be due, in part, to π-anions at the α positions being more stabilized by the positive sulfur atom than when located farther away at the β positions such as in F3.2b,e.

Another important fundamental aspect of thiophene reactivity is that the σ-heteroaryl C—H bonds can be deprotonated by strong bases such as alkyl lithiums or lithium dialkylamides to give the corresponding anions at the α and β positions. Table 3.1 lists approximate pK_a values for alkanes, benzene, dialkylamines, and thiophene in H_2O. Note that the presence of the electronegative sulfur atom increases the acidity of thiophene by 10 pK_a units compared to that

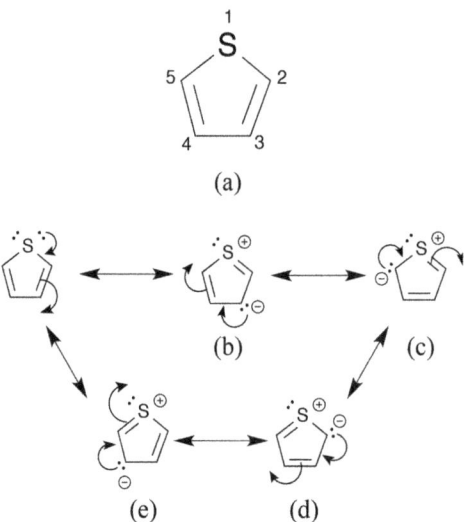

(a)

(b)   (c)

(e)   (d)

Figure 3.2   Structure and numbering of the thiophene ring system with major resonance forms.

Table 3.1   The acidity (pK_a) of organic functional groups having strong conjugate bases that are used in organic synthesis.

Entry	1	2	3	4
Structure	R—C(H)(H)—H	⟨benzene⟩—H	R—N(H)—R	⟨thiophene⟩
~pK_a	50	43	38	33

of benzene itself (Entry 2 *vs.* 4). Within thiophene, the α positions are more acidic than the β positions due largely to the α position proximity to the inductively withdrawing sulfur atom, which stabilizes the anion. As will presented below in Section 3.2.1 and in other chapters, the stability of the anion at the α positions along with the 2,5 regioselectivity of electrophilic aromatic substitution can be effectively used in the synthesis of a variety of thiophene derivatives with predictable regiochemistry between all the α and β positions to provide almost any substitution pattern at the 2, 3, 4, and 5 positions.

3.2.1 Thiophene Metallation

Metallation of thiophene and reaction of the thienyl anion with electrophiles is widely used in the synthesis of organic electronic and photonic materials. This is mostly because metallation can occur selectively at any one of the carbons of the thiophene ring depending on the substitution of the thiophene and the metallating reagent used. The three basic means of metallating thiophene compounds (S3.1a) are shown in Scheme 3.1. The first is deprotonation of a C−H bond by an organometallic or dialkylamide base to give the metallated species (S3.1b). According to the pK_a values in Table 3.1, the conjugate bases of alkanes, benzene, and amines are able to deprotonate the α and β positions of thiophene; however, deprotonation is regiospecific to the 2 and 5 positions in comparison to the 3 and 4 positions since the α anions are more stable than the β anions. The most common reagents used to deprotonate thiophene are alkyl lithiums (or in some cases alkyl Grignards) and lithium amides of sterically hindered secondary amines (such as LDA). The by-product of deprotonation is the

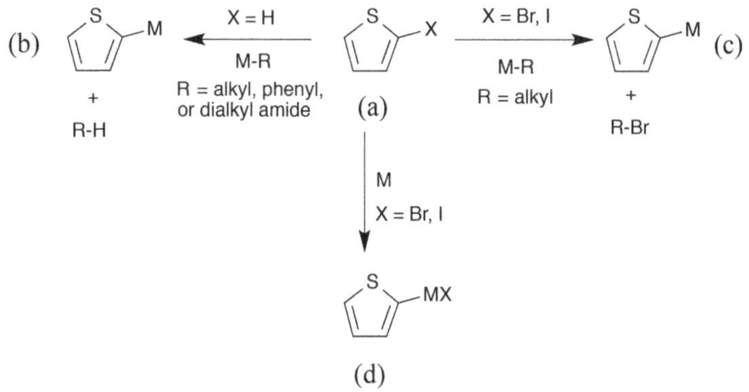

Scheme 3.1 Overview of thiophene metallation conditions.

conjugate acid of the base, which for alkyl lithiums are low boiling alkanes and for lithium amides are sterically hindered amines, which typically do not interfere with the metallated thiophene or with the reaction. A second method to metallate thiophene is metal–halogen exchange (S3.1c), which is typically accomplished with alkyl lithiums, although a "turbo" Grignard[2] may also be used. A third method to generate the metallated thiophene is by direct reaction with a metal such as magnesium or lithium (S3.1d), which may require activation of the metal by addition of a trace amount of reductant or by using a highly active form of the metal such as with Rieke metals.[3] Typically, either deprotonation or metal–halogen exchange is used for metallating thiophene since the reagents are readily available from commercial vendors as solutions that are used directly; however, alkyl lithium solutions that are used in deprotonation and lithium–halogen exchange are *pyrophoric and therefore can be very hazardous* to use, and require specific training regarding their safe handling. Although alkyl lithium hazards can be managed on the smaller scale with proper training and techniques, larger scale reactions of lithiated thiophenes may be less hazardous using the direct reaction of lithium with the bromo- or iodothiophene.

One potential drawback to metallation using metal–halogen exchange is that the alkylhalide by-product (R–Br in S3.1c) may undergo reaction with the thiophene, other reactants, or the product. The by-products of lithium–halogen exchange with *n*-butyl lithium and *t*-butyllithium are *n*-butylbromide and *t*-butylbromide, respectively, and common deleterious side-reactions that may occur with these by-products are shown in Scheme 3.2. The lithiated thiophene can react with *n*-butylbromide *via* a nucleophilic substitution reaction (S3.2a) to give alkylated 2-butylthiophene as a side-product. On the other hand, lithiated thiophene can react with *t*-butylbromide *via* an elimination mechanism (S3.2b) to give the protonated thiophene and isobutylene. The substitution side-products can usually be avoided by keeping the reaction cold (*e.g.*, −78 °C) and the elimination side-product may be avoided by adding an extra equivalent of *t*-butyllithium to act as a base in the elimination reaction to give isobutylene and 2-methylpropane (S3.2c), which is disadvantageous because an extra equivalent of the hazardous *t*-butyllithium must be used; however, the increased reactivity of *t*-butyllithium is sometimes required, so a due increase in caution must be exercised when an extra equivalent of *t*-butyllithium is used, which may be especially relevant when reaction scales are borderline too large for safe use of commercial alkyl lithium solutions.

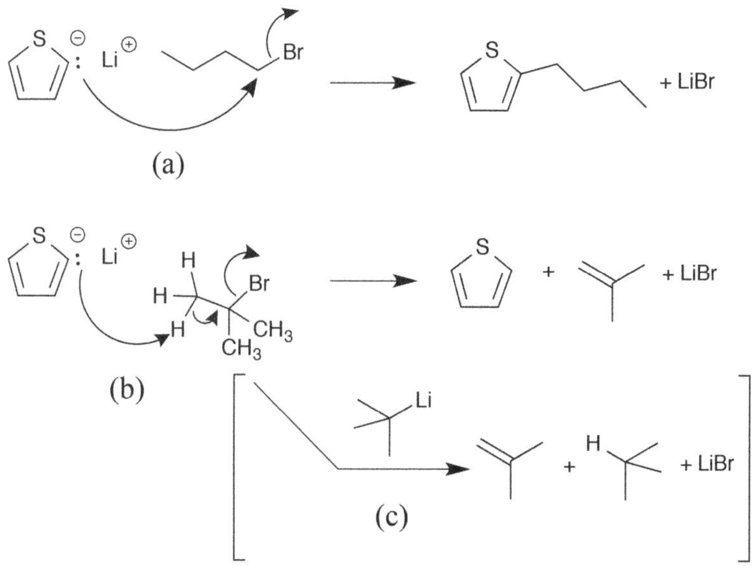

(a)

(b)

(c)

Scheme 3.2 Side-reactions of alkylhalide by-products of alkyllithium–halogen exchange.

All the methods to metallate thiophene occur regioselectively, with deprotonation occurring preferably at the 2 and 5 positions and metal–halogen exchange occurring at the halide position. One caveat is that anions formed at the β position may rearrange to give the more stable anion at the α position. The *regiostability* of thienyl anions between the α and β positions has been examined for a variety of reagents. As shown Scheme 3.3, the 3-thienyl Mg (**1**), Zn (**2**), and Mn (**3**) reagents are regiostable at room temperature but 3-lithiothiophene **4** rearranges to 2-lithiothiophene **5** at temperatures above −25 °C.[4] This lithium exchange from the β to the α position may be slower in less coordinating solvents such as diethyl ether. For example, lithiation of 3-bromothiophene **6** by *n*-buLi in ether at −70 °C gives the regiostable 3-thienyllithium **7**, but isomerization still occurs to the 2 isomer at room temperate (**8**). Even less coordinating conditions such as using hexanes as the main solvent reportedly results in 3-lithiothiophene **9**, which is stable at room temperate, and can be trapped as 3-iodothiophene **10** by reaction with 1,2-diidoethane.[5] However, many π-conjugated substrates have poor solubility in hexanes, so practically this often means reactions utilizing β-lithiothiophenes are limited to temperatures below −50 °C *unless the α positions are blocked* with groups such as trimethylsilyl (TMS). Fortunately, formation of lithiated thiophenes and reaction of the thienyl anion is typically

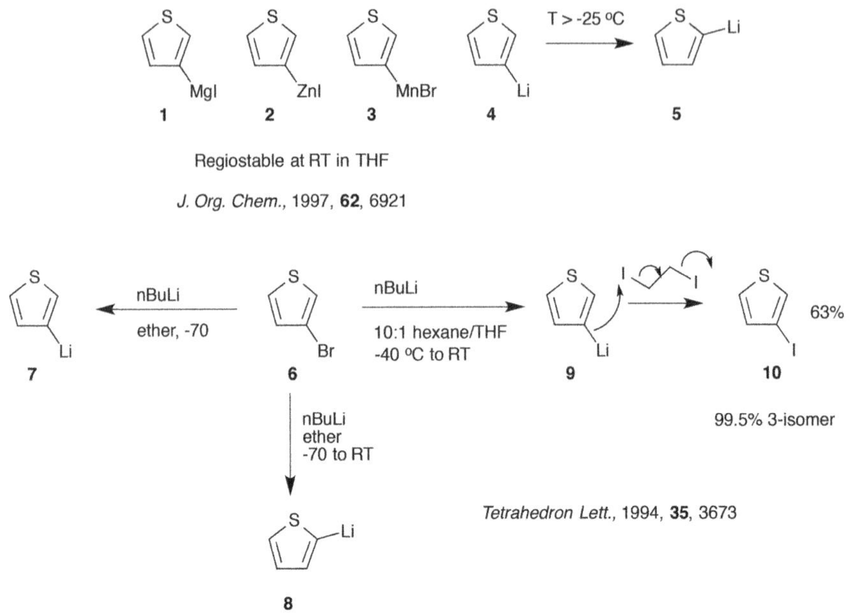

Regiostable at RT in THF

J. Org. Chem., 1997, **62**, 6921

Scheme 3.3 Regiostability overview and formation of α- and β-metallated thiophenes.[4,5]

rapid even at −78 °C and colder. Indeed, many of the metallated thiophenes used in organic synthesis are lithiothiophenes because they are formed readily, they have generally high reactivity, and reagents used to effect lithiation are widely commercially available.

3.2.2 Thiophene Lithiation by Exchange

Generally, lithium–halogen exchange occurs *before* deprotonation when alkyl lithiums are used. Examples of lithium–halogen exchange on various thiophenes are shown in Scheme 3.4 and illustrate some general principles of reactivity. Lithiation typically occurs to give more stable α anions before β anions, which is demonstrated by the lithiation of 2,3-dibromothiophene **11** followed by addition of iodine and warming to room temperature to give 3-bromo-2-iodobenzene **12** in high yield.[6] Lithiation between α positions also tends to give the most stable anion, which may be determined by adjacent electron withdrawing groups. This is demonstrated by the lithiation of the α carbon adjacent to the electron withdrawing bromine atom in 2,3,5-tribromothiophene **13** to give 2,4-dibromothiophene **14** after protonation in 79% yield.[7] Generally, both α positions are more

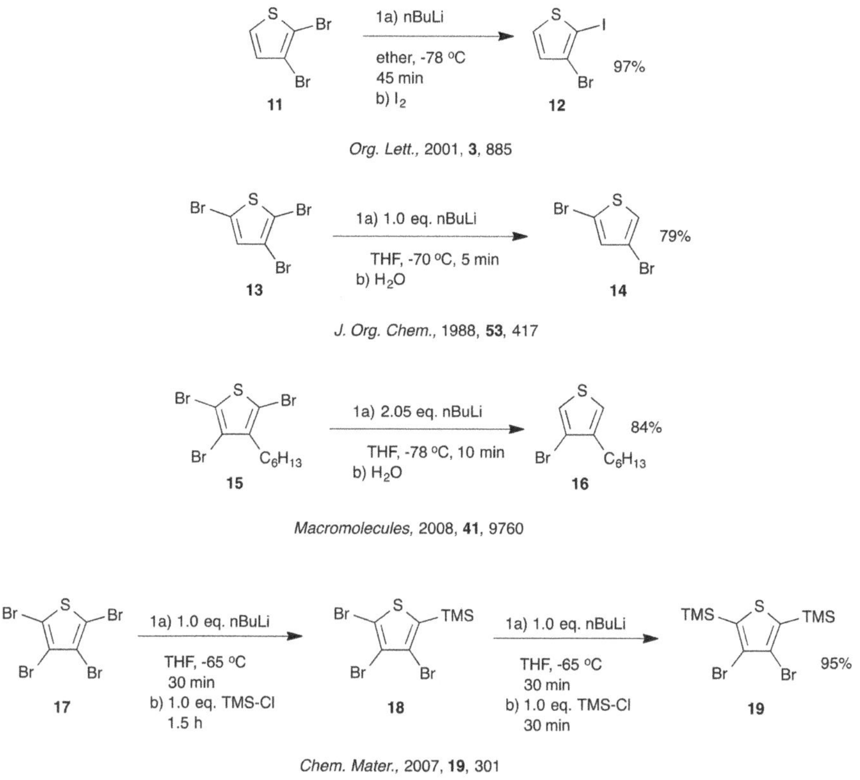

Org. Lett., 2001, **3**, 885

J. Org. Chem., 1988, **53**, 417

Macromolecules, 2008, **41**, 9760

Chem. Mater., 2007, **19**, 301

Scheme 3.4 Synthetic examples of thiophene lithiation by bromine–lithium exchange.[6–9]

reactive than the β positions, as demonstrated by double lithiation of 2,3,5-tribromo-4-hexylthiophene **15** by 2 equivalents of *n*-BuLi to give 3-bromo-4-hexylthiophene **16** in high yield.[8] Double α lithiation also may work when there are two β bromines, although a two-step lithium–halogen exchange/alkylation reaction in one-pot may result in a higher yield, such as mono lithiation of tetrabromothiophene **17** and reaction with TMS-Cl to give **18** and then lithiation of the second α position and reaction with TMS-Cl to give **19** in high yield.[9] Such a scheme may be particularly useful if the doubly lithiated thiophene is not soluble in the reaction solvent.

3.2.3 Thiophene Lithiation by Deprotonation

Lithiation by direct deprotonation of the thiophene can be accomplished by a variety of strong carbon- and nitrogen-centered bases such as alkyl lithiums, phenyllithiums, and lithium dialkylamides.

Typically, two equivalents of strong base will deprotonate both α protons of the thiophene ring before the β protons. One important question is the 2 *vs.* 5 regioselectivity of α deprotonation when there is a substituent in the 3 position since groups that increase solubility are often installed in the 3 position. Table 3.2 summarizes factors that affect the 2/5 regioselectivity of 3-substituted thiophenes. In Table 3.2, Entries 1–4 examine the effect of base size on the 2/5 regioselectivity for 3-methylthiophene **20**.[10] For the smallest base, MeLi, steric effects are large enough to favor 5-deprotonation (**21**) although 25% of 2-deprotonation (**22**) is also observed. Regioselectivity of 5-deprotonation can be improved to 99% by increasing the size of the base to LiTMP (C3.2–4). Thus, generally, deprotonation occurs regioselectively in varying degrees at the 5-position when the 3-substituent is an alkyl group, but can be increased to almost complete 5-deprotonation by using a bulky base. However, 2-deprotonation may be favored when there is an electronegative, σ-withdrawing substituent in the 3-position, which increases acidity of the proximal 2-proton compared to the 5-proton, as illustrated by Entries 5–7. In some cases, lone pairs

Table 3.2 Regioselectivity of thiophene α-deprotonation in the presence of β-functional groups.

Entry	R¹	Base	E	R²	21 (%)	22 (%)
1	−Me	MeLi	MeI	−Me	75	25
2	−Me	*n*-BuLi	MeI	−Me	81	19
3	−Me	(dicyclohexylamide, Li)	MeI	−Me	94	6
4	−Me	(tetramethylpiperidide, Li)	MeI	−Me	99	1
5	−OMe	*n*-BuLi	SO₂(OMe)₂	−Me	5	95
6	−OPh	PhLi	CO₂	−CO₂H	2	98
7	−Br	LDA	TMS-Cl	−TMS	Not reported	85 (isolated)

of electrons on the 3-substituent may help direct the deprotonation by chelating the lithium as well. The selectivity is demonstrated by the 95% regioselectivity of 2-deprotonation for 3-methoxythiophene by *n*-BuLi (T3.2–5),[11] which is essentially the same even with a sterically larger 3-substituent and base (T3.2–6).[12] The 2-regioselectivity has also been seen with 3-bromothiophene, which gives a high isolated yield of the 2-TMS substituted product (T3.2–7),[13] although in this case if any 5-deprotonation was observed it was not reported (also note that LDA is necessary for this deprotonation since an alkyl lithium would likely result in lithium–halogen exchange).

Some examples of lithiation with 3-hexyl substituted thiophenes are shown in Scheme 3.5. Deprotonation of 3-hexylthiophene **23** with LDA at −78 °C to give the lithiated intermediate **24** and subsequent reaction with tributyltin chloride gives the 5-stannane regioisomer **25** in high isolated yield.[14] Note that the stannylation part of the reaction is run (and presumably substantially finished) at −78 °C, which helps prevent isomerization of the 5-lithio intermediate to the 2-lithio intermediate. However, when reactions or substrate solubility may not allow lithiation and/or subsequent reaction at lower temperatures,

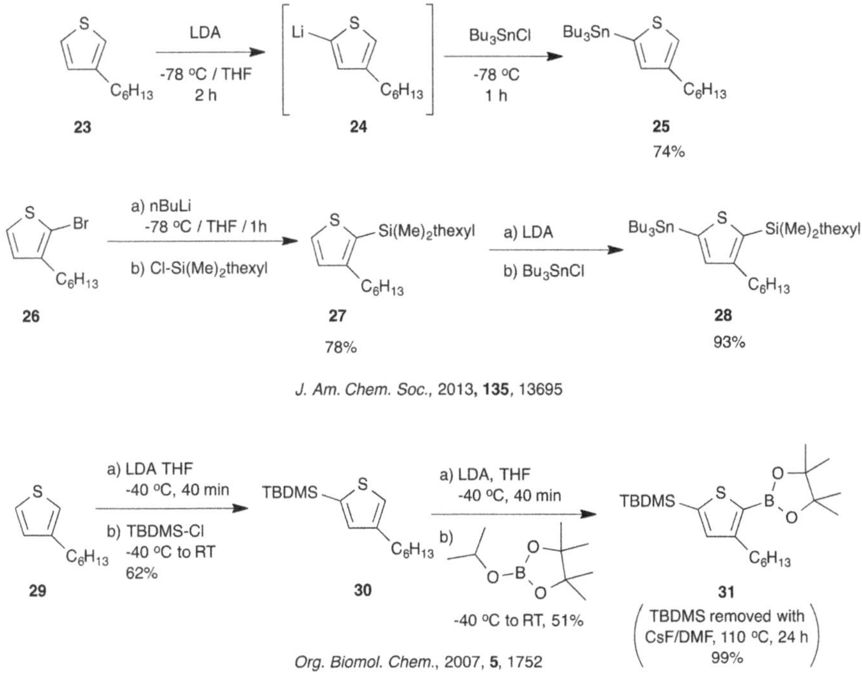

J. Am. Chem. Soc., 2013, **135**, 13695

Org. Biomol. Chem., 2007, **5**, 1752

Scheme 3.5 Synthetic examples of thiophene lithiation by deprotonation.[14,15]

then 2-isomerization can be prevented by installing a silyl protecting group at the 2-position. This is demonstrated by the regiospecific lithium–halogen exchange of 2-bromo-3-hexylthiophene 26 and subsequent silylation to give intermediate 27 which can then be deprotonated with LDA and stannylated to give only the 5-regiostannane 28 in high yield.[14] Silyl protecting groups can also be used to achieve 2-deprotonation as shown by deprotonation and silylation of 29 with LDA at −40 °C followed by TBDMS-Cl to give 30 in moderate yield. Subsequent deprotonation and borylation gave the 2-substituted boronate ester 31. In this case, the TBDMS group was reportedly much more stable to hydrolysis by acetic acid than the TMS group.[15] Note that, along with protecting particular reactive sites, the silyl groups can be used to increase the solubility of synthetic intermediates or final products, especially with the thexyldimethylsilyl group shown in 28.

3.2.4   Thiophene Halogen Dance

Another reaction of lithiated thiophenes that can be used to set regioselectivity is the so-called "halogen dance" reaction, which is known to occur in a variety of lithiated aryl and heteroaryl halides.[16] The halogen dance is effectively a lithium/halogen "self-exchange" of a β-lithiated thiophene having an α-halogen. Scheme 3.7 shows a conceptual mechanism for the halogen dance where the lithiated thiophene exchanges once (S3.7a) to give a doubly lithiated species and a dibromothiophene (S3.7b) that then undergoes another exchange to give the final thiophene lithiated in the α position (the more stable anion) and halogenated in the β-position (S3.7c). The actual mechanism may involve simultaneous lithium/halogen exchanges or may involve electron transfer depending on the substrate. Nevertheless, the halogen dance has been used to synthesize a variety of thiophenes with differing substitution, with examples illustrated in Scheme 3.6. When treated with LDA, 2,3-dibromothiophene 32 deprotonates at the 5-position and rearranges to the 2-lithiated species via lithium–halogen exchange to give 33,[17] which has the most stable α anion next to the electron withdrawing 3-Br. The lithiated species 33 can also be generated by deprotonation of the β-proton of 2,5-dibromothiophene 34 in LDA as evidenced by subsequent reaction with TMS-Cl to give 3,5-dibromo-2-trimethylsilylthiophene 35.[18] Deprotonation of the β protons has also been used in the reaction of 5-TMS substituted 36 to give the lithiated intermediate 37[19] and from the reaction of 2-bromo-5-n-butylthiophene 38 with LDA and subsequent halogen dance and iodination to give 3-bromo-2-iodo thiophene 39.[20]

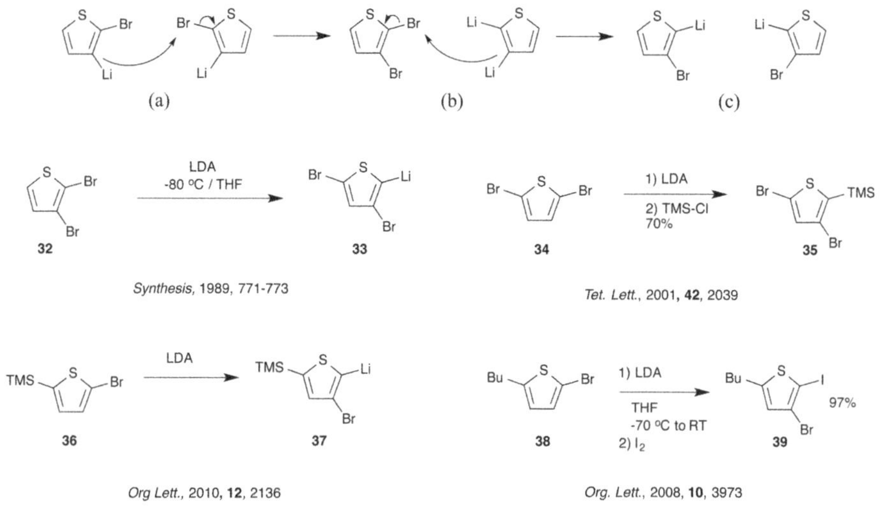

Scheme structures labeled (a), (b), (c); compounds 32, 33, 34, 35, 36, 37, 38, 39 with reaction conditions.

Synthesis, 1989, 771-773

Tet. Lett., 2001, **42**, 2039

Org Lett., 2010, **12**, 2136

Org. Lett., 2008, **10**, 3973

Scheme 3.6 Synthetic examples of the thiophene halogen dance rearrangement.[17-20]

A number of different halogen dance reactions will be shown in synthetic sequences throughout this book.

3.2.5 Donor Synthesis *via* Lithiations

A number of the common donors listed in Figure 3.1 have been synthesized using various aspects of lithiation chemistry in key steps. Scheme 3.7 illustrates the synthesis of dithienothiophene (DTT)[21] and dithienosilole (DTS)[22] donors utilizing thiophene lithiations in key steps. Lithiation of 2,3-dibromothiophene results in the 2-lithiated thiophene that is coupled with CuCl to give bithiophene **40** in high yield. Double lithiation of **40** with *n*-BuLi to give dual β-lithiated species **41** followed by reaction with benzenesulfonic acid thioanhydride gave DTT **42** in 70% yield. The 3,3′-dibromo thiophene **40** was also used in the synthesis of silylated intermediate **43** that can be doubly lithiated and reacted with a di-*n*-octyldichlorosilane to give DTS **44** in high yield. Note that the TMS protecting groups of **43** would allow higher temperature reactions by blocking isomerization of the dual β-lithiated intermediate to α-isomers. Intermediate **43** has also been prepared in a one-pot reaction from commercially available 2-bromo-thiophene.[19] The first step involves deprotonation of the 5-position with LDA and silylation to give 2-bromo-5-(trimethylsilyl)thiophene, and then the halogen dance illustrated in Scheme 3.6 (**36** → **37**) is induced by deprotonation with LDA at the 3-position, which is likely

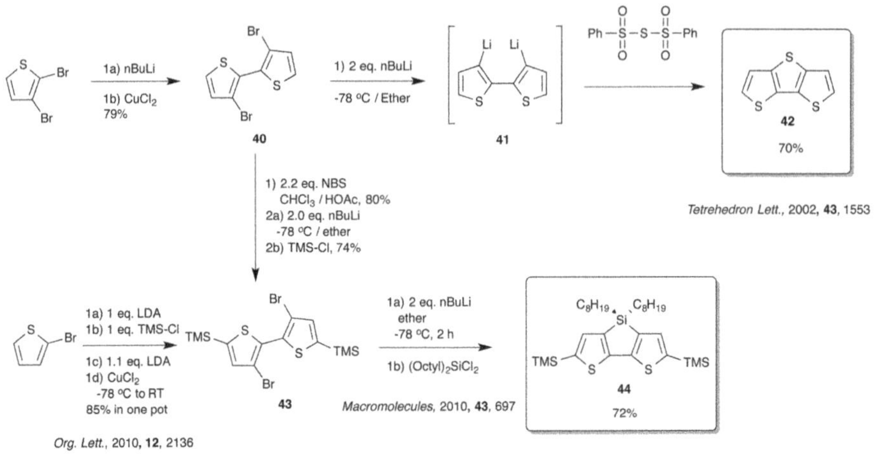

Scheme 3.7 Synthesis of donors dithienothiophene (DTT) and dithienosilole (DTS) utilizing lithiation in key synthetic steps.[19,21,22]

most acidic due to the σ-withdrawing bromine as well as being less sterically encumbered, followed by the halogen dance to give the 2-lithio-3-bromo intermediate that is coupled with $CuCl_2$ to give **43** in 85% yield.

Cyclopenta[2,1-*b*:3,4-*b'*]dithiophene (CPDT, **51**, Scheme 3.8) has been used as a donor in a number of polymers and the synthetic routes developed involve lithiations in key steps as shown in Scheme 3.8. Recently, key intermediate **49** was synthesized from intermediate **43** described in Scheme 3.7 by double lithiation and trapping with an acylating agent to give the bis-TMS protected **45** in high yield[23] that could then be deprotected with triflic acid (protodesilylated) to give **49** in 95% yield.[24] Previously, intermediate **49** had been synthesized by lithiating 3-bromothiophene in ether and then reaction with thiophene-2-carboxaldehyde to give the dithienylmethanol **46** as a critical intermediate.[22] Lithiation of **46** is a special circumstance that (1) requires three equivalents of base, the first of which deprotonates the alcohol and (2) is an example of *ortho* direction of lithiation of thiophene[25] by the alcohol oxygen to give the doubly lithiated **47** at the 3-positions instead of the 5-positions, which are normally more acidic. Such direction is known to occur generally in deprotonation of aromatics.[26] *Ortho*-directed 3-deprotonation over 5-deprotonation may be highly dependent on competition for coordination of the lithium atom, and more highly coordinating bases and conditions may shift deprotonation away from the 3-position.[27] In the case of **46**, *ortho*-directed deprotonation occurred with *n*-BuLi in ether to give

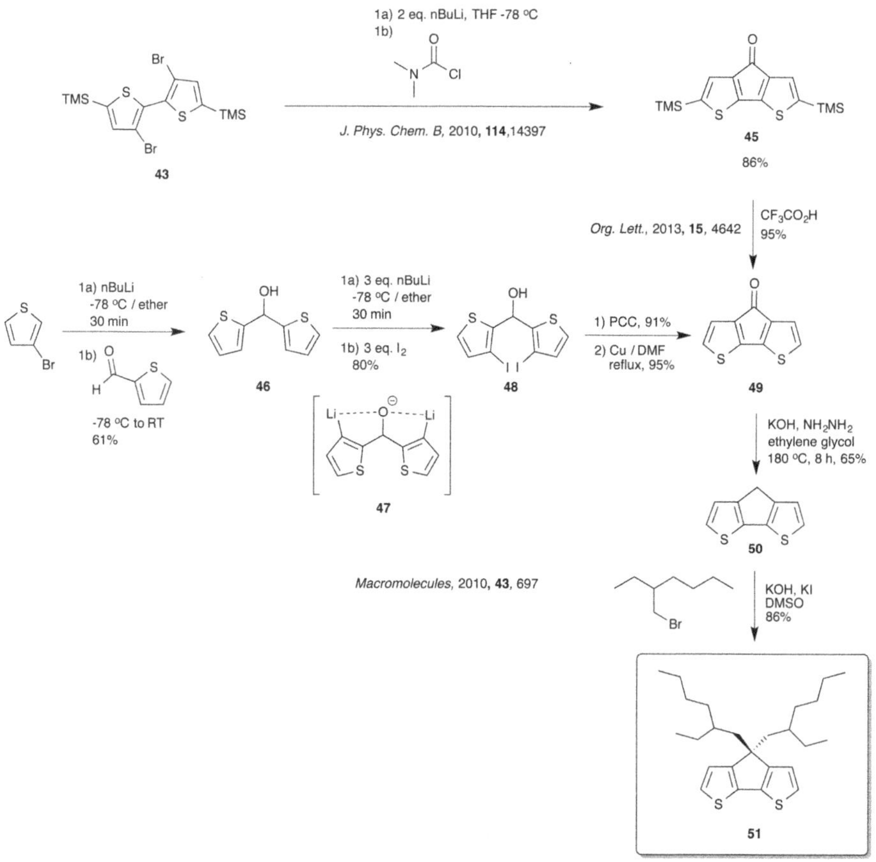

Scheme 3.8 Synthesis of the donor cyclopenta[2,1-*b*:3,4-*b'*]dithiophene (CPDT) utilizing lithiation in key synthetic steps.[22-24]

intermediate **48** in high yield after iodination with iodine. Oxidation of the alcohol and Cu coupling of the 3-positions then gave intermediate **49** in high yield. The CPDT acceptor can then be synthesized from **49** by Wolff–Kishner reduction of the ketone to **50** and subsequent alkylation of the cyclopentadienyl-like ring to give **51**. Many different alkyl groups may be used in the alkylation of **50**, and the final alkylation generally occurs in high yield.

3.2.6 Thiophene Lithiation Summary

Table 3.3 attempts to summarize some of the important points of lithiating thiophenes. Note that in almost every case, both direct deprotonation or lithium–halogen exchange chemistry are viable alternatives. In general, deprotonation is favored when the alkyl halide by-product of lithium–halogen exchange might react with any of the reaction mixture constituents. In addition, any undesired isomerizations between α positions or between β and α positions may require low temperatures throughout the reaction or blocking the undesired α position with a silyl protecting group. When there is a 3-alkyl group, 5-regioselectivity can be highly favored by using a bulky base (LiTMP) or 2-regioselectivity can be favored by lithium–halogen exchange. Both the 2 and 5 positions can be blocked with a silyl protecting group by various means to enforce regioselectivity.

Table 3.3   Summary of lithiation of various thiophenes.

Entry	R¹	Substrate	Base	Notes[a]
1		Thiophene 2-Bromothiophene	R–Li, R₂N–Li R–Li	(1)
2		3-Bromothiophene	R–Li	(1), (2) and (6)
3		3-Alkylthiophene	R₂N–Li	(2) and (3)
4		2-Bromo-3-alkylthiophene	R–Li	(1) and (2)
5		3-Alkoxythiophene	R–Li, R₂N–Li	
6		3-Bromothiophene	R₂N–Li	(4) and (6)
7		2,5-Dibromothiophene 2,3-Dibromothiophene	R₂N–Li	(4), (5) and (6)
8		2-DG-thiophene DG = carboxy, amide, –CH₂–OH, *etc.*	R–Li	(6)

[a](1) Alkyl bromide by-product may react with lithiated species or other constituents in the reaction; (2) isomerization to another α-isomer possible at elevated temperatures; (3) bulky bases increase regioselectivity; (4) lithium–halogen exchange would occur with R–Li; (5) halogen dance; (6) lower coordination media/reagents preferable to decrease isomerization.

3.2.7 Thiophene Grignard Formation

Although lithiated thiophenes are used in a variety of synthetic schemes for organic electronic and photonic materials, Grignards are also used in special cases, which includes the Kumada coupling of a Grignard and an aryl-bromide[28] and when 3-regiostability at higher temperatures is required. Examples of Grignard metallation of thiophene are shown in Scheme 3.9. Formation of the Grignard of 5-butyl-2-bromothiophene can occur from reaction of the lithiated species with magnesium(II) bromide diethyletherate to give the corresponding magnesiothiophene **52** that can then be coupled with another equivalent of 5-butyl-2-bromothiophene to give dithiophene **53** in high yield.[29] This formation of Grignards from lithiated species is general across a variety of aromatic and heteroaromatic substrates, but one complication is that $MgBr_2$ is

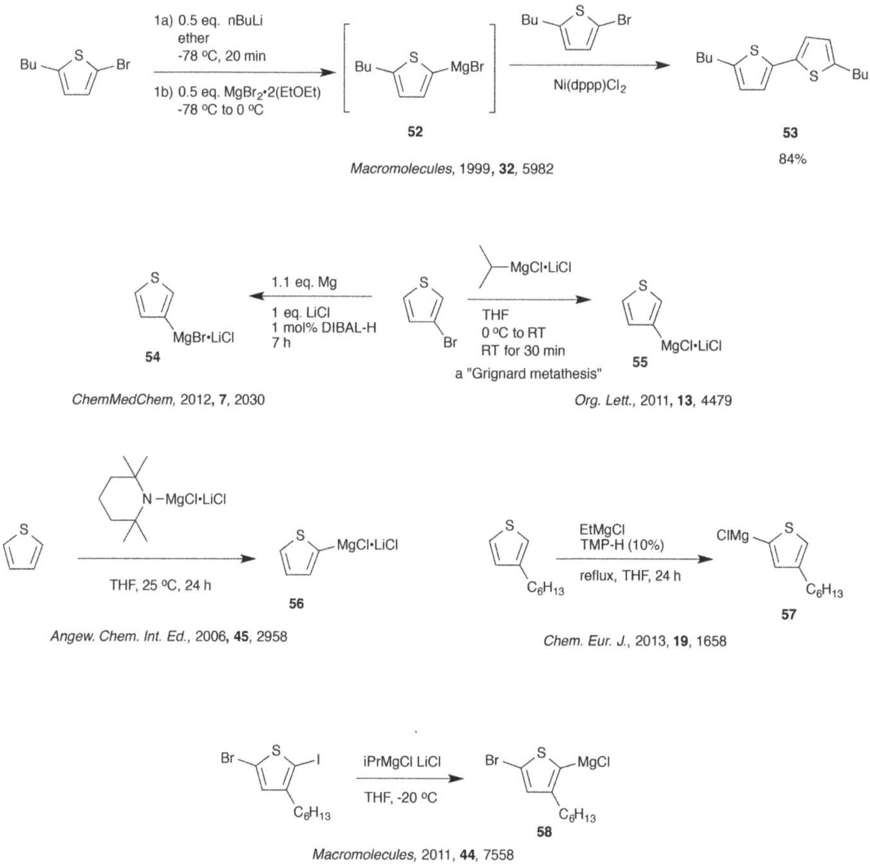

Macromolecules, 1999, **32**, 5982

ChemMedChem, 2012, **7**, 2030

Org. Lett., 2011, **13**, 4479

Angew. Chem. Int. Ed., 2006, **45**, 2958

Chem. Eur. J., 2013, **19**, 1658

Macromolecules, 2011, **44**, 7558

Scheme 3.9 Synthetic examples of formation of thiophene Grignard reagents.[29,30,32–35]

highly hygroscopic and may absorb water that can protonate any lithiated species and reduce the yield of Grignard. Other methods to form the Grignard include the classic reaction of a halogenated thiophene with magnesium metal to give that can be done with highly active Rieke magnesium[4] or by activating the magnesium with a trace amount of iodine, 1,2-dibromoethane, or a reducing agent. Such activation is demonstrated by the formation of the 3-Grignard/lithium chloride complex **54** from 3-bromothiophene using 1% DIBAL-H to activate the magnesium.[30] Lithium chloride is known to increase the rate of Grignard formation and impart special reactivity to the resulting Grignard.[31] A Grignard also can be prepared from a so-called "super" or "turbo" Grignard[2] such as in the case of formation of **55** from isopropylmagnesiumchloride-LiCl.[32] The formation of **55** in Scheme 3.9 is an example of a "Grignard metathesis," which in general is the formation of an aryl Grignard by halogen–Mg exchange with another Grignard. Magnesium dialkylamides can also be used to directly metallate thiophene through deprotonation in an analogous manner to lithium dialkylamides such as with MgTMP to give the 2-Grignard **56**,[33] which also gives 5-regioselectivity when a 3 alkyl group is present such as with the formation of **57**, influenced by the size of the base.[34] In addition, super Grignards may preferentially exchange with iodine relative to bromine such as with 5-bromo-3-hexyl-2-iodothiophene to give the corresponding 2-Grignard **58**.[35] This type of high regioselectivity is particularly important in synthesizing highly regioregular poly 3-hexylthiophene by the Grignard metathesis polymerization.[36,37]

3.2.8   Thiophene Alkylation

The alkylation of thiophenes is important as one of the main methods to install solubilizing and/or functional groups on π-conjugated materials. To allow π-conjugation through the thiophene, alkyl groups must be installed in the β positions. Some examples of β alkylation of thiophenes are shown in Scheme 3.10. By far, the most widely used method to install alkyl groups in the 3- and 4-positions is Kumada coupling of alkyl Grignards with β-bromothiophenes. One example of this is formation of the Grignard of 1-bromo-2-methylbutane with dibromoethane as a magnesium activating agent and then Kumada coupling with 3-bromothiophene to give the corresponding 3-alkyl thiophene **59** in moderate yield.[38] The Kumada coupling as well as other organometallic coupling reactions will be discussed in Chapter 5. Another example is the double coupling of octylmagnesium bromide with 3,3′-dibromo-2,2′-bithiophene to give the corresponding

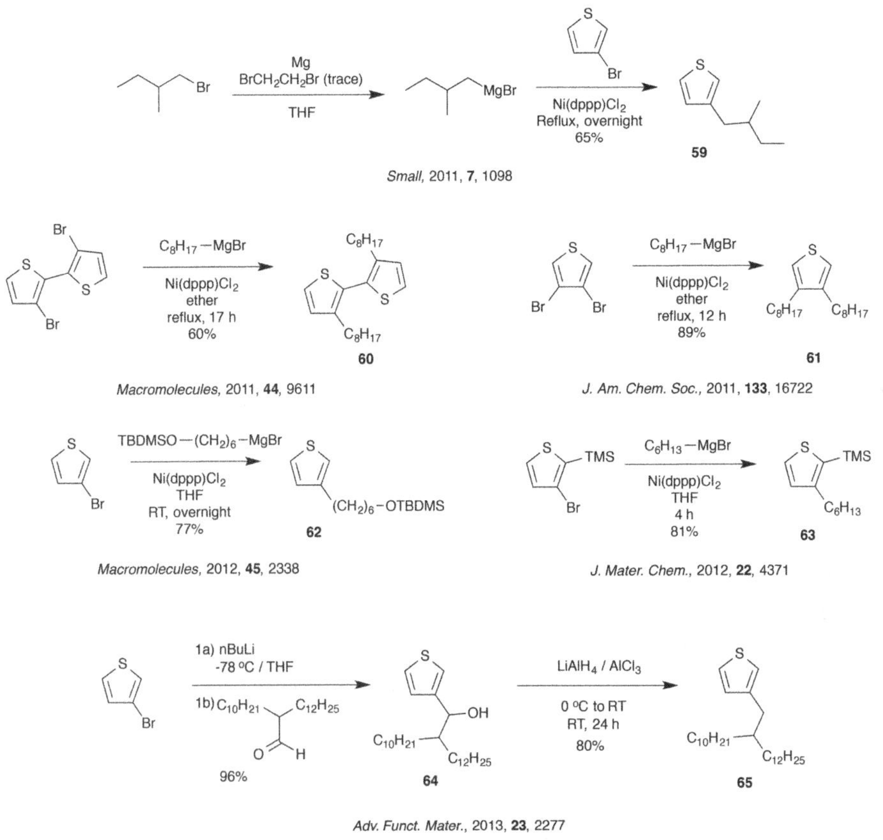

Scheme 3.10   Alkylation of thiophene at the β position.[13,38–42]

alkylated thiophene **60** in moderate yield.[39] Both the 3- and 4-positions of thiophene can be alkylated in this manner as reported for the dioctylthiophene **61**.[40] This type of coupling has also been used with a TBDMS-protected Grignard to give the corresponding thiophene **62** in high yield.[41] The reaction also has been reported to occur in the presence of a 2-TMS-thiophene such as with the synthesis of **63**.[13] Finally, another method to alkylate thiophenes has been reported that utilizes an addition of 3-lithiothiophene to an aldehyde to give an intermediate alcohol **64** that is then reduced with LiAlH$_4$ and AlCl$_3$ to give the corresponding alkyl thiophene **65** in high yield.[42] This method may be an attractive alternative to Kumada coupling when alkyl halides of certain branched alkyl groups are difficult to obtain synthetically and where the aldehydes are more readily available.

Alkylation of thiophene at the α positions is less common than at the β positions since within a π-conjugated material an alkyl group in the α position would block conjugation. However, thiophenes alkylated at an α position have been used for end groups and occasionally as side groups in polymers and small molecules. Scheme 3.11 shows a few examples of synthetic methods to install alkyl groups at an α position in thiophene. The first is perhaps the most straightforward as an alkylation of a lithiated thiophene formed by deprotonation to give the corresponding alkyl thiophene **66** in high yield.[43] Using a strong base to affect deprotonation in this case is highly preferred over using a halogenated thiophene with lithium–halogen exchange since the alkyl halide by-product of the exchange may itself alkylate the thiophene and therefore must be eliminated with an extra equivalent of alkyl lithium (Schemes 3.1 and 3.2). Another complication is that thiophene is often doubly deprotonated under the reaction conditions, which would result in lower yields of the monoalkylated product. The amount of doubly lithiated species may be lower when a large excess of thiophene is used and the alkyl lithium is slowly added to a solution of the thiophene so that the concentration of thiophene is high relative to the mono-lithiated species. The excess thiophene can be removed during purification by distillation; however, one problem that is more difficult to overcome is that separation of the mono- and dialkylated species may be difficult by column chromatography if the molecular weight of the monoalkylated species is too high for distillation without decomposition; however, these alkylation issues can be avoided using the other methods illustrated in Scheme 3.11. One such method to install the α alkyl groups in thiophene is by reduction of the corresponding ketone. Such an example is the Wolff–Kishner

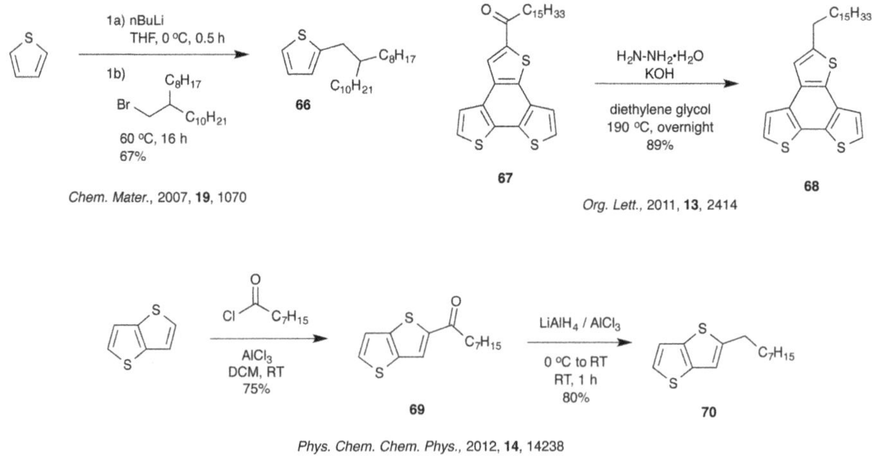

Chem. Mater., 2007, **19**, 1070

Org. Lett., 2011, **13**, 2414

Phys. Chem. Chem. Phys., 2012, **14**, 14238

Scheme 3.11   Alkylation of thiophene at the α position by various methods.[43–45] reduction of the ketone of **67** to the corresponding alkyl group in **68** in high yield.[44] The ketone can be installed *via* a Friedel–Crafts acylation of the thiophene derivative such as for thienothiophene to the corresponding ketone **69** and then an alternative (to Wolff–Kishner) reduction of the ketone to an alkyl group in **70** using LiAlH$_4$ and AlCl$_3$ in high yield.[45]

3.2.9 Thiophene Ring Synthesis

The synthesis of most thiophene derivatives used in organic electronics and photonics begins with commercially available thiophene derivatives. However, certain thiophene derivatives may be more readily available through synthesis of the thiophene ring itself. Scheme 3.12 shows some examples of thiophene forming reactions that have been used in the context of organic electronics and photonics, but other methods are also known.[1] One common method is the Paal–Knorr synthesis of thiophenes from 1,4-diketones using a thiation reagent. Some examples are the reaction of 1,4-di(thienyl)ketone **71** with Lawesson's reagent to give terthiophene **72** in high yield.[46]

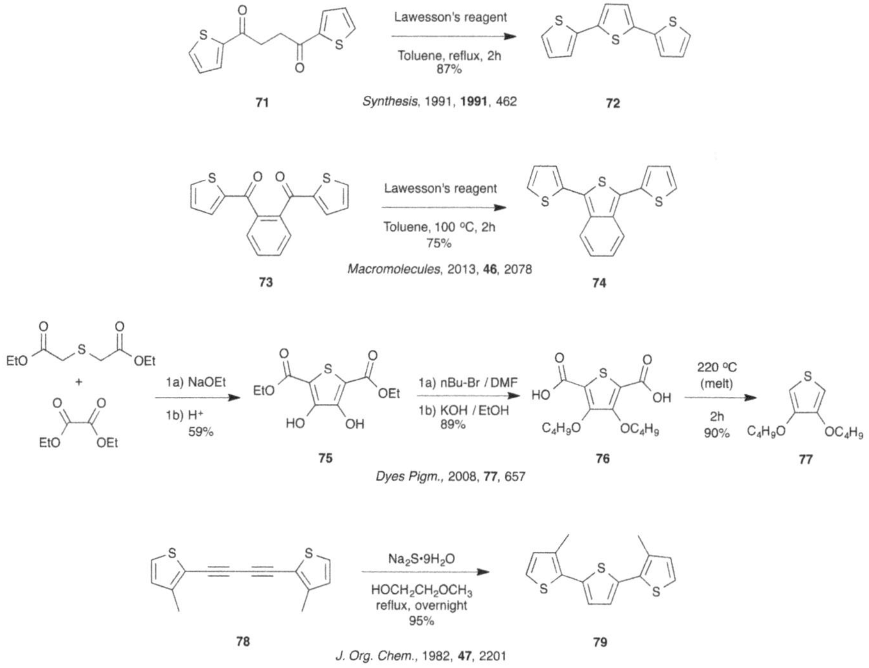

Scheme 3.12 Synthesis of the thiophene ring by various methods.[46–48,51]

Lawesson's reagent is a convenient commercially available thiation reagent, but has a strong odor and should be used strictly in a fume hood. Lawesson's reagent has also been used to synthesize isobenzothiophenes from *o*-phenyldiketones such as from **73** to give the corresponding isobenzothiophene **74** in high yield.[47] Another method is the Hinsberg synthesis of thiophenes from a thiodiacetate diester and a 1,2-dicarbonyl compound. An interesting example of the Hinsberg synthesis is the formation of the 3,4-dihydroxy-2,5-dicarboxylic ester **75** from diethyl thiodiacetate and diethyloxalate in moderate yield. Diester **75** can then by alkylated at both hydroxy groups and hydrolyzed to give diacid **76** that can be decarboxylated to give 3,4-dialkoxythiophene **77** in high yield.[48] Several 3,4-dialkoxythiophenes have been synthesized in this manner, including EDOT[49] and functionalized EDOT.[50] Another synthesis of the thiophene ring involves the reaction of sodium sulfide with a butadiyne such as the reaction of **78** with Na_2S in refluxing methoxyethanol to give the corresponding terthiophene **79** in high yield.[51] Although the methods illustrated in Scheme 3.12 are less common in the synthesis of thiophenes in the context of organic electronics and photonics, there may be situations, such as with the synthesis of 3,4-dialkoxythiophenes, where starting from commercially available thiophene rings is less practical.

3.2.10   Thieno Heteroannulation

Heteroannulation of a thieno group from an aryl or heteroaryl has been used in the synthesis of organic electronic and photonic materials to influence π-conjugation, change donor strength, and/or increase surface area of π-conjugated moieties in polymers and small molecules. Scheme 3.13 illustrates some examples of heteroannulation to form the thieno group. One of the more widely used methods is reaction of ethyl thioglycolate with an aryl bromide such as 3-bromothiophene-2-carboxaldehyde under basic conditions to give the corresponding thieno heteroannulated ester **80** that can be hydrolyzed with lithium hydroxide and decarboxylated to give the thienothiophene (TT) **81** in high yield.[47] The ethyl thioglycolate heteroannulation also works with ketones to heteroannulate *and* install a solubilizing alkyl group in the same sequence. An example of this is the deprotonation of thienothiophene **82** with LDA (at the more acidic α position proximal to the electronegative bromine) followed by reaction with a long chain aldehyde to give the corresponding alcohol that was oxidized to ketone **83** in high yield overall. Reaction of ketone **83** with ethyl thioglycolate, hydrolysis, and

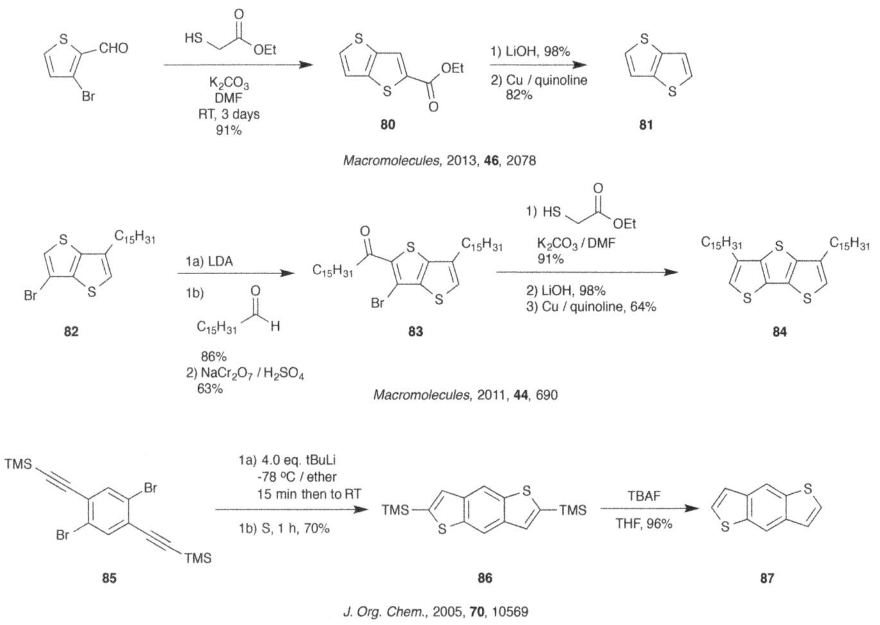

Scheme 3.13 Synthetic examples thieno group synthesis *via* heteroannulation.[47,52,53]

decarboxylation in separate steps gave dialkyl dithienothiophene **84** in moderate yield overall.[52] The installation of the solubilizing alkyl group as part of the thieno group annulation is particularly useful. Another method for thieno heteroannulation is the reaction of *ortho*-lithiated arylalkynes with elemental sulfur. An example of this is a double lithiation of **85** and subsequent reaction with S to give the corresponding thienobenzothiophene **86** in high yield, which can be easily desilylated with tetrabutylammonium fluoride (TBAF) to give **87**.[53] Again, note the use of four equivalents of *t*-BuLi in the lithiation, which is likely used so that the two equivalents of *t*-BuBr generated as a by-product of the lithiation can be eliminated with the remaining two equivalents of *t*-BuLi to avoid reaction of *t*-BuBr with intermediates (Schemes 3.1 and 3.2).

3.3 ARYL AMINE CHEMISTRY

Like thiophenes, aryl amines also have been used widely as donors in π-conjugated materials (Figure 3.1). Tri- and diarylamines have been used in organic electronics as hole transporting polymers and small molecules and *N*,*N*-dialkyl arylamines have been used extensively as donors in nonlinear optical materials. Given the widespread use of arylamines, many synthetic methodologies have been applied to materials synthesis of mono-, di-, and triarylamines. The chemistries typically used include alkylation, nucleophilic substitution, and, in particular, organometallic coupling. As illustrated throughout this chapter and other chapters, there a number of different substrates and reactions that can be used and combined in the synthesis of arylamine donors.

3.3.1   Alkylation of Arylamines

The N−H bonds of monoarylamines (*i.e.*, anilines) have a $pK_a \sim 30$ and diarylamines have a $pK_a \sim 25$ and both can be alkylated in a manner that is similar to aliphatic alcohols. Numerous examples of alkylation of both mono- and diarylamines exist in the literature. In general, the reactions can be run without solvent or many different solvents can be used, but dipolar aprotic solvents may be preferred over polar aprotic solvents to increase the rate of S_N2 reaction. Likewise, many different bases can be used including amines and bases that fully deprotonate the arylamine including NaH, but alkali metal bases such as K_2CO_3, Na_2CO_3, NaOH, *etc.*, are often employed as well. Typically, the reaction yield is high, but the reaction can often take days since delocalization of the amine lone pairs into the aryl group/s tends to decrease nucleophilicity of aryl amine compared to that of alkyl amines. In part because of the reduced nucleophilicity, over alkylation to the quaternary ammonium salt, which is generally seen with alkyl amines, is typically not seen with alkylation of aryl amines, although in certain circumstances quaternization may occur such as when the reaction is run in the presence of excess alkylating agent without solvent (a "neat" reaction where the solvent is effectively the alkylating agent).

Scheme 3.14 illustrates some examples of alkylation of anilines with various conditions and alkylating agents. The first few conditions show the alkylation of *N*-ethylaniline. In the first case, an alkyliodide is effectively used as the solvent of the reaction to give the corresponding dialkylaniline **88** in high yield after reflux for 6 h.[54] The second example illustrates that the reaction can occur in protic solvents in the presence of a hydroxyl group on the alkylhalide to give the corresponding dialkylaniline **89** in high yield.[55] Note that the KI is added to the reaction as a catalyst to displace the chloride of the alkylating agent to produce the more highly reactive alkyliodide *in situ* (this activation of the chloride to give the more reactive iodide is known as the Finkelstein reaction). Generally, protic solvents can be used since anilines tend to be more nucleophilic than alcohols. The final example in Scheme 3.14 illustrates the use of a dipolar aprotic solvent with

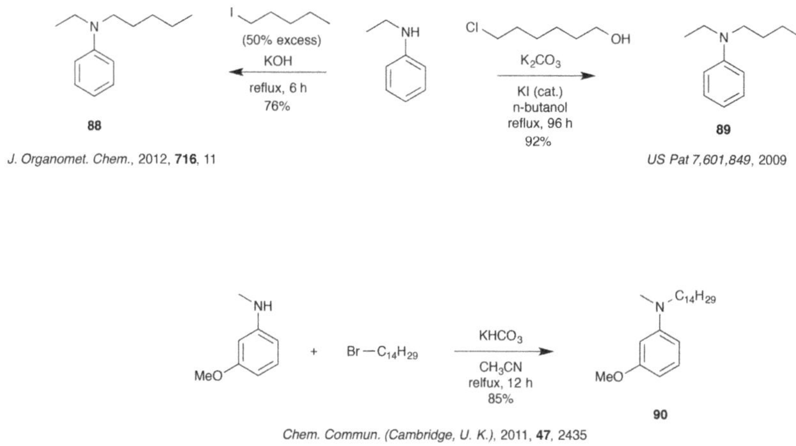

Scheme 3.14 Synthetic examples of *N*-alkylation of anilines.[54–56]

an *N*-methyl-3-methoxyaniline and an alkylbromide to give the dial-kylaniline **90** in high yield.[56]

Examples of the alkylation of diarylamines are shown in Scheme 3.15. The first example is a tandem alkylation with a dihaloalkane that first uses NaH as the base for diphenylamine in the chelating solvent 1,2-dimethoxyethane (DME) with an excess of 1,6-dibromohexane to give bromoalkylated diarylamine **91** in moderate yield. Subsequent reaction of **91** with phenoxazine and KOH in DMSO gives the corresponding alkylated phenoxazine-diphenylamine product **92** in high yield.[57] The lower yield of **91** compared to **92** is likely due in part to the general reduced nucleophilicity of acyclic diphenylamines compared to cyclic analogues since in cyclic analogues the lone pair on the nitrogen generally has less steric hindrance, although in the specific case of phenoxazine, the electron-donating oxygen also increases electron density in the aromatic rings and thereby increases nucleophilicity of the nitrogen as well. Because of the relative ease of alkylating cyclic dia-rylamines, alkylation of carbazoles and carbazole derivatives has been used widely in the synthesis of π-conjugated materials. Two examples of this are the alkylation of carbazole with 2-ethylhexyl bromide in DMF to give alkylated carbazole **93** in high yield.[58] The reaction also works with non-halogenated alkylating agents such alkyl tosylates. An example of tosylate alkylation is illustrated by the double alkylation of the carbazole derivative **94** with a branched chain secondary tosylate to give the doubly alkylated **95** in high yield.[59] Tosylates may be preferable in certain cases since tosylate is a better leaving group than bromide or iodide, tosylates generally are less susceptible to electron

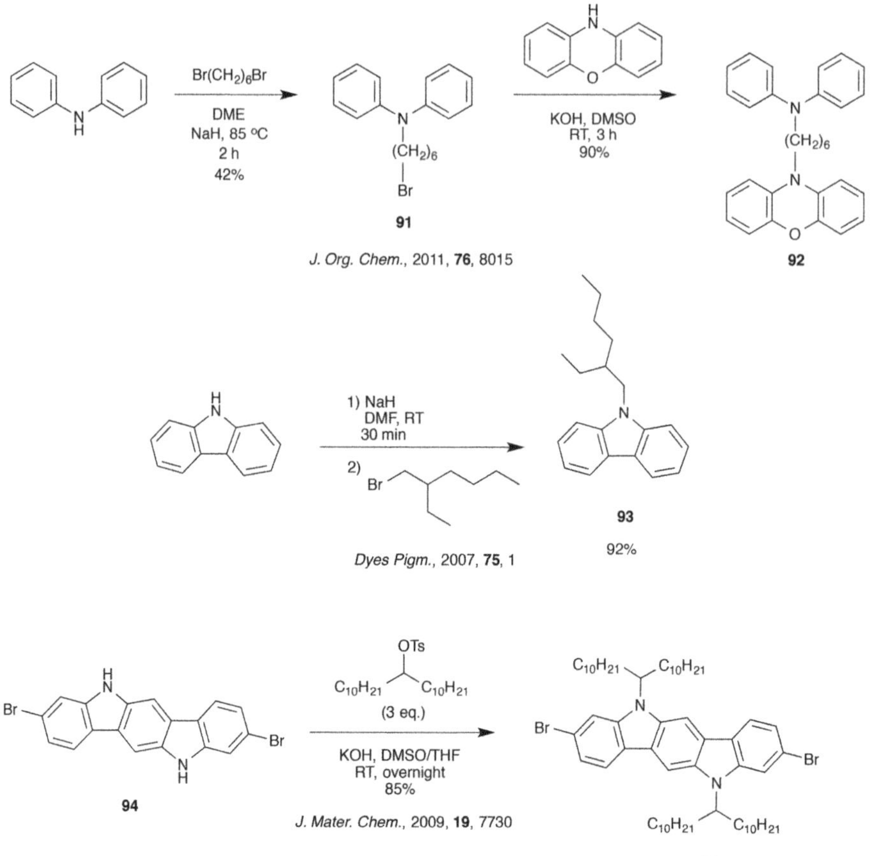

Scheme 3.15   Synthetic examples of *N*-alkylation of diarylamines.[57-59]

transfer reactions that may cause significant side-products, and specific tosylates may be more synthetically accessible or convenient than the analogous halide.

Other methods of alkylating amines include amidation and reduction and reductive amination. Although these methods have been used less far less frequently in the synthesis of π-conjugated materials, their use may be preferred when certain substrates are not stable to alkylating conditions or a suitable alkylating agent is not conveniently accessible. Scheme 3.16 illustrates two examples of amine alkylation involving a reductive step. The first involves reacting an arylamine with an acid chloride (or an equivalent reagent such as an *N*-hydroxysuccinimide (NHS) ester) to give amide **96** in high yield followed by BH$_3$ reduction of the amide to the amine **97** in high yield.[60] Another example is reductive amination with an aldehyde to give an intermediate imine that can then be reduced with NaBH(OAc)$_3$ (or an equivalent

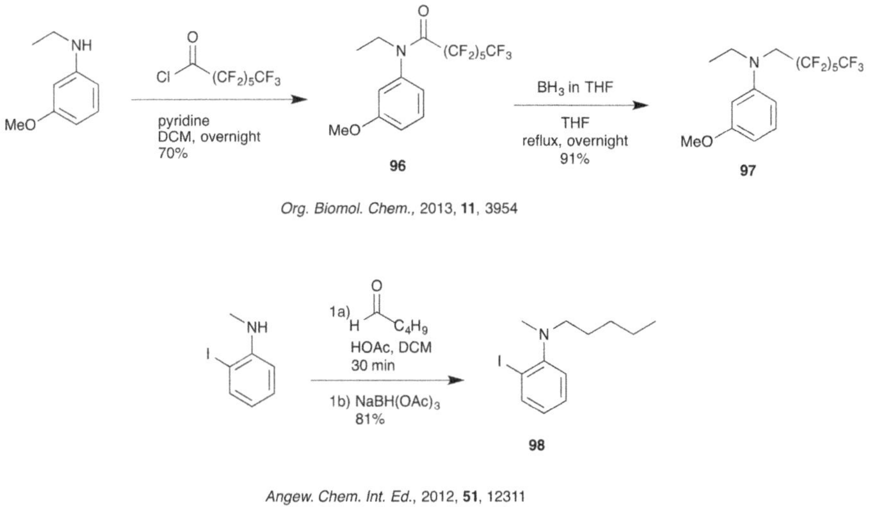

Org. Biomol. Chem., 2013, **11**, 3954

Angew. Chem. Int. Ed., 2012, **51**, 12311

Scheme 3.16 Synthesis of *N*-alkyl anilines using amide and imine reductions.[60,61]

such as NaBH$_3$CN) such as the iminization of 2-iodo-*N*-methylaniline with pentanal and subsequent reduction in one pot to give the alkylated amine **98** in high yield.[61]

3.3.2 Nucleophilic Aromatic Substitution

Nucleophilic substitution of a leaving group on an aromatic ring having an electron-withdrawing group (EWG) may also be used to synthesize donors. A number of leaving groups (X) may be used, and as illustrated in Figure 3.3, the mechanism generally proceeds by nucleophilic attack on the aromatic ring of the carbon bearing the leaving group (F3.16a) to give a Meisenheimer complex (F3.16b) and stabilization of the electrons on the EWG, and then finally elimination of X to give the final product (F3.16c). The rate of the reaction generally is higher when the EWG is a stronger π-withdrawing group and when X is a stronger σ-withdrawing group. An important consequence of this is that fluoride is typically a much better leaving group than other halides, and generally the leaving group reaction rate trend is F ≫ Cl > Br > I, which parallels electronegativity. A number of other leaving groups may also be used, including the diazonium group, but then other mechanisms may be operative.[62]

Some examples of nucleophilic aromatic substitution in the preparation of donors in π-conjugated materials are illustrated in Scheme 3.17. Typically, these reactions have been used with electron deficient aromatic halides (usually with fluoride leaving groups) where the EWG is an aldehyde that can be used in later steps in the π-conjugated material synthesis. One example is the reaction of an N-methyl aminoethanol with 4-fluorobenzaldehyde in DMSO to give the N-hydroxyethyl functionalized **99** in high yield.[63] The substrate 4-fluorobenzaldehyde also reacts with carbazole in DMF to give the corresponding N-phenylcarbazole **100** in moderate yield.[64] The reaction may also work on thiophene substrates such as with the reaction of 5-bromothiophene carboxaldehyde with piperidine to give aminothiophene **101** in high yield.[65] In this case, the reaction is accelerated by acid-catalyzed formation of an iminium salt **102** between the piperidine and the aldehyde to give the stronger π-withdrawing iminium group *in situ*. Note that certain substrates may be sensitive to the

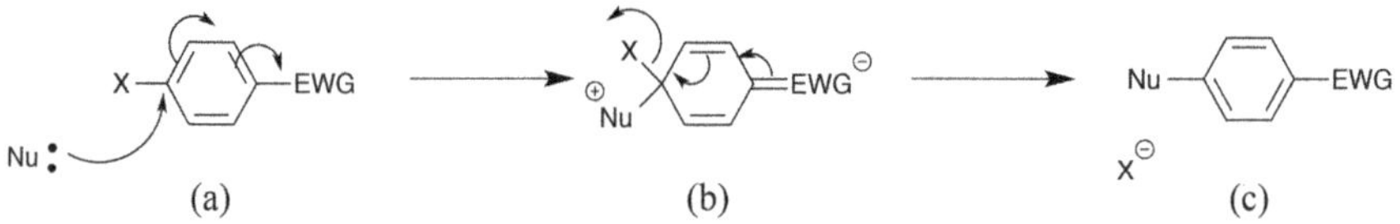

Figure 3.3 General mechanism of nucleophilic aromatic substitution showing (a) attack; (b) rearomatization; and (c) the final product.

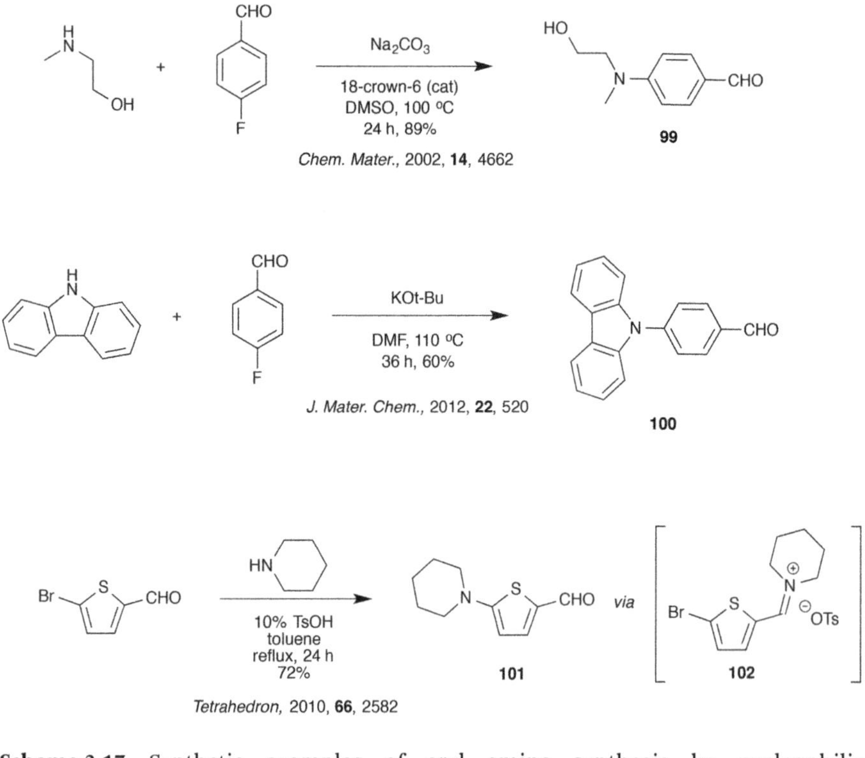

Chem. Mater., 2002, **14**, 4662

J. Mater. Chem., 2012, **22**, 520

Tetrahedron, 2010, **66**, 2582

Scheme 3.17 Synthetic examples of aryl amine synthesis by nucleophilic substitution.[63–65]

high temperatures, dipolar aprotic solvents, and long reaction times that often are needed to complete the reaction.

3.3.3 Ullmann *N*-Aryl Coupling

The Ullmann *N*-aryl coupling is another method for synthesizing aryl amine donors. The reaction typically couples an aryl halide to an aryl amine using a cuprous salt catalyst with the support of a chelating agent, which is usually a diamine or a polycyclic heteroaromatic compound with two chelating sp^2 nitrogens. The reaction has been known for over 100 years and has been used to successfully couple a variety of aryl amines and nitrogen heterocycles to mostly aryl iodides.[66] Some examples of the *N*-aryl coupling applicable to amine donor synthesis are illustrated in Scheme 3.18. Reaction of carbazole with an aryl

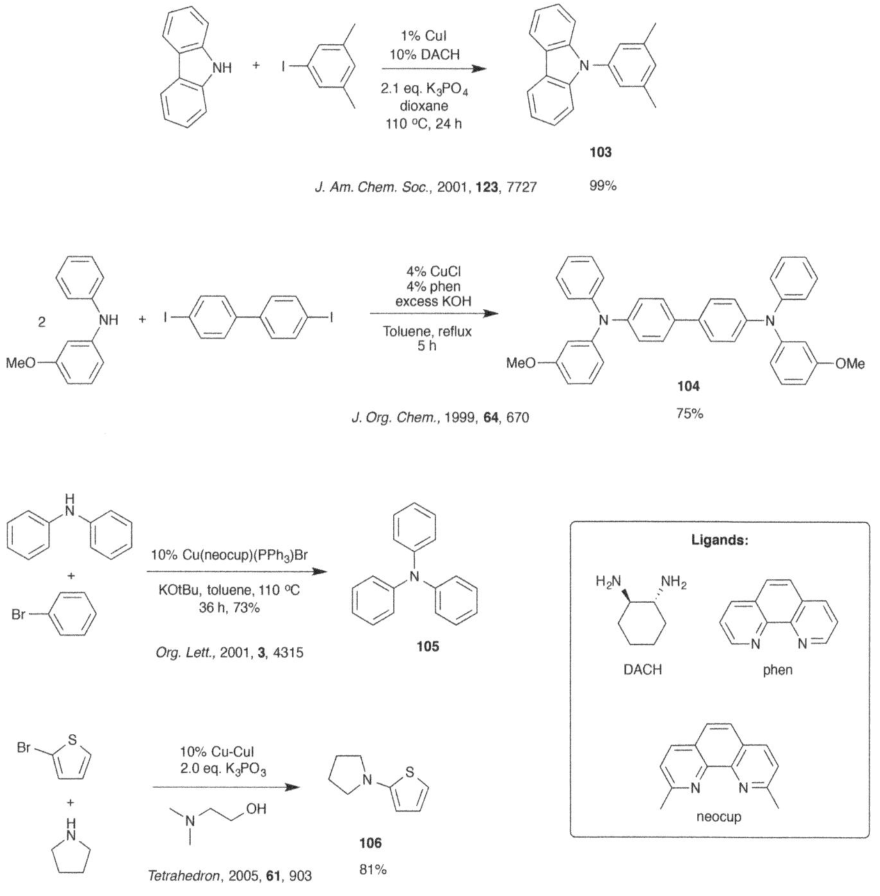

Scheme 3.18 Synthetic examples of aryl amine synthesis by the *N*-Ullmann coupling.[67–70]

halide using cuprous iodide and the 1,2-diaminocyclohexane (DACH) ligand gave the *N*-phenyl carbazole **103** in high yield.[67] In another example, double coupling of a diphenyl amine to 4,4′-diiodobenzene using cuprous chloride and the phen ligand gave the corresponding TPD derivative **104** in high yield.[68] Although less common, aryl bromides also react to give the corresponding aryl amine, such as **105**, in moderate to high yield, although the reaction times are typically much longer than for the corresponding aryl iodides.[69] A final example uses a metallic copper–cuprous iodide mixture in a coordinating solvent to couple the cyclic amine pyrrolidine and 2-bromothiophene to give **106** in high yield.[70] Some advantages of the *N*-aryl Ullmann couplings are the relatively inexpensive copper catalyst and reasonable reaction times and temperatures for aryl iodides. However, the lower reactivity of aryl bromides may require long reaction times at higher temperatures that are incompatible with certain substrates. In addition, the inability to efficiently couple with a wide range of aryl chlorides is also a limitation.

3.3.4 Buchwald–Hartwig *N*-Aryl Coupling

An alternative to the Ullmann coupling is the Buchwald–Hartwig Pd-catalyzed coupling reaction.[71–73] The reaction is highly useful in many fields of chemistry and has many variants that have been developed over the last twenty years. Figure 3.4 summarizes the main features of the Buchwald–Hartwig coupling that are typically relevant to

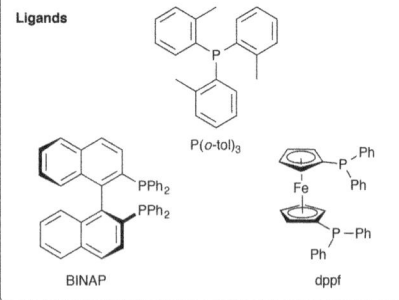

Figure 3.4   Overview of Buchwald–Hartwig coupling applied to the synthesis of organic electronics and photonic materials.

the synthesis of donors for π-conjugated materials. The fundamental reaction involves use of an aryl substrate having a halogen or pseudo-halogen (*e.g.*, OTf, OTs) for Pd insertion and an amine to act as the nucleophile that displaces the halogen or pseudohalogen on the Pd followed by reductive elimination to the aryl amine product (organo-metallic mechanisms will be discussed in Chapter 5). The scope of the reaction in terms of the substrate and the amine is very broad, and unlike most cases of the Ullmann coupling, both aryl chlorides and bromides are suitable. Typically, the Pd source is relatively air-stable such as $PdCl_2$, $Pd(OAc)_2$, or $Pd_2(dba)_3$. A variety of ligands have been used, especially in the case of specialized substrates, but often a steri-cally hindered triaryl phosphine or bidentate bis(triarylphosphine) is employed such as $P(o\text{-tol})_3$, dppf, or BINAP. The "workhorse" base used in most cases has been NaO*t*-Bu, but other bases have been employed in more specialized conditions. In general, the reaction may be very sensitive to the choice of bases, and a reaction that is high yielding using NaO*t*-Bu may not work at all when using KO*t*-Bu. The reaction is often run in toluene, but DME and dioxane are known to work as well. At this point, catalyst/ligand systems and conditions have been developed for a very wide variety of substrates and amines, and check-ing the literature for substrates and amines similar to those in a par-ticular synthetic plan to get conditions that may be more applicable to specific substrates and amines is advisable.

Some examples of the Buchwald–Hartwig coupling are shown in Scheme 3.19 for the synthesis of various aryl amines. The first exam-ple uses "classic" conditions to give the *N*-arylmorpholine **107** in high yield from the corresponding arylbromide.[74] These conditions can be used for a variety of substrates and amines. One interesting feature of the reaction is that the coupling of diarylamines is significantly slower than for anilines, which allows one pot assembly of triarylamines from anilines by sequential addition of aryl halides. An example of this is the reaction of 4-butylaniline and bromobenzene to give the interme-diate diarylamine **108**, which need not be isolated, and then the triaryl amine **109** by addition of an excess of 1,4-dibromobenzene and an additional equivalent of NaO*t*-Bu.[75] The overall yield of the reaction is high and a number of different aryl halides can be added sequentially, although complete formation of the intermediate diarylamine should be confirmed by some method, such as TLC, GC/MS, or HPLC, before addition of the second aryl halide to avoid unwanted side-products. The reaction can also be used on multi-halogenated aryls such as the stepwise coupling of 9,9-dihexyl-2,7-diiodofluorene with 4-(hex-yloxy)aniline to give the intermediate dicoupled product (not shown)

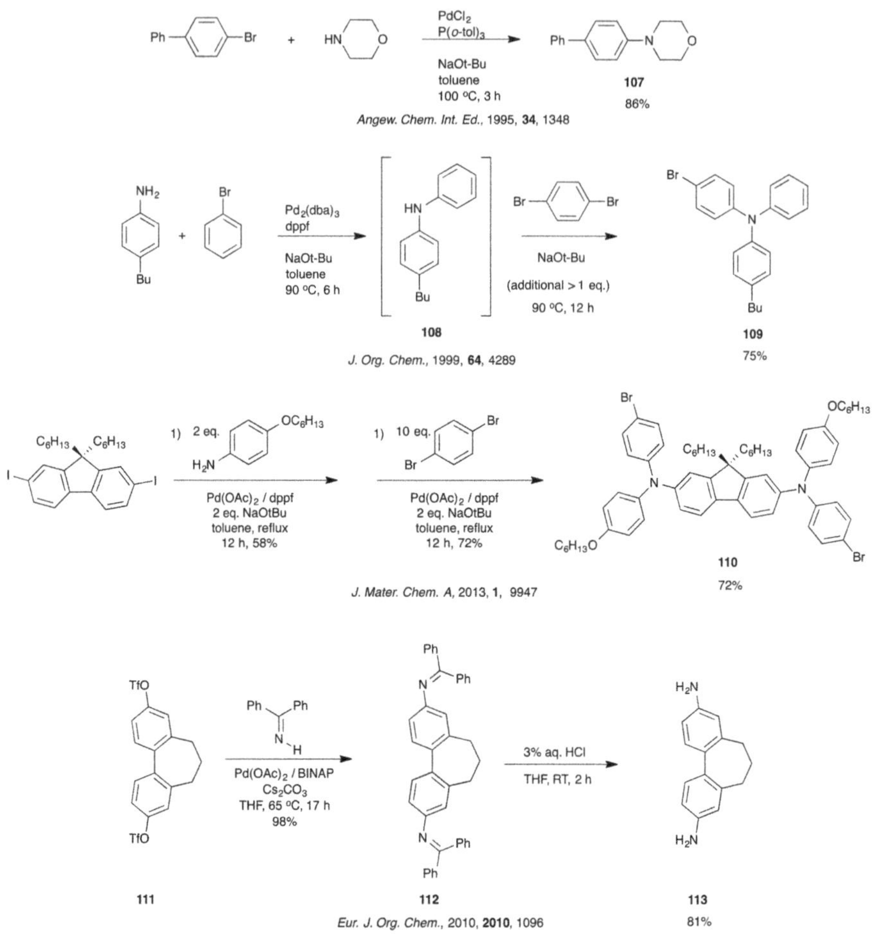

Scheme 3.19   Synthetic examples of the Buchwald–Hartwig coupling in the synthesis of aryl amines.[74–77]

in 58% isolated yield and then subsequent reaction with an excess
1,4-dibromobenzene to give the 2,7-diamino fluorene derivative **110**
in high yield.[76] A similar example is the reaction benzophenone imine
with bistriflate **111** to give the intermediate diimine **112** in high yield
followed by hydrolysis to give the diarylamine **113** in high yield.[77] Note
that each of the products in the bottom three examples (**109, 110,** and
112) are synthesized from either arylbromides (**109, 110**) or amines
(**113**) that can in turn be used in with Buchwald–Hartwig coupling to
give still other products.

The Buchwald–Hartwig coupling also works with halogenated heteroarenes, although the scope may be narrower than with halogenated
arenes depending on the substrate; however, among halogenated heteroaryl substrates, the coupling of amines with bromothiophenes

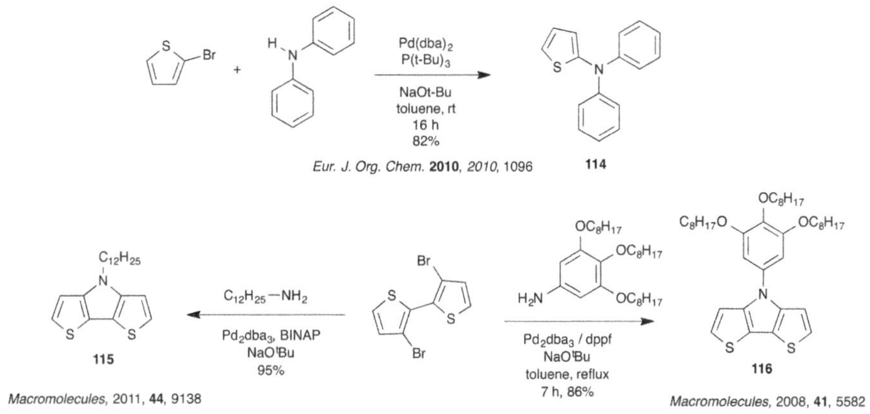

Scheme 3.20 Synthetic examples of the Buchwald–Hartwig coupling of amines and thiophenes.[77–79]

is particularly relevant to the synthesis of donors for π-conjugated materials. Scheme 3.20 illustrates a few examples of bromothiophene coupling to amines. Typical Buchwald–Hartwig conditions with the trialkylphosphine P(*t*-Bu)$_3$ has been used to couple 2-bromothiophene to diphenylamine to give the aminothiophene **114** in high yield.[77] An interesting variation of this reaction occurs when 3,3′-dibromo-2,2′-bithiophene is used as the substrate in reaction with an aliphatic amine to give the *N*-alkyl DTP donor **115** in high yield[78] while reaction with an aniline derivative gives an *N*-phenyl DTP donor **116** in high yield as well.[79]

3.4 CONCLUSION

This chapter has covered the synthesis of donor units having thiophene and arylamine moieties that are typical in π-conjugated materials. In particular, the synthesis of various thiophene regioisomers, modification of the thiophene with solubilizing alkyl groups, synthesis of thiophene rings themselves, and extension of the thiophene unit through heteroannulation was discussed. A number of different synthetic routes to mono-, di-, and triaryl amines that can enable different substituents on both the nitrogen and/or aryl groups were also discussed. However, the incorporation of these units into larger small molecule and polymeric materials typically requires that the thiophenes and aryl amines be functionalized further to extend the conjugation of the material. The chemistry used to functionalize the donors and extend conjugation in intermediates and the final materials is the subject of the next chapters.

REFERENCES

1. J. Joule and K. Mills, in *Heterocyclic Chemistry*, John Wiley and Sons, Chichester, UK, 5th edn, 2010, ch. 17, pp. 325–345.
2. A. Krasovskiy and P. Knochel, *Angew. Chem., Int. Ed.*, 2004, **43**, 3333.
3. R. D. Rieke, *Science*, 1989, **246**, 1260.
4. R. D. Rieke, S.-H. Kim and X. Wu, *J. Org. Chem.*, 1997, **62**, 6921.
5. X. Wu, T.-A. Chen, L. Zhu and R. D. Rieke, *Tetrahedron Lett.*, 1994, **35**, 3673.
6. M. J. Marsella, Z.-Q. Wang, R. J. Reid and K. Yoon, *Org. Lett.*, 2001, **3**, 885.
7. D. L. Ladd, P. B. Harrsch and L. I. Kruse, *J. Org. Chem.*, 1988, **53**, 417.
8. Y. He, W. Wu, G. Zhao, Y. Liu and Y. Li, *Macromolecules*, 2008, **41**, 9760.
9. H. Pang, P. J. Skabara, S. Gordeyev, J. J. W. McDouall, S. J. Coles and M. B. Hursthouse, *Chem. Mater.*, 2007, **19**, 301.
10. K. Smith and M. L. Barratt, *J. Org. Chem.*, 2007, **72**, 1031.
11. L. L. Miller and Y. Yu, *J. Org. Chem.*, 1995, **60**, 6813.
12. J. W. H. Watthey and M. Desai, *J. Org. Chem.*, 1982, **47**, 1755.
13. Y. A. Getmanenko, J. M. Hales, M. Balu, J. Fu, E. Zojer, O. Kwon, J. Mendez, S. Thayumanavan, G. Walker, Q. Zhang, S. D. Bunge, J.-L. Brédas, D. J. Hagan, E. W. Van Stryland, S. Barlow and S. R. Marder, *J. Mater. Chem.*, 2012, **22**, 2.
14. F. P. Koch, P. Smith and M. Heeney, *J. Am. Chem. Soc.*, 2013, **135**, 13695.
15. D. J. Turner, R. Anemian, P. R. Mackie, D. C. Cupertino, S. G. Yeates, M. L. Turner and A. C. Spivey, *Org. Biomol. Chem.*, 2007, **5**, 1752.
16. M. Schnurch, M. Spina, A. F. Khan, M. D. Mihovilovic and P. Stanetty, *Chem. Soc. Rev.*, 2007, **36**, 1046.
17. F. Sauter, H. Fröhlich and W. Kalt, *Synthesis*, 1989, **1989**, 771.
18. E. Lukevics, P. Arsenyan, S. Belyakov, J. Popelis and O. Pudova, *Tetrahedron Lett.*, 2001, **42**, 2039.
19. Y. A. Getmanenko, P. Tongwa, T. V. Timofeeva and S. R. Marder, *Org. Lett.*, 2010, **12**, 2136.
20. M. J. O'Connor and M. M. Haley, *Org. Lett.*, 2008, **10**, 3973.
21. F. Allared, J. Hellberg and T. Remonen, *Tetrahedron Lett.*, 2002, **43**, 1553.
22. C.-H. Chen, C.-H. Hsieh, M. Dubosc, Y.-J. Cheng and C.-S. Hsu, *Macromolecules*, 2010, **43**, 697.
23. S. Barlow, S. A. Odom, K. Lancaster, Y. A. Getmanenko, R. Mason, V. Coropceanu, J. L. Bredas and S. R. Marder, *J. Phys. Chem. B*, 2010, **114**, 14397.

24. G. Pozzi, S. Orlandi, M. Cavazzini, D. Minudri, L. Macor, L. Otero and F. Fungo, *Org. Lett.*, 2013, **15**, 4642.

25. A. Carpenter and D. Chadwick, *J. Chem. Soc., Perkin Trans. 1*, 1985, 173.

26. V. Snieckus, *Chem. Rev.*, 1990, **90**, 879.

27. A. J. Carpenter and D. J. Chadwick, *Tetrahedron Lett.*, 1985, **26**, 1777.

28. M. Kumada, *Pure Appl. Chem.*, 1980, **52**, 669.

29. M. J. Marsella and R. J. Reid, *Macromolecules*, 1999, **32**, 5982.

30. Z. Su, A. A. Yeagley, R. Su, L. Peng and C. Melander, *ChemMedChem*, 2012, **7**, 2030.

31. E. Hevia and R. E. Mulvey, *Angew. Chem., Int. Ed.*, 2011, **50**, 6448.

32. T. Leermann, F. R. Leroux and F. Colobert, *Org. Lett.*, 2011, **13**, 4479.

33. A. Krasovskiy, V. Krasovskaya and P. Knochel, *Angew. Chem., Int. Ed.*, 2006, **45**, 2958.

34. S. Tanaka, D. Tanaka, G. Tatsuta, K. Murakami, S. Tamba, A. Sugie and A. Mori, *Chem.–Eur. J.*, 2013, **19**, 1658.

35. S. Wu, L. Huang, H. Tian, Y. Geng and F. Wang, *Macromolecules*, 2011, **44**, 7558.

36. M. C. Stefan, M. P. Bhatt, P. Sista and H. D. Magurudeniya, *Polym. Chem.*, 2012, **3**, 1693.

37. M. C. Stefan, A. E. Javier, I. Osaka and R. D. McCullough, *Macromolecules*, 2009, **42**, 30.

38. H. C. Chen, I. C. Wu, J. H. Hung, F. J. Chen, I. W. Chen, Y. K. Peng, C. S. Lin, C. H. Chen, Y. J. Sheng, H. K. Tsao and P. T. Chou, *Small*, 2011, **7**, 1098.

39. A. Bhuwalka, J. F. Mike, M. He, J. J. Intemann, T. Nelson, M. D. Ewan, R. A. Roggers, Z. Lin and M. Jeffries-El, *Macromolecules*, 2011, **44**, 9611.

40. S. Ko, E. Verploegen, S. Hong, R. Mondal, E. T. Hoke, M. F. Toney, M. D. McGehee and Z. Bao, *J. Am. Chem. Soc.*, 2011, **133**, 16722.

41. C.-Y. Nam, Y. Qin, Y. S. Park, H. Hlaing, X. Lu, B. M. Ocko, C. T. Black and R. B. Grubbs, *Macromolecules*, 2012, **45**, 2338.

42. Q. Wu, S. Ren, M. Wang, X. Qiao, H. Li, X. Gao, X. Yang and D. Zhu, *Adv. Funct. Mater.*, 2013, **23**, 2277.

43. S. Ellinger, U. Ziener, U. Thewalt, K. Landfester and M. Möller, *Chem. Mater.*, 2007, **19**, 1070.

44. C. B. Nielsen, J. M. Fraser, B. C. Schroeder, J. Du, A. J. White, W. Zhang and I. McCulloch, *Org. Lett.*, 2011, **13**, 2414.

45. J. Huang, H. Jia, L. Li, Z. Lu, W. Zhang, W. He, B. Jiang, A. Tang, Z. Tan, C. Zhan, Y. Li and J. Yao, *Phys. Chem. Chem. Phys.*, 2012, **14**, 14238.

46. A. Merz and F. Ellinger, *Synthesis*, 1991, **1991**, 462.
47. K. Kawabata, M. Takeguchi and H. Goto, *Macromolecules*, 2013, **46**, 2078.
48. K. Guo, J. Hao, T. Zhang, F. Zu, J. Zhai, L. Qiu, Z. Zhen, X. Liu and Y. Shen, *Dyes Pigm.*, 2008, **77**, 657.
49. H. Zhang, C. Qian and X.-Z. Chen, *J. Chem. Res.*, 2011, **35**, 339.
50. M. Yamada, N. Ohnishi, M. Watanabe and Y. Hino, *Chem. Commun. (cambridge, U. K.)*, 2009, 7203.
51. J. P. Beny, S. N. Dhawan, J. Kagan and S. Sundlass, *J. Org. Chem.*, 1982, **47**, 2201.
52. J. Li, H.-S. Tan, Z.-K. Chen, W.-P. Goh, H.-K. Wong, K.-H. Ong, W. Liu, C. M. Li and B. S. Ong, *Macromolecules*, 2011, **44**, 690.
53. K. Takimiya, Y. Konda, H. Ebata, N. Niihara and T. Otsubo, *J. Org. Chem.*, 2005, **70**, 10569.
54. P. J. W. Elder, J. C. Landry, A. F. Cozzolino, A. E. A. Chapman and I. Vargas-Baca, *J. Organomet. Chem.*, 2012, **716**, 11.
55. A. K.-Y. Jen, L. R. Dalton and H. Ma, *US Pat.*, 7,601,849, 2009.
56. Y. Hai, J. J. Chen, P. Zhao, H. Lv, Y. Yu, P. Xu and J. L. Zhang, *Chem. Commun. (cambridge, U. K.)*, 2011, **47**, 2435.
57. Y. Hong, J. Y. Liao, D. Cao, X. Zang, D. B. Kuang, L. Wang, H. Meier and C. Y. Su, *J. Org. Chem.*, 2011, **76**, 8015.
58. A. K. Mishra, J. Jacob and K. Müllen, *Dyes Pigm.*, 2007, **75**, 1.
59. E. Zhou, S. Yamakawa, Y. Zhang, K. Tajima, C. Yang and K. Hashimoto, *J. Mater. Chem.*, 2009, **19**, 7730.
60. D. K. Kolmel, B. Rudat, D. M. Braun, C. Bednarek, U. Schepers and S. Brase, *Org. Biomol. Chem.*, 2013, **11**, 3954.
61. B. Yao, Q. Wang and J. Zhu, *Angew. Chem., Int. Ed.*, 2012, **51**, 12311.
62. M. Smith and J. March, in *March's Advanced Organic Chemistry: Reactions, Mechanisms, and Structure*, John Wiley & Sons, New York, 5th edn, 2001, ch. 13, pp. 850–893.
63. M. He, T. M. Leslie and J. A. Sinicropi, *Chem. Mater.*, 2002, **14**, 4662.
64. B. Hu, F. Zhuge, X. Zhu, S. Peng, X. Chen, L. Pan, Q. Yan and R.-W. Li, *J. Mater. Chem.*, 2012, **22**, 520.
65. J. Sutharsan, D. Lichlyter, N. E. Wright, M. Dakanali, M. A. Haidekker and E. A. Theodorakis, *Tetrahedron*, 2010, **66**, 2582.
66. K. Kunz, U. Scholz and D. Ganzer, *Synlett*, 2003, 2428.
67. A. Klapars, J. C. Antilla, X. Huang and S. L. Buchwald, *J. Am. Chem. Soc.*, 2001, **123**, 7727.
68. H. B. Goodbrand and N.-X. Hu, *J. Org. Chem.*, 1999, **64**, 670.
69. R. K. Gujadhur, C. G. Bates and D. Venkataraman, *Org. Lett.*, 2001, **3**, 4315.

70. Z. Lu and R. J. Twieg, *Tetrahedron*, 2005, **61**, 903.
71. A. Muci and S. Buchwald, *Top. Curr. Chem.*, 2002, **219**, 131.
72. J. F. Hartwig, *Acc. Chem. Res.*, 2008, **41**, 1534.
73. J. P. Wolfe, S. Wagaw, J.-F. Marcoux and S. L. Buchwald, *Acc. Chem. Res.*, 1998, **31**, 805.
74. A. S. Guram, R. A. Rennels and S. L. Buchwald, *Angew. Chem., Int. Ed.*, 1995, **34**, 1348.
75. S. Thayumanavan, J. Mendez and S. R. Marder, *J. Org. Chem.*, 1999, **64**, 4289.
76. K. D. Seo, B. S. You, I. T. Choi, M. J. Ju, M. You, H. S. Kang and H. K. Kim, *J. Mater. Chem. A*, 2013, **1**, 9947.
77. J. Rotzler, D. Vonlanthen, A. Barsella, A. Boeglin, A. Fort and M. Mayor, *Eur. J. Org. Chem.*, 2010, **2010**, 1096.
78. P. Imin, M. Imit and A. Adronov, *Macromolecules*, 2011, **44**, 9138.
79. W. Vanormelingen, K. Van den Bergh, T. Verbiest and G. Koeckelberghs, *Macromolecules*, 2008, **41**, 5582.

CHAPTER 4

Creating Conjugation: Functionalization and Alkenes

4.1 INTRODUCTION

Creating conjugated linkages between aromatic or heteroaromatic intermediates is a key step in the synthesis of π-conjugated materials. Typically, this is performed by creating an alkene linkage, and alkyne linkage, or by direct aryl–aryl coupling between aromatic or heteroaromatic intermediates. Choosing a specific linkage will often depend on what specific set of properties are desired, or whether one type of linkage is more synthetically feasible while not adversely affecting other material properties. One significant issue is what effect linkages have on the overall π-conjugation of the system. Many of the factors that effect π-conjugation between subunits were discussed in various aspects in Section 2.3. In summary, aryl–alkene–aryl linkages promote more efficient conjugation between the aryl groups than aryl–alkyne–aryl linkages, or direct aryl–aryl linkages. In some cases, direct aryl–aryl linkages may offer increased π-conjugation relative to aryl–alkyne–aryl linkages if there is no significant hindrance to planarization between the aryl groups around the aryl–aryl bond due to steric torsional twisting (some steric twisting is often the case). Fortunately, many synthetic methods exist to cover most combinations of alkene-, alkyne-, and direct aryl–aryl linked π-conjugated materials.

Synthetic Methods in Organic Electronic and Photonic Materials: A Practical Guide
By Timothy C. Parker and Seth R. Marder
© Timothy C. Parker and Seth R. Marder, 2015
Published by the Royal Society of Chemistry, www.rsc.org

Due to the breadth of the synthetic topics, the methods for creating alkene, alkyne, and direct aryl–aryl π-conjugated linkages have been divided into two chapters: Chapter 4 focuses on functionalization of heteroaromatic systems and the creation of alkene linkages *via* phosphorous reagents; and Chapter 5 focuses on organometallic coupling reactions to create direct aryl–aryl, alkene, and alkyne linkages.

4.2 FUNCTIONALIZING INTERMEDIATES

Extending conjugation of a π-system through aryl–aryl bonds, alkenes, or alkynes requires functionalized starting materials or synthetic intermediates. Since many materials contain aryl moieties with structural variations that are more complex than those in commercially available starting materials, functionalizing intermediates is an important part of the synthesis of π-conjugated materials. Although any one of the many available functional groups may be used in the synthesis of any specific material, the most common functional groups used in the synthesis of π-conjugated are aryl halides and aryl aldehydes. As is shown in Chapters 4 and 5, reactions that form aryl–aryl linkages, alkenes, and alkynes typically involve at least one reaction partner that has either an aryl halide (especially in cases of organometallic coupling) or an aldehyde (especially in the cases of Wittig-type couplings to form alkenes). Additionally, both aryl halides and aldehydes are useful intermediates for the installation of acceptors in π-conjugated materials. Indeed, as illustrated throughout this book, aryl halides and aldehydes play a central role in the synthesis of donors, extending π-conjugation, and introducing acceptors into materials. As such, functionalizing aromatic substrates to aryl halides and aldehydes is of primary importance.

4.2.1 Halogenation

Halogenation of aromatic substrates has been widely used in all areas of organic chemistry and there are many conditions that have been developed. In Larock's *Comprehensive Organic Transformations* (1999), there are 36 methods listed for aromatic chlorination, 57 methods for aromatic bromination, and 65 methods for aromatic iodination.[1] The transformations and related interconversions that are often used in the synthesis of π-conjugated materials are summarized schematically in Scheme 4.1. Electrophilic aromatic substitution (S4.1a) where an arene (Ar—H) or a TMS–arene (TMS—Ar) is reacted under conditions where a halogen can react as an electrophile ("X$^+$", S4.1a) is a method that has been used widely for the halogenation of π-conjugated intermediates. Since the electrophilic reaction is most facile for materials that have a low ionization energy (IE), it is most easily done on electron donors such as thiophenes or arylamines, although "forcing" conditions (*e.g.*, heat, acid, catalyst addition and amount, longer time) can sometimes be employed to halogenate electron acceptor intermediates. Another option employed frequently is nucleophilic attack of a metallated arene on a halogen bonded to a leaving group (S4.1b). Examples of this type of halogenation have been illustrated previously in Schemes 3.3 and 3.4 in the sections on metallation of thiophene. Note that Ar−H potentially can be used for electrophilic halogenation or halogenation of a metallated species formed by direct deprotonation (S4.1c) such as with thiophenes. One caveat with halogenation of a metallated species is that with acceptors having a high electron affinity (EA), the metallated carbon nucleophile may attack other metallated acceptors in the reaction and decompose to unwanted products and/or polymers through electron transfer reactions or nucleophilic addition reactions. Thus, the functionalization of stronger electron acceptors and intermediates is often practically limited to electrophilic halogenation.

Another important aspect of halogenation in π-conjugated materials synthesis is the interconversion of halogenated species. This is often seen when one aryl halide is metallated and then halogenated with a second halide to give another aryl halide (the S4.1d to S4.1b cycle in Scheme 4.1). This is most often encountered in the exchange of bromine for either iodine or, much less often, chlorine or fluorine. Another useful interconversion is the silylation (S4.1e) of a metallated arene to give a TMS−arene that can be used as a blocking/protecting group and can then either be desilylated (S4.1f) with acid or fluoride

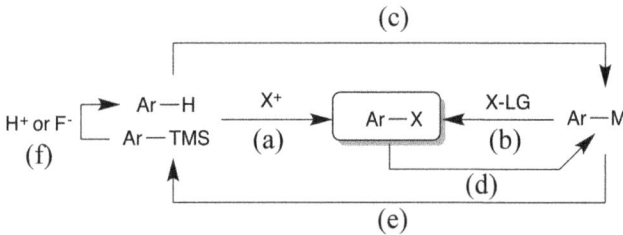

X = halogen; M = metal; LG = leaving group

Scheme 4.1 Summary of interconversion between aryl halides, organometallics, and arenes.

or reacted directly with an electrophilic halogenation reagent to give Ar—X (in this case, the silyl group acts as a "bulky proton"). This TMS protection/deprotection strategy is most often employed on thiophene rings.

Several conditions that have been used to electrophilically halogenate 3-alkyl thiophenes are illustrated in Scheme 4.2 on 3-alkylthiophene **1**. Clean monobromination to give **2** can often be accomplished with *N*-bromosuccinimide in a polar solvent such as DMF typically in high yields in less than four hours.[2] Note that the most electrophilically active position is adjacent to both the sulfur and the electron-releasing alkyl group. Other high-yielding conditions that have been used for monobromination with NBS include halogenated solvent/acetic acid mixtures,[3] acetic acid as the solvent,[4] and lower temperatures with THF as the solvent.[5] Brønsted acids in general tend to accelerate halogenation reactions, presumably by providing a more polar medium to stabilize carbocation intermediates and by increasing electrophilicity of the bromine atom through protonation of a lone pair of electrons on an atom adjacent to the halogen to weaken the Br—X bond and increase "bromenium" ion (Br⁺) character, such as

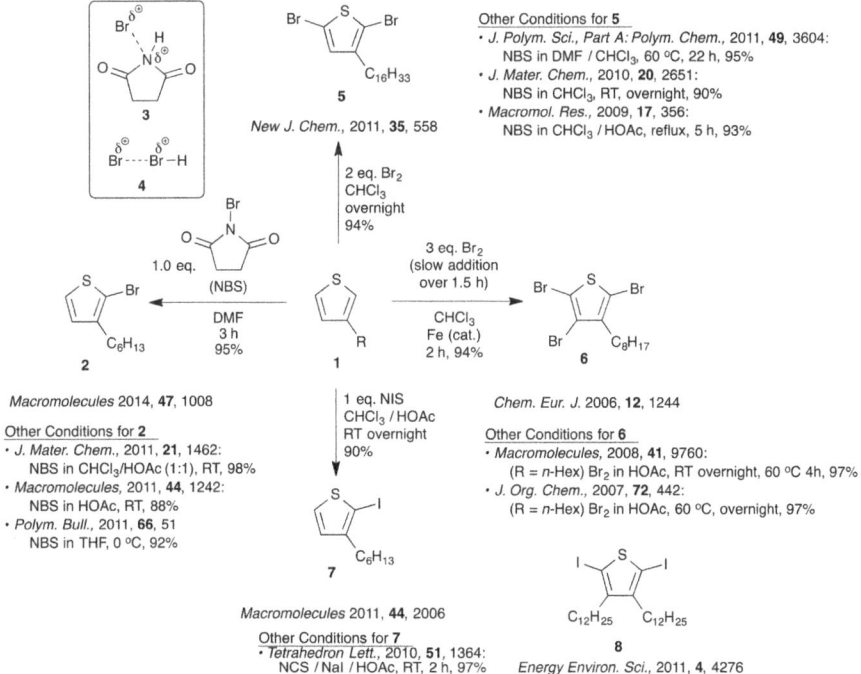

Scheme 4.2 Mono-, di-, and trihalogenation of 3-alkylthiophene.[2–15]

shown in Scheme 4.2 inset for NBS (**3**) and Br_2 (**4**). Two brominations can occur with high yields if two or more equivalents of a brominating reagent and slightly more forcing conditions are employed such as for **5** with more reactive molecular Br_2 as the brominating agent,[6] NBS with heat,[7] longer reaction times,[8] acid,[9] or combinations thereof. Tribromination to **6** is known as well, and again the conditions are more forcing, such as with using $FeBr_3$ as a catalyst[10] (which forms *in situ* from Br_2 and Fe powder) and with Br_2 in acetic acid at 60 °C.[11,12] Similar to NBS, *N*-iodosuccinimide (NIS) can be used as an iodinating reagent to give monoiodo **7**[13] and diiodo **8**[14] (from 3,4-didodecylth-iophene) in high yields. Another method utilizes *N*-chlorosuccinim-ide (NCS) and NaI to generate iodine monochloride (I-Cl) *in situ* as an iodinating agent to give **7** in high yield in two hours.[15] Many other variations on these conditions and reagents exist, so when a planned reaction with a certain substrate fails, a thorough literature search for similar substrates and conditions can be beneficial.

One important aspect of halogenation of thiophenes is the regi-oselective bromination/iodination of 3-alkylthiophenes, in particular as monomers for highly regioregular poly(3-hexylthiophene) (P3HT), which will be discussed further in Section 7.4.3 on Grignard metath-esis polymerization of thiophenes.[16,17] Methods that have been used for such halogenations are summarized in Scheme 4.3 for 3-hexylth-iophene. Most straightforwardly, NBS can be used in a first step fol-lowed by a hypervalent iodinating agent I_2/Ph–$I(OAc)_2$[18] or NIS[19] to give 2-bromo-3-hexyl-5-iodothiophene **9** in high overall yield. This

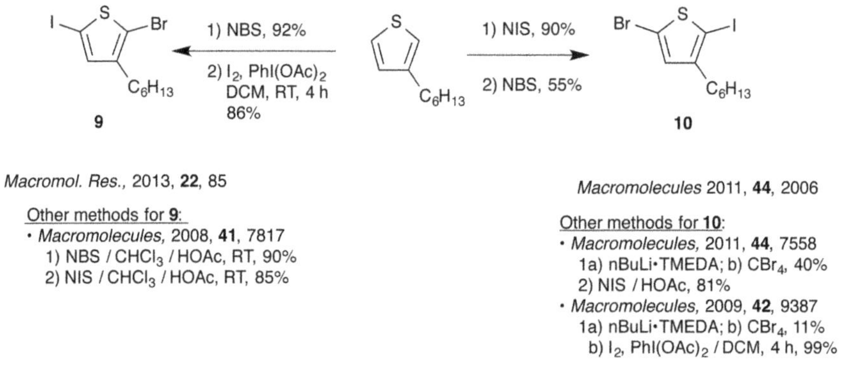

9	1) NBS, 92% 2) I_2, PhI$(OAc)_2$ DCM, RT, 4 h 86%	1) NIS, 90% 2) NBS, 55%

Macromol. Res., 2013, **22**, 85

Other methods for **9**:
• *Macromolecules*, 2008, **41**, 7817
1) NBS / CHCl$_3$ / HOAc, RT, 90%
2) NIS / CHCl$_3$ / HOAc, RT, 85%

Macromolecules 2011, **44**, 2006

Other methods for **10**:
• *Macromolecules*, 2011, **44**, 7558
1a) nBuLi·TMEDA; b) CBr$_4$, 40%
2) NIS / HOAc, 81%
• *Macromolecules*, 2009, **42**, 9387
1a) nBuLi·TMEDA; b) CBr$_4$, 11%
b) I$_2$, PhI$(OAc)_2$ / DCM, 4 h, 99%

Scheme 4.3   Differential dihalogenation of 3-alkylthiophene.[13,18-21] Reproduced from ref. 18 (formation of **9**) and ref. 13 (formation of **10**). Other meth-ods for **9**: (1) NBS/CHCl$_3$/HOAc, RT, 90%, (2) NIS/CHCl$_3$/HOAc, RT, 85% (ref. 19). Other methods for **10**: (1a) *n*-BuLi·TMEDA; (b) CBr$_4$, 40%, (2) NIS/HOAc, 81% (ref. 20); 1(a) *n*-BuLi·TMEDA; (b) CBr$_4$, 11% (2) I$_2$, PhI$(OAc)_2$/DCM, 4 h, 99% (ref. 21).

strategy can be reversed to give 5-bromo-3-hexyl-2-iodothiophene **10** in moderate overall yield, with some Br for I exchange occurring during the reaction.[13] A potential work around for bromination in the second step for **10**, with conditions given in Scheme 4.3, is to deprotonate at the sterically favored 5-position and brominate the lithiated thiophene with CBr_4; however, this first lithiation/bromination step reportedly results in low to modest yields.[20,21]

Scheme 4.4 shows other substrates and methods of halogenation that should be noted. The first is tetrabromination of bithiophene with reactive conditions to give **11** in high yield, which can be debrominated with Zn and acid at the α positions (4,4′ positions) to give 2,2′-dibromobithiophene **12** in moderate yield.[22] This halogenation/ dehalogenation is perhaps the main strategy used for the synthesis of β-halothiophenes, and has been used to prepare 3-bromothiophene from 2,3,5-tribromothiophene[23] and 3,4-dibromothiophene from tetrabromothiophene.[24] Also illustrated is TMS-protection of the more active α-positions of 3,3′-dibromobithiophene **13** with deprotonation– lithiation and TMS–Cl, followed by halogen-lithiation and quenching with I_2 to give a 3,3′-diiodobithiophene **14**, followed by bromination to replace the TMS groups to give **15** in high yield.[25] Another example of bromine–TMS exchange is the reaction of bis-TMS-dithienosilole **16** with approximately one equivalent of NBS to give monobrominated **17** with dropwise addition of the NBS solution to keep the concentration of NBS low; then, trifluoroacetic acid (TFA) was used to remove the remaining TMS group to give monobromo **18** in high yield.[26] The TMS group also reacts with iodine monochloride (I-Cl) to give iodination typically in high yield as with the conversion of **19** to **20**.[27] It should be noted that when mixed donor materials are present in a molecule, such as with the carbazole and terminal thiophene in **21**, any potential issue with lack of chemospecificity during electrophilic halogenation between the electron donor carbazole and thiophene segments can be avoided by lithiating the thiophene with high chemospecificity and iodination with I_2 such as with the conversion of **21** to **22** in high yield.[28] Finally, electrophilic halogenations may even work well on thiophenes when there are electron withdrawing/acceptor moieties such as with the reaction of **23** with NIS to give **24** in high yield.[29]

Halogenation of arylamines also can be accomplished with NBS and NIS to give *p*-halogenated products in generally high yields; however, to avoid over halogenation of the aryl amine, the NBS and NIS reagents should be added as a solution in a suitable solvent to the reaction mixture containing the amine. Generally, if a solid reagent is added to a reaction mixture and dissolves, there is a high local concentration of

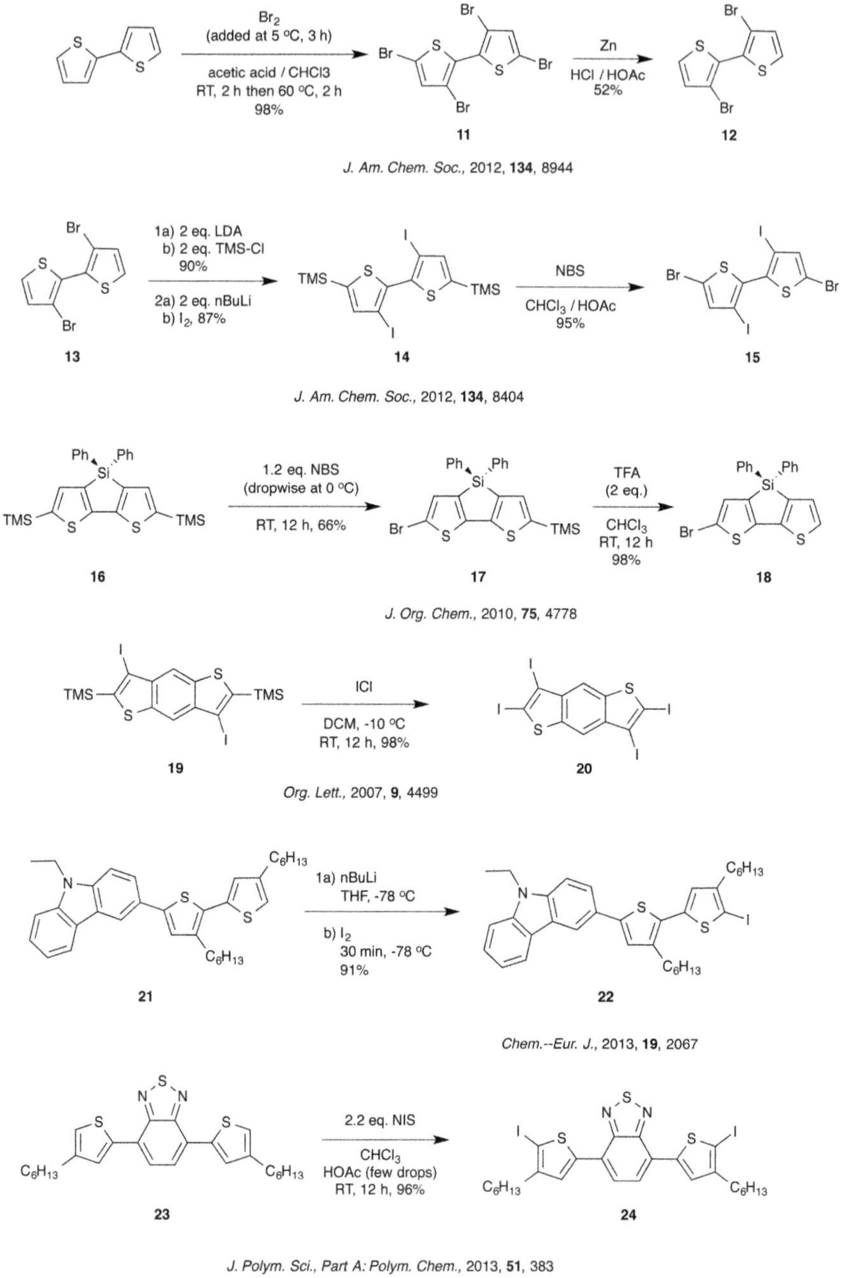

J. Am. Chem. Soc., 2012, **134**, 8944

J. Am. Chem. Soc., 2012, **134**, 8404

J. Org. Chem., 2010, **75**, 4778

Org. Lett., 2007, **9**, 4499

Chem.--Eur. J., 2013, **19**, 2067

J. Polym. Sci., Part A: Polym. Chem., 2013, **51**, 383

Scheme 4.4   Halogenation of several thiophene and thieno substrates.[22,25–29]

the reagent near the dissolving solid. If there is a high local concentration of NBS or NIS near the solid as it dissolves, any aryl amine also in solution near the solid can be overhalogenated. Scheme 4.5 demonstrates some important points of bromination of arylamines. First, the halogenation can occur with relatively mild conditions that does not require acids, which may be critical when acid sensitive groups are present, such as in the case of **25** giving bromo **26**[30] and iodo **27**,[31] where acidic conditions might cleave the TBDMS–O ethers.[32] When triarylamines are substrates without blocking groups in *p*-positions, monobromination (**28**),[33] dibromination (**29**),[34] and tribromination (**30**)[35] may occur in high yields by increasing the number of NBS equivalents; however, when *N*-phenyl carbazole **31** is the substrate, bromination may occur selectively at the carbazole *p*-positions to give monobromo **32** and dibromo **33** as products in high yields, but not bromination on the *N*-phenyl group.[36] This is most likely because

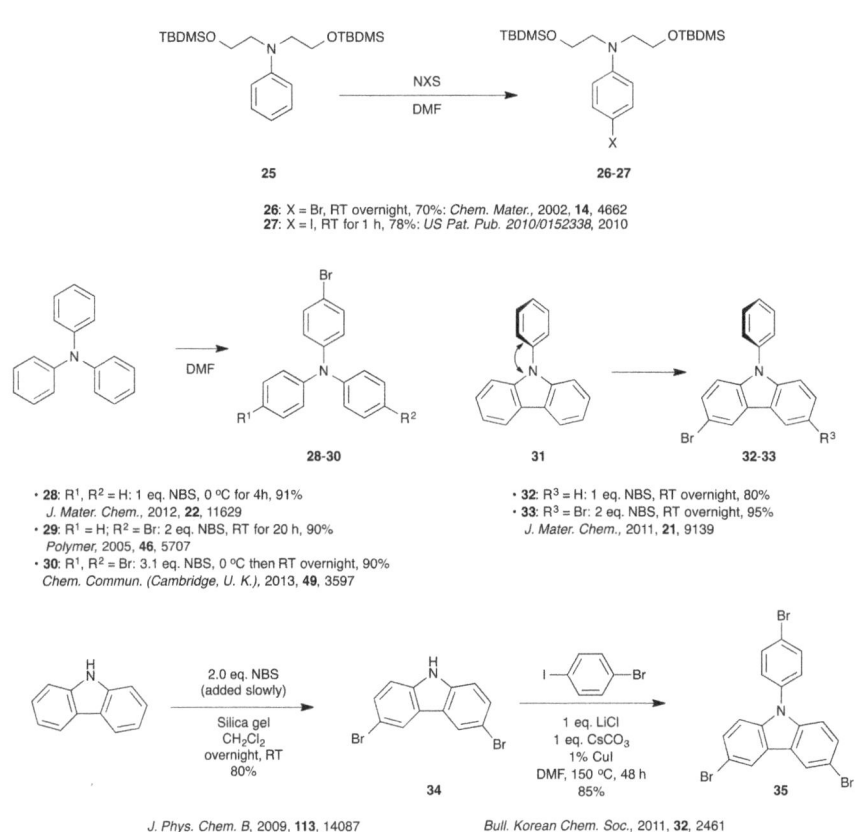

26: X = Br, RT overnight, 70%: *Chem. Mater.*, 2002, **14**, 4662
27: X = I, RT for 1 h, 78%: *US Pat. Pub. 2010/0152338*, 2010

• **28**: R¹, R² = H: 1 eq. NBS, 0 °C for 4h, 91%
  J. Mater. Chem., 2012, **22**, 11629
• **29**: R¹ = H; R² = Br: 2 eq. NBS, RT for 20 h, 90%
  Polymer, 2005, **46**, 5707
• **30**: R¹, R² = Br: 3.1 eq. NBS, 0 °C then RT overnight, 90%
  Chem. Commun. (Cambridge, U. K.), 2013, **49**, 3597

• **32**: R³ = H: 1 eq. NBS, RT overnight, 80%
• **33**: R³ = Br: 2 eq. NBS, RT overnight, 95%
  J. Mater. Chem., 2011, **21**, 9139

J. Phys. Chem. B, 2009, **113**, 14087

Bull. Korean Chem. Soc., 2011, **32**, 2461

Scheme 4.5 *para* Halogenation of aryl amines.[29–31,33–38]

the *o*-phenyl hydrogens of the *N*-phenyl group sterically repulse the hydrogens on the 1 and 8 positions of the carbazole (*o* to the carbazole nitrogen), thereby causing a torsional twist around the N−C phenyl bond, which breaks conjugation with the nitrogen lone pair and reduces the electron density in the ring, making it more difficult to electrophilically halogenate. In fact, the reported method to have a halogen in the *p*-*N*-phenyl position is to halogenate 9-*H*-carbazole to give **34** followed by Ullmann coupling (Section 3.3.3) with iodobromobenzene to give **35** in high yield.[37,38]

Although this section has presented halogenation of donor substrates, halogenation also has been used on electron acceptor intermediates; however, the reaction conditions are typically forcing and tend to be very substrate specific. Because of these relatively narrow substrate conditions, halogenation of electron acceptors will be presented in Chapter 6.

4.2.2 Aldehydes

Aldehydes are used extensively in the synthesis of π-conjugated materials as a functional group for the synthesis of alkene π-linkages and for the attachment of acceptor groups. Although there are a number of methods to synthesize aldehydes, in π-conjugated material synthesis there are two prominent methods: (1) the formylation of lithiated aromatics and (2) the Vilsmeier reaction. Each of these reactions has their own limitations for substrates and functional groups. Lithiation–formylation must have substrates that can tolerate the highly nucleophilic lithiation reaction, whereas substrates for the Vilsmeier reaction must tolerate the acidic conditions that develop during the reaction. Another important aspect of the Vilsmeier reaction is that it is an electrophilic aromatic substitution, and as such, it is limited practically to reactions where electron density is highest (*e.g.*, *para* to electron-donating groups in benzene rings and adjacent to sulfur in thiophene rings), whereas lithiation–formylation can occur wherever there is a halogen for lithium–halogen exchange or an acidic proton for deprotonation–lithiation. Thus, lithiation–formylation may have more flexible regioselectivity than the Vilsmeier reaction.

4.2.3 Aldehydes: Lithiation–Formylation

Scheme 4.6 gives a general mechanism for the formylation of organolithiums. Typically this proceeds by nucleophilic attack (S4.6a) of the aryllithium (Ar^1−Li) on the formyl group of a formamide to give a hemiaminal

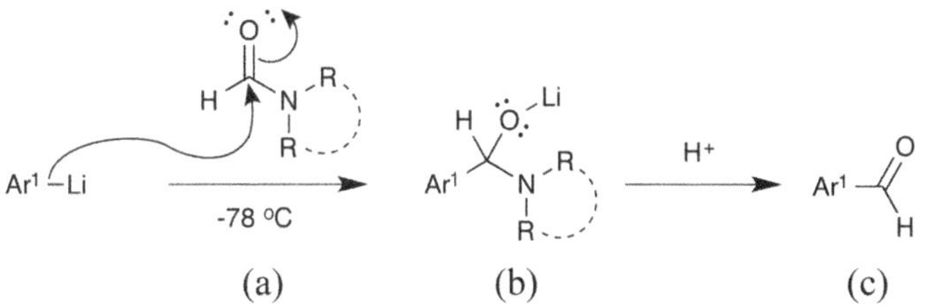

Scheme 4.6 General illustration of formylation of aryllithium reagents.

lithium salt (S4.6b), which hydrolyzes with water or acidic workup to give corresponding aldehyde (S4.6c). Typically, water workup will suffice or a "buffered" acid such as aqueous ammonium chloride may be necessary. If the substrate or product is not acid sensitive, then stronger acids such as dilute aqueous HCl are very effective. DMF is most commonly used as the formylating agent; however, alternatives such a *N*-formylmorpholine and *N*-formylpiperidine sometimes give higher yields since they do not absorb water as readily as DMF and the intermediate lithiated hemiaminal tends to be less reactive. The main limitation as stated earlier is that the substrate must be able to withstand being an organolithium reagent, which can be generated by any of the methods mentioned in Sections 3.2.1–3.2.6. Since the substrate must be able to form a stable organolithium reagent, acceptor groups that have high EA are often not suitable for the reaction.

Scheme 4.7 illustrates some examples of lithiation–formylation with various substrates and conditions. The first example of thieno-thiophene **36** employs classic conditions of deprotonation followed by quenching with DMF to give aldehyde **37** in high yield.[39] Note that deprotonation occurs at the least sterically hindered α-position that is distal from the propyl group. Another example with thienothiophene is the double lithiation of the tetrabromide **38** and quenching with *N*-formylpiperidine **39** to give **40** in high yield.[40] Double lithiation has also been used on 2,7-dibromocarbazole **41** followed by quenching with DMF to give dialdehyde **42** in high yield.[41] Note that the normal electrophilic aromatic substitution positions of **41**, which are *para* to the nitrogen, can be reacted as shown for the bromination of **42** to give **43** in high yield. Another example in Scheme 4.7 illustrates a deprotonation–lithiation where there are other acidic groups such as with the conversion of **44** using three equivalents of *n*-BuLi, two of which deprotonate the most-acidic alcohols, to give aldehyde **45** after quenching with DMF.[30] A final example shows the lithiation of a relative electron accepting dithiadiazole **46** and quenching with *N*-formylmorpholine **47** to give corresponding aldehyde **48** in high yield.[42]

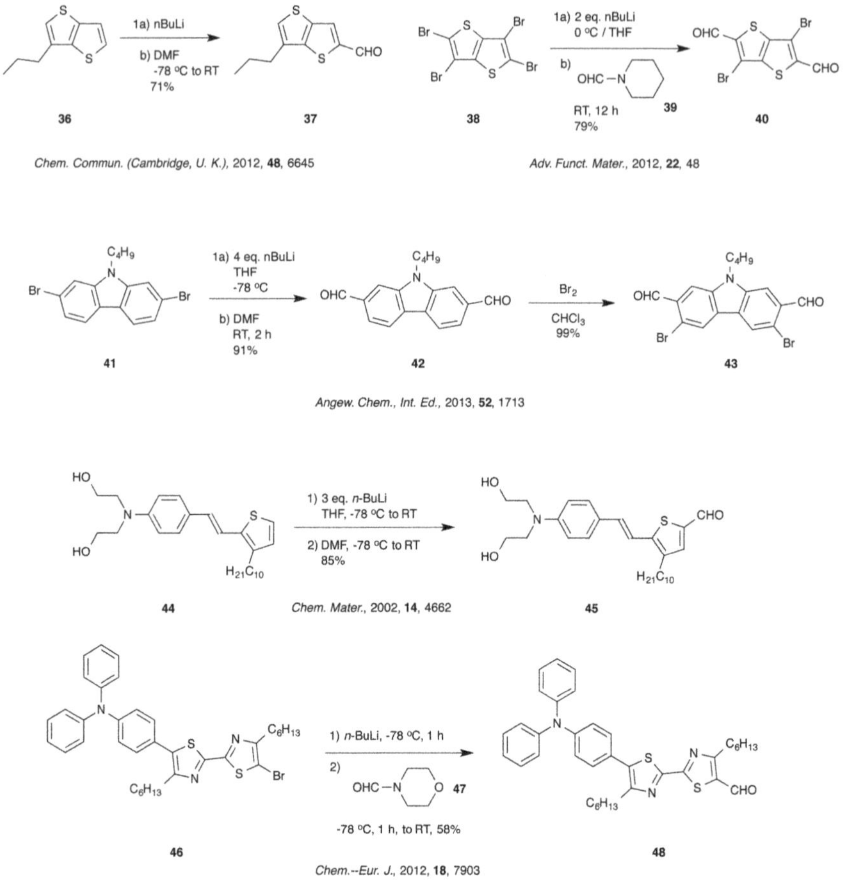

Chem. Commun. (Cambridge, U. K.), 2012, **48**, 6645

Adv. Funct. Mater., 2012, **22**, 48

Angew. Chem., Int. Ed., 2013, **52**, 1713

Chem. Mater., 2002, **14**, 4662

Chem.--Eur. J., 2012, **18**, 7903

Scheme 4.7   Formylation of several aryllithium substrates.[30,38,39,41,42]

4.2.4   Carbonyl-π Extension *via* Transposition

One useful variation of the lithiation–formylation is the reaction of lithiated arenes with vinylogous formamides, which is a specific example of carbonyl transposition. The reaction mechanism is illustrated in Scheme 4.8. Addition of the lithiated arene to the vinylogous formamide (S4.8a) gives the intermediate aminoalkoxide (S4.8b) that when protonated can expel water (S4.8c) to give an imine (S4.8d) that undergoes attack by water to give a hemiaminal (S4.8e) and then hydrolysis to give an aldehyde (S4.8f), where the aldehyde is extended from the aryl group by one alkene. This reaction can be particularly useful for installing longer polyene bridges in π-conjugated materials. Two examples of such "aldehyde extension" are shown in the bottom half of Scheme 4.8, where aniline **49** is lithiated and reacted with

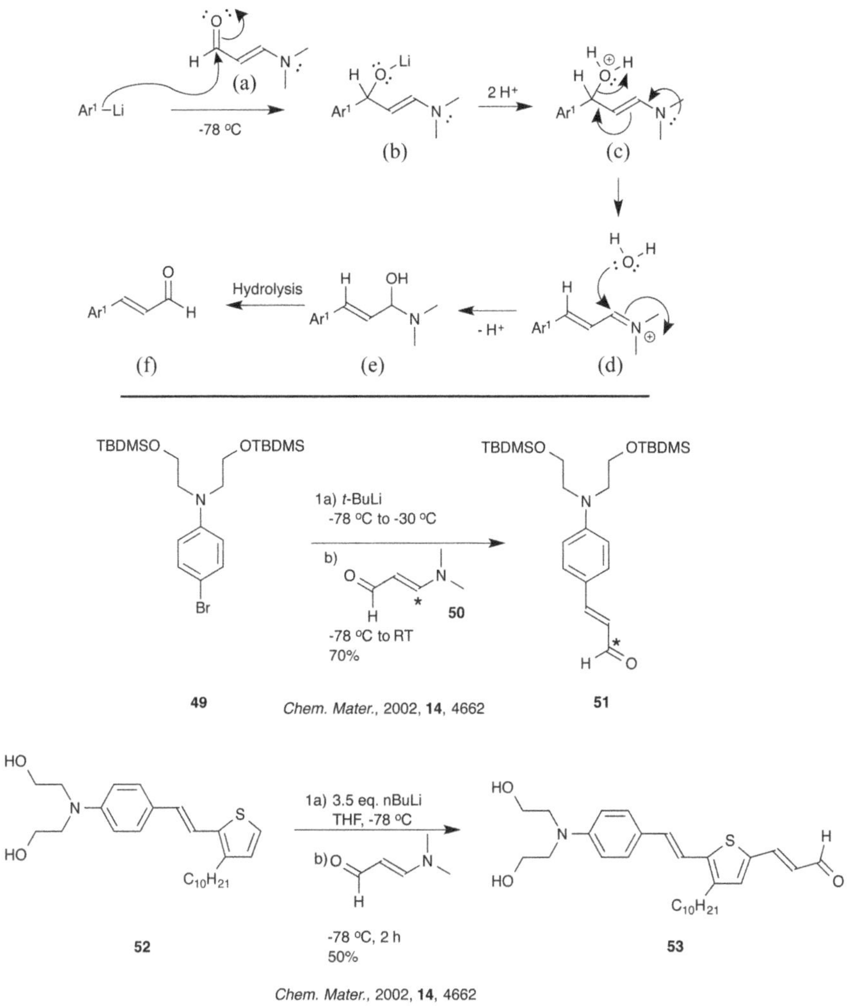

Scheme 4.8   Aldehyde π-extension of aryllithiums by vinylogous formamides.[30]

3-(*N,N*-dimethylamino)-acrolein **50** to give the corresponding extended aldehyde **51** in high yield.[30] Note the starred carbon positions in **50** and **51**. Also note that the somewhat acid-sensitive –OTBDMS ether protecting groups[32] do not cleave in this basic to mildly acidic reaction. Another example again uses a substrate with two hydroxyl groups in **52** that gives the extended aldehyde **53** in moderate yield.[30] The lower yield for **53** compared to **51** may be due to complications arising from having to deprotonate the hydroxyl groups, such as low solubility, excess lithiating *n*-BuLi, and/or Michael-addition reaction of the nucleophilic lithium alkoxides to dimethylaminoacrolein.

Carbonyl transposition has also been used to give extended ketones. Although ketones have been used far less than aldehydes in the synthesis of π-conjugated materials, they can be useful in the synthesis of ring-locked polyene bridges. The mechanism is similar to the aldehyde extension in Scheme 4.8 and typically uses vinylogous esters, outlined in Scheme 4.9 for an example with a 3-(alkoxy)cyclopent-2-en-1-one (S4.9a). Addition of the lithiated arene to the vinylogous ester gives the intermediate alkoxide-ester (S4.9b) that when protonated expels water (S4.9c) to give an intermediate (S4.9d)

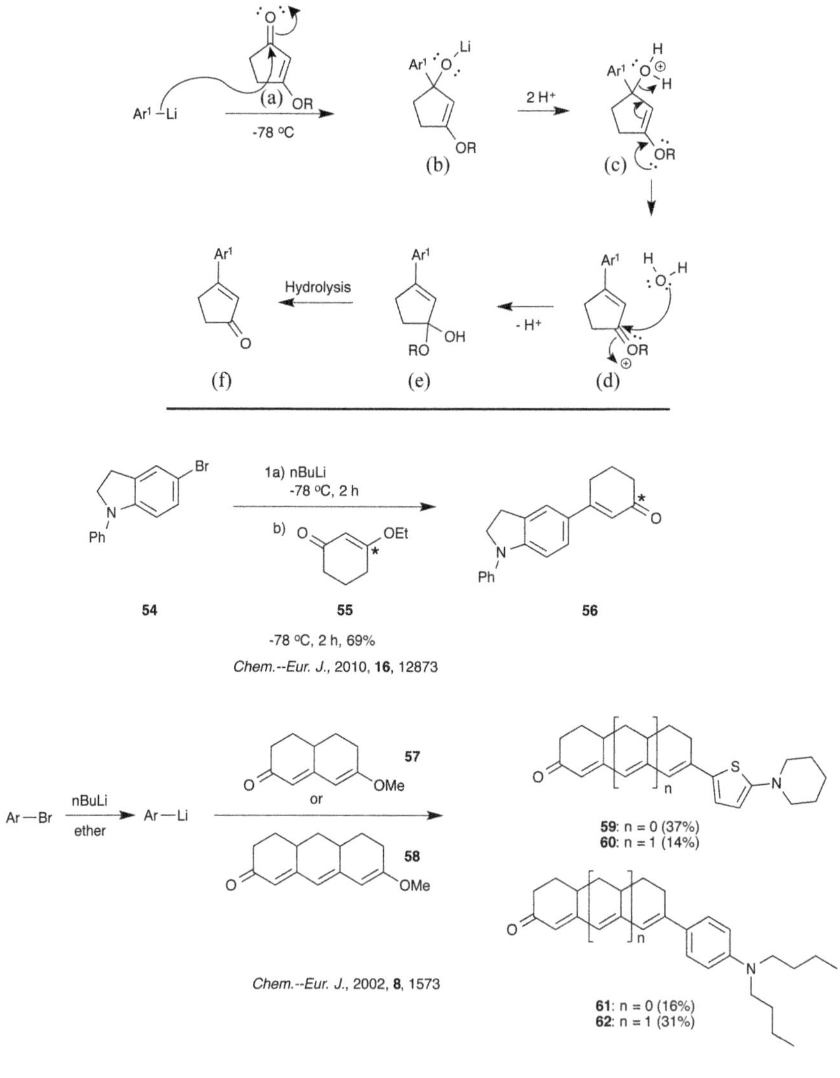

Scheme 4.9   Ketone π-extension of aryllithiums by carbonyl transposition.[43,44]

that undergoes attack by water to give the hemiketal (S4.9e) that hydrolyzes to give a ketone (S4.9f), which has a ketone extended from the aryl group by one ring locked alkene. Two examples of such carbonyl transposition also are shown in Scheme 4.9 with the lithiation of aniline **54** followed by quenching with the vinylogous ester **55** to give the extended ketone **56** in moderate yield.[43] The method can also be used in the synthesis of more extended systems, such as the reaction of the extended vinylogous esters **57** and **58** with either lithiated thiophenes or lithiated anilines to give the corresponding extended systems **59–60** and **61–62**, respectively, in low yields.[44] Note that, although the yield of this step is low in this case, synthesizing the polyene ring systems by annulating from the aromatic substrates would be more time intensive and likely may end up having a lower overall yield.

4.2.5 Aldehydes: Vilsmeier Formylation

The Vilsmeier reaction, also known as the Vilsmeier–Haack reaction, is outlined in Scheme 4.10. The method involves the reaction of DMF with POCl$_3$ to form the chloroiminium ion (S4.10a, sometimes referred to as the "Vilsmeier complex"), which then can electrophilically substitute an aromatic or heterocyclic ring to give an imine (S4.10b), which in turn can be hydrolyzed to an aldehyde on aqueous workup (S4.10c). Formation of the chloroiminium ion can occur with or without the presence of the aromatic substrate, although adding the aromatic substrate after formation of the Vilsmeier complex can lead to decomposition of the complex in certain circumstances, especially when external cooling fails.[45] Being an electrophilic substitution, the Vilsmeier reaction typically works well only for aromatic and heteroaromatic rings that have relatively low ionization potential (IP), which is often the case for donor intermediates in synthesis of π-conjugated materials. One very

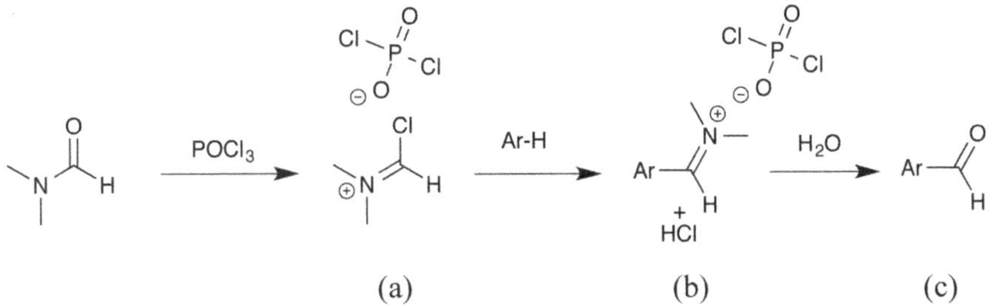

Scheme 4.10 General illustration of the Vilsmeier reaction.

important general note is that the reaction mixture is typically very acidic, and thus substrates that are acid sensitive should be used with due caution. For example, an acetate ester protecting group for an alcohol may survive, but –OTBDMS ether protecting groups most likely would not.[32] Additionally, the chloroiminium ion can react with amines (to give formamides) or alcohols (to give formates), and $POCl_3$ itself can react with alcohols to give alkyl chlorides. Thus, alcohols and amines as well as any other potentially nucleophilic groups should be protected with an acid stable group before Vilsmeier reaction. It also should be pointed out that the chloroiminium ion reacts with a number of different substrates,[46] so caution should be exercised when planning selective formylations of complex substrates.

Scheme 4.11 illustrates several examples of the Vilsmeier reaction on thiophene substrates. The first examples are of common electron donor/π-bridges that are formylated by classic Vilsmeier conditions with DMF or DMF/1,2-dichloroethane (DCE) as solvents to give the corresponding monoaldehydes for DTS (**63**),[47] DTT (**64**),[48]

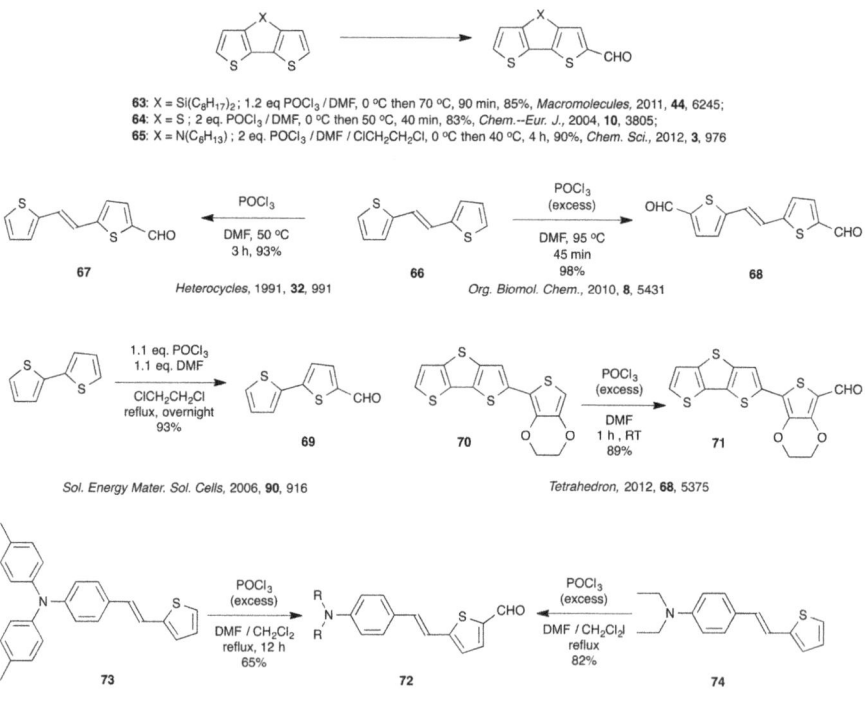

63: X = Si(C_8H_{17})$_2$; 1.2 eq $POCl_3$ / DMF, 0 °C then 70 °C, 90 min, 85%, *Macromolecules*, 2011, **44**, 6245;
64: X = S ; 2 eq. $POCl_3$ / DMF, 0 °C then 50 °C, 40 min, 83%, *Chem.--Eur. J.*, 2004, **10**, 3805;
65: X = N(C_6H_{13}) ; 2 eq. $POCl_3$ / DMF / ClCH_2CH_2Cl, 0 °C then 40 °C, 4 h, 90%, *Chem. Sci.*, 2012, **3**, 976

67 · POCl$_3$ / DMF, 50 °C / 3 h, 93% · *Heterocycles*, 1991, **32**, 991

66 · POCl$_3$ (excess) / DMF, 95 °C / 45 min / 98% · *Org. Biomol. Chem.*, 2010, **8**, 5431 · **68**

69 · 1.1 eq. POCl$_3$ / 1.1 eq. DMF / ClCH$_2$CH$_2$Cl, reflux, overnight / 93% · *Sol. Energy Mater. Sol. Cells*, 2006, **90**, 916

70 · POCl$_3$ (excess) / DMF / 1 h, RT / 89% · **71** · *Tetrahedron*, 2012, **68**, 5375

73 · POCl$_3$ (excess) / DMF / CH$_2$Cl$_2$, reflux, 12 h / 65% · **72** · *New J. Chem.*, 2009, **33**, 868

72 · POCl$_3$ (excess) / DMF / CH$_2$Cl$_2$, reflux / 82% · **74** · *Chem. Commun. (Cambridge, U. K.)*, 2006, 2792

Scheme 4.11   Vilsmeier formylation of several thiophene and thieno substrates.[47–55]

and DTP $(65)^{49}$ all in high yield. Since the reaction is an electrophilic substitution, and the imine intermediate is an electron withdrawing group $(R^- \sim 0.20)$, reaction conditions can often be chosen to easily avoid over-formylation. This is demonstrated by the formylation of dithienyl alkene **66** to give the monoaldehyde **67** in high yield[50] or the dialdehyde **68** also in high yield[51] with an excess of DMF and POCl$_3$ and at higher temperature. Clean monoformylation can also occur when DCE is used as the solvent with longer reaction times such as with the conversion of bithiophene to monoaldehyde **69** in high yield.[52] This relative ease of synthesizing monoformylated products in the Vilsmeier reaction contrasts to lithiation–formylation, since selectively deprotonating compounds that have one or more equivalent α-protons in a thiophene, such as in **63–65, 66**, and bithiophene, can be challenging, which often results in mixtures of monoformylated and diformylated products. Since the Vilsmeier reaction is an electrophilic aromatic substitution, more electron donating parts of a molecule (and therefore more reactive toward electrophilic substitution) may be selectively formylated, such as with the selective formylation of **70** on the EDOT thiophene to give **71** in high yield.[53] The reaction will also give thiophene formylation of arylamine containing compounds like **72** if the *p*-positions on the triaryl amines are blocked such as with the *p*-methyl groups on product **73**,[54] or if the arylamine is a dialkylaniline to give moderate to high yield of the corresponding aldehyde on the thiophene moiety such as in **74**.[55]

Scheme 4.12 illustrates some examples using trisubstituted arylamine substrates in the Vilsmeier reaction. First note that monoformylation is typically as possible with carbazoles as it is with thiophenes to give, for example, **75** in high yield.[56] Also, diformylation can be accomplished in both the active 3- and 6-positions of carbazole with an excess of DMF and POCl$_3$ to give **76** in high yield.[57] Note the contrast in regioselectivity between the 3,6 positions in carbazole for the Vilsmeier reaction and the 2,7 selectivity possible for lithiation–formylation as in the reaction of **42** to **43** in Scheme 4.7. Another important aspect of the Vilsmeier reaction is that it can be used to formylate and halogenate in one pot, such as with the reaction of an *N*-(3-hydroxypropyl)carbazole to give the corresponding chloride-aldehyde **77** in high yield.[58] In contrast, the alcohol group can be protected with a relatively acid-stable protecting group such as acetate in **78** to give the aldehyde-alcohol **79** in high overall yield after Vilsmeier reaction and deprotection (KOH/H$_2$O).[59] Finally, one variation of the Vilsmeier that may be required is the use of POBr$_3$ instead of POCl$_3$ to

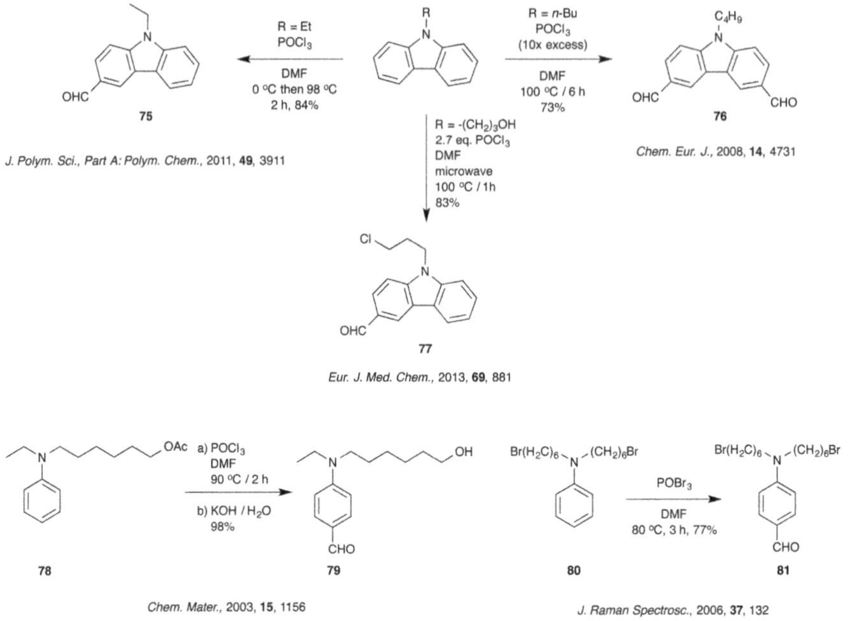
R = Et
POCl3
DMF
0 °C then 98 °C
2 h, 84%
R = n-Bu
POCl3
(10x excess)
DMF
100 °C / 6 h
73%
R = -(CH2)3OH
2.7 eq. POCl3
DMF
microwave
100 °C / 1h
83%
OHC
CHO
75
76
77
J. Polym. Sci., Part A: Polym. Chem., 2011, 49, 3911
Chem. Eur. J., 2008, 14, 4731
Eur. J. Med. Chem., 2013, 69, 881
a) POCl3
DMF
90 °C / 2 h
b) KOH / H2O
98%
POBr3
DMF
80 °C, 3 h, 77%
78
79
80
81
Chem. Mater., 2003, 15, 1156
J. Raman Spectrosc., 2006, 37, 132

Scheme 4.12 Vilsmeier formylation of several aryl amine substrates including carbazole.[56–60]

reduce any unwanted chlorine exchange products, such as with the reaction of the di(bromoalkyl) aniline **80** to **81** in high yield,[60] which might give a mixture of bromoalkyl and chloroalkyl products if $POCl_3$ were used, possibly though HCl acid catalyzed Cl/Br exchange.

4.3 ALKENES *VIA* PHOSPHOROUS REAGENTS

Alkene linkages are widely utilized in the synthesis of π-conjugated materials. In many cases, double bonds are formed *via* reaction of aldehydes with phosphorous reagents such as in the Wittig reaction and the Horner–Wadsworth–Emmons (HWE) modification. The common features of many olefination reactions include: (1) the 1,2-nucleophilic addition of the phosphorous bearing component to the carbonyl and then (2) elimination of the oxygen in a by-product that contains an oxygen–phosphorous (P—O) bond, which acts as a thermodynamic driving force. Two of the fundamental differences between the Wittig and HWE reactions are: (1) that the HWE reagents are more nucleophilic and typically more reactive and (2) that the HWE P—O by-products are typically soluble in water, which facilitates reaction workup and purification. Many other olefination reactions are known, including the Petersen olefination,[61] which uses the oxygen–silicon bond as a thermodynamic driving force; however, in the synthesis of π-conjugated materials, the Wittig and HWE reactions have proven to be very useful and as such are the focus of Sections 4.3.1 to 4.3.2 below.

4.3.1 Wittig Reaction

Georg Wittig was awarded the 1979 Nobel Prize mainly for developing the Wittig reaction, which has been extensively reviewed.[62] Scheme 4.13 illustrates some of the important aspects of the reaction. Generally, the Wittig reaction involves deprotonation of a phosphonium salt (S4.13a, a "Wittig salt") to form an ylide (S4.13b, a "Wittig reagent") that has a stabilized resonance form called an "ylene" (S4.13c). The positively charged phosphorous of the phosphonium salt and the resonance stabilization of the anion result in an acidity of the phosphonium salt of around $pK_a \sim 22$, which can be more or less depending on the other substituents on the carbon (R^2, R^3). Conceptually, the ylide can be viewed as attacking the aldehyde nucleophilically to give a 1,2 carbonyl addition adduct (S4.13d) referred to as a "betaine," which then quickly gives oxaphosphetane intermediates (S4.13e and/or S4.13f) that decompose to either a *trans* alkene (S4.13g from S4.13e) or a *cis* alkene (S4.13h from S4.13f), and a phosphine oxide by-product (S4.13i). There has been a tremendous amount of mechanistic work done on the Wittig reaction, and generally the betaine has not been

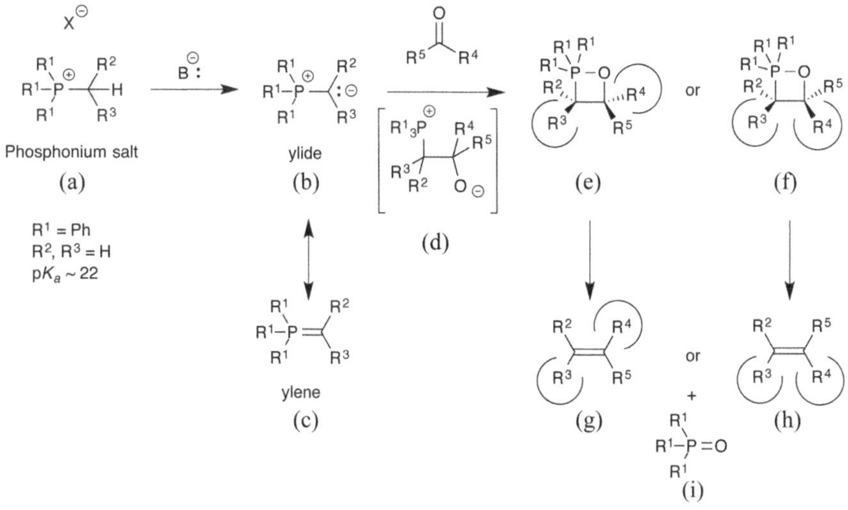

Scheme 4.13 General illustration of the Wittig Reaction.

experimentally observed, with the oxaphosphetane intermediates having been observed in some cases.[63] Some of the implications of the generally accepted mechanism will be explained below.

Some of the general features of the reaction are that alkoxides can be used to deprotonate many phosphonium salts, especially when an electron withdrawing group at R^2 or R^3 increases the proton acidity; however, stronger bases such as NaH, LiN^iPr_2 (LDA), n-BuLi, or PhLi may be preferred when R^2 and R^3 are H or alkyl and the phosphonium salt is less acidic. Generally, aldehydes are more reactive towards the ylide than ketones, as both the electron releasing alkyl group and more steric hindrance reduces reactivity in ketones. R^1 groups on the phosphorous atom that favor the ylide resonance form over the ylene resonance form, such as electron releasing alkyl groups that stabilize the positive charge, tend to increase the *reactivity* of the ylide, but do not necessarily result in higher reaction yields, with more side-products sometimes resulting. Substituent R^2, R^3 groups that stabilize the carbanion through resonance and/or are electron withdrawing decrease the reactivity of the ylide. Ylides generally can be classified as reactive (R^2, R^3 = H or alkyl), moderate (R^2 = H; R^3 = allyl, phenyl), or stable (R^2 = H, R^3 = strong electron withdrawing group). The reactivity of these ylides are illustrated in Figure 4.1, where reactive ylides may react with both ketones and aldehydes and moderate ylides may react mainly with aldehydes, but stable ylides (*e.g.*, R^2 = H; R^3 = $-CO_2R$, $-COR$, $-CN$) mainly react with aldehydes or may react with ketones only under forcing conditions such as high heat and polar solvents and/or with certain salt additives. It should also be noted that reactive ylides tend be sensitive to oxygen and moisture, so care must be taken in their preparation and storage.

The *trans/cis* stereoselectivity around the double bond is important in π-conjugated materials synthesis since steric hindrance between aryl groups that are *cis* on the alkene will likely cause twisting and reduction of π-conjugation. Some of the general principals for

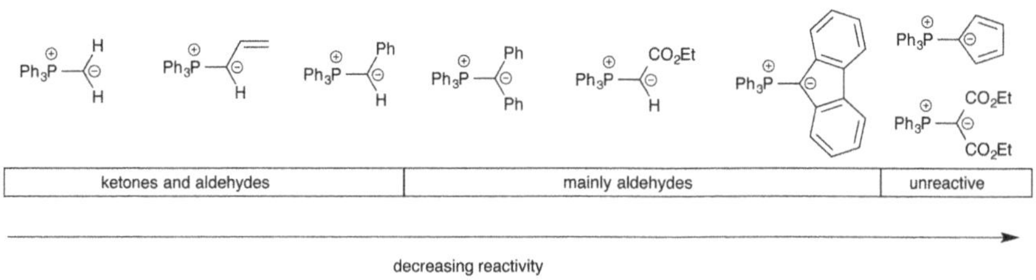

| ketones and aldehydes | mainly aldehydes | unreactive |

decreasing reactivity

Figure 4.1 Reactivity trends of various Wittig reagents.

stereoselectivity in the Wittig reaction are summarized in Table 4.1.[62] Reactive ylides tend to favor formation of the *cis* isomer (T4.1, Entry 1), moderate ylides tend to give mixtures of around 50% each of *trans* and *cis* isomers (T4.1, Entry 2), and stable ylides tend to increase the *trans* isomer (T4.1, Entry 3) with alkyl groups on the phosphorous of stable ylides tending to increase the *trans* isomer the most (T4.1, Entry 4). Generally, solvent conditions, temperature, additives, *etc.* also can influence *trans/cis* stereochemistry; although the effects are sometimes unpredictable.[63] Fortunately, for π-conjugated materials, where typically the maximum amount of π-conjugation is desired, mixtures of *trans* and *cis* alkenes may be converted to all *trans* by post reaction isomerization, such as adding a trace amount of I_2 to the crude *trans/cis* mixture and refluxing for a brief period in toluene.

Some methods for the synthesis of Wittig salts useful in π-conjugated materials synthesis are illustrated in Scheme 4.14. One of the main methods is the nucleophilic substitution between a triaryl or trialkyl phosphine and an alkyl halide (S4.14a), but other leaving groups and Michael addition to activated multiple bonds are known as well.[64] It should be noted that some trialkylphosphines *may present significant safety hazards* including being pyrophoric, such as tri-*n*-butyl phosphine, which has a H250 hazard statement, amongst others. Another reagent that converts alcohols into phosphonium salts is $Ph_3P \cdot HBr$ **82** (S4.14b), which is a shelf stable reagent that is easily prepared from Ph_3P and aqueous HBr, and is effective at converting electron donor alcohols to Wittig reagents. The $Ph_3P \cdot HBr$ reagent is particularly useful for substrates where aryl halides might be unstable or not as synthetically available.[65] Another related reagent is $Ph_3P/NaI/$paraformaldehyde **83** (S4.14c), which in one pot adds the methyl phosphonium fragment for electron donor substrates, such as in the preparation of **84** from julolidine.[66] The reaction mechanism likely involves electrophilic aromatic substitution.

Table 4.1 Wittig reagent structure and its affect on *cis–trans* stereoselectivity.

Entry	R^1	R^2	% *trans*	% *cis*
1	—Ph	—Et	5	95
2	—Ph	—CH=CH$_2$	50	50
3	—Ph	—CO$_2$Et	85	15
4	—*n*-Hexyl	—CO$_2$Et	98	2

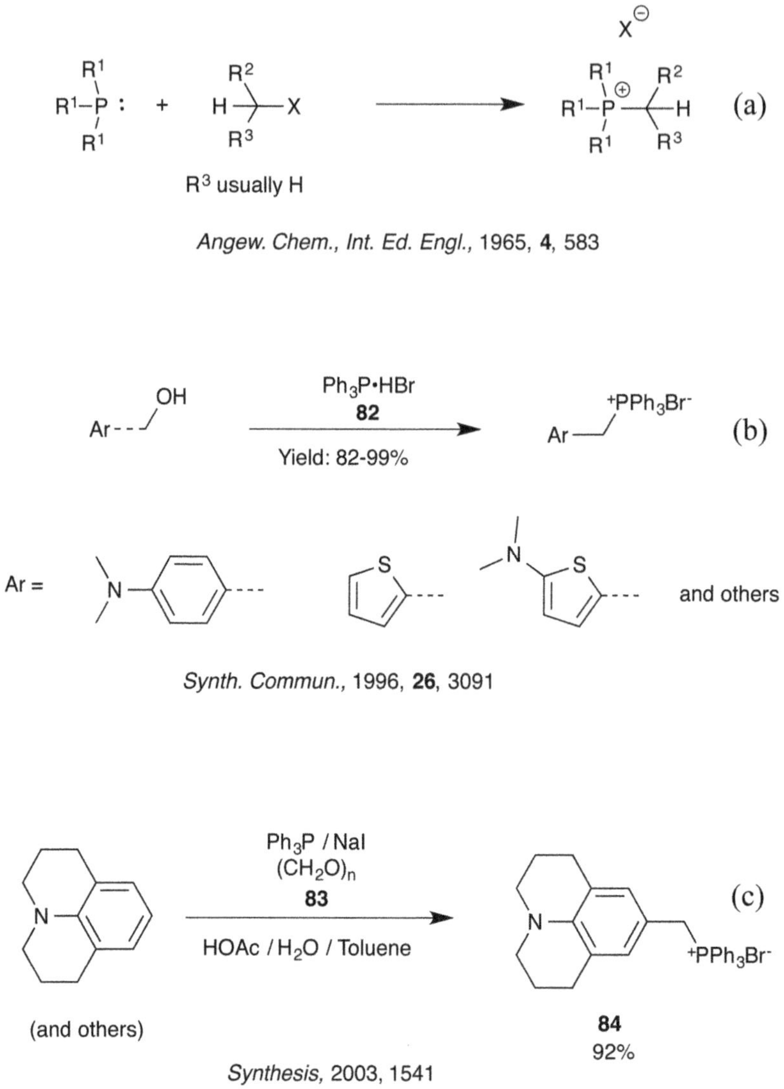
R³ usually H
Angew. Chem., Int. Ed. Engl., 1965, 4, 583
(a)
Ph3P•HBr
82
Yield: 82-99%
(b)
Ar =
and others
Synth. Commun., 1996, 26, 3091
Ph3P / NaI
(CH2O)n
83
HOAc / H2O / Toluene
(c)
(and others)
84
92%
Synthesis, 2003, 1541

Scheme 4.14 Methods of preparing Wittig Salts.[64–66]

Some examples of Wittig reactions are illustrated in Scheme 4.15. Perhaps the "simplest" Wittig salt, methyl triphenylphosphonium bromide, can be used to prepare the vinyl group in high yields, such as with the conversion of *N*-(octyl)carbazole 3-carboxaldehyde to vinyl carbazole **85**,[67] which uses an alkoxide base (KO*t*-Bu) for deprotonation of the phosphonium salt. Another example that converts aldehyde **86** to alkene **87** uses phenyl lithium in THF to first deprotonate the Wittig salt, and then adds aldehyde **86** to the Wittig reagent in a second step to give **87** in high yield as a mixture of isomers.[68]

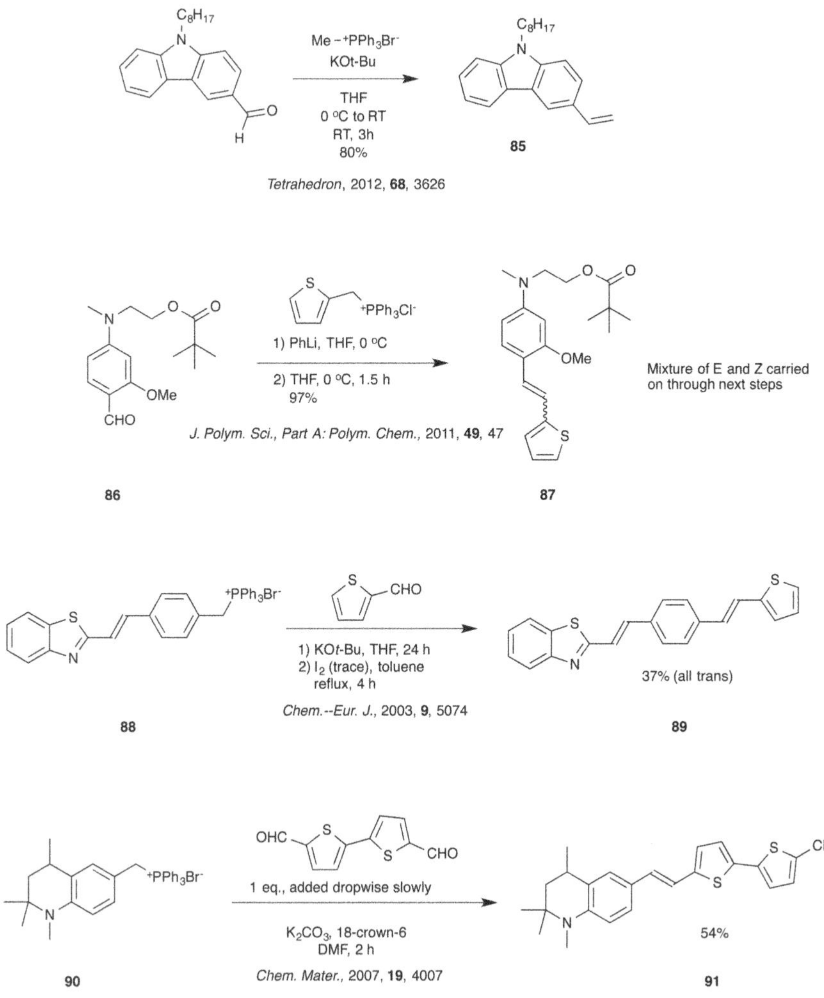

Scheme 4.15 Single Wittig reaction of several substrates.[67–70]

These conditions can be useful if the substrate or Wittig salt/reagent is unstable to alcohols, which are the by-product of deprotonation with alkoxides, whereas the by-product of phenyl lithium deprotonation (*i.e.*, benzene) is inert under the reaction conditions. Note that adding phenyl lithium to a solution with the aldehyde and Wittig salt may lead to 1,2 nucleophilic addition of phenyl lithium to the aldehyde, so the Wittig reagent should be formed in the absence of aldehyde **86**. Another example in Scheme 4.15 uses I_2 to isomerize the crude reaction mixture from Wittig salt **88** and thiophene carboxaldehyde to give the all-*trans* alkene **89** in moderate-to-low yield.[69] Another useful general feature of the Wittig reaction is that, since electron donation into an aldehyde reduces reactivity towards nucleophiles, a controlled addition of a dialdehyde to an electron donor Wittig reagent can lead to formation of the monocoupled product in moderate yields. An example of this is reaction of the Wittig salt **90** with 4,4′-bithiophene dicarboxaldehyde to give the monoalkene **91** in moderate yield.[70] Also note that the conversion of **90** to **91** is affected by K_2CO_3 in DMF with 18-crown-6, which is a less common base than alkoxides, but demonstrates that a number of bases may be used.

Double Wittig reactions have been used widely in the synthesis of many π-conjugated materials to rapidly extend conjugation. Scheme 4.16 illustrates some double Wittig reactions. The first is the reaction of the double Wittig salt **92** with 4-(*N,N*-dibutylamino)benzaldehyde to give the corresponding double addition product **93** in low yield.[71] Note that the second step isomerization in the conversion of **92** to **93** is not the cause of the relatively low yield of **93**, and yields of the I_2 isomerization step based on crude Wittig coupling mixtures are typically

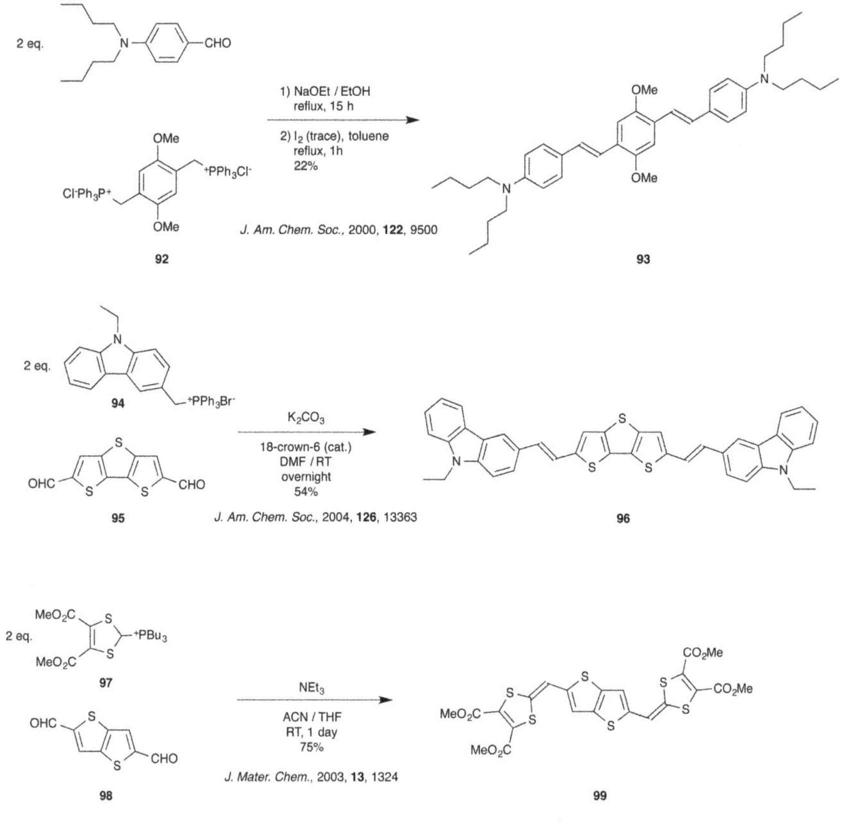

Scheme 4.16   Double Wittig reaction on dialdehyde substrates.[71–73]

quantitative. Another example uses Wittig salt **94** to add to a dialdehyde **95**, again with K_2CO_3/18-crown-6, to give the corresponding double addition product **96** in moderate yield.[72] A final example in Scheme 4.16 illustrates a specialized tributyl phosphonium Wittig salt of a thioketal (**97**) that is relatively active and reacts with dialdehyde **98** to give the corresponding double alkene **99** in high yield using an amine base.[73]

4.3.2 Horner–Wadsworth–Emmons Reaction

The Horner–Wadsworth–Emmons coupling (HWE), also referred to as the "Wittig–Horner" or "Horner–Emmons" reaction, is a modification of the Wittig reaction illustrated in Scheme 4.17 that utilizes a neutral phosphonate (S4.17a) in place of a phosphonium salt.[74] Since the phosphorous is not as electropositive in phosphonates as it is in phosphonium salts, phosphonates tend to be less acidic, and have pK_a values ~10 units higher than analogous phosphonium salts. This also makes the anion of phosphonates (S4.17b), which are stabilized by the resonance structure S4.17c, generally more nucleophilic and more reactive than Wittig reagents. The difference in reactivity is summarized in Table 4.2 for some ketones with varying steric hindrance, and the reaction yields reflect the generally higher reactivity of HWE reagents over Wittig reagents.[74] Another significant advantage of the HWE coupling is that the resulting alkene tends to be the *trans* isomer in high percentage or exclusively. *Trans* isomers especially are favored when one of the α substituents on the phosphonate and one of the substituents on the aldehyde are conjugated to the alkene in the final product, which is often the case in π-conjugated materials synthesis.[63]

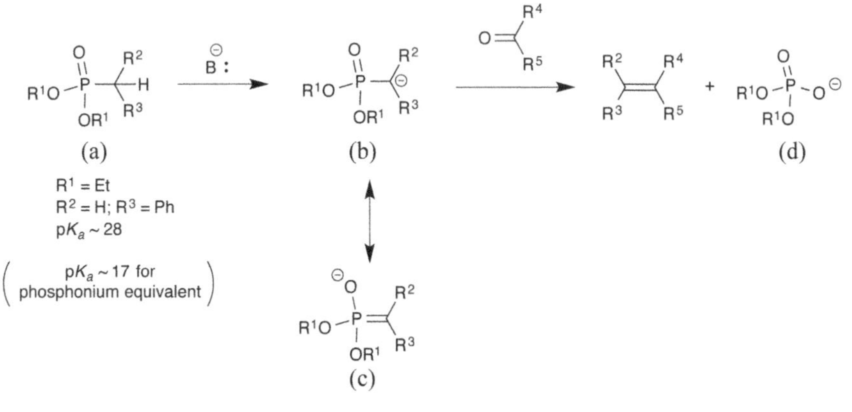

Scheme 4.17 General illustration of the Horner–Wadsworth–Emmons (HWE) reaction.

Regarding the practical process of the reaction, HWE also has significant advantages over the Wittig reaction, including: (1) the trialkyl phosphites used to make phosphonates are typically less expensive than the trialkyl or triaryl phosphines used to make phosphonium salts (and trialkylphosphines may be pyrophoric, such as tri-*n*-butyl phosphine, which has a H250 hazard statement, amongst others); (2) phosphonates, though polar, can typically be chromatographed on normal phase silica (often with acetone/CH_2Cl_2 mobile phase mixtures), whereas phosphonium salts typically cannot, which provides more flexibility in purifying the starting materials; and (3) the dialkyl phosphate by-product (S4.17d) of HWE reactions is very often water soluble, whereas the triaryl or trialkyl phosphine oxide by-product of the Wittig reaction is not, which generally results in HWE by-products being easier to separate from the desired products. One potential caveat with the HWE coupling that is potentially important in the synthesis of π-conjugated materials is that reaction of the phosphonate anion with α,β-unsaturated ketones and aldehydes may give Michael addition products as significant side-products or as the main product.

Some methods for the preparation of phosphonates are summarized in Scheme 4.18. Phosphonates are typically prepared by the Arbuzov reaction of a trivalent trialkyl phosphite and an alkyl halide (S4.18a) to give an intermediate salt that rapidly decomposes (S4.18b) to give the

Table 4.2 Reactivity comparison of Wittig reagents and HWE reagents.

Entry	Ketone	$Ph_3P^{\oplus}\text{—}CH^{\ominus}CO_2Et$ Reaction conditions (yield of alkene)	$(EtO)_2P(O)\text{—}CH_2CO_2Et$ Reaction conditions (yield of alkene)
1	(methyl ethyl ketone)	Xylene, reflux (0%)	NaOEt–DMF, 20 °C (~65%)
2	(cyclohexanone)	Benzene, reflux (25%)	NaOEt–DMF, 20 °C (~65%)
3	(β-ionone)	Toluene, reflux, 2 days (<50%)	NaOMe–MeOH, 10 h, 40 °C (~90%)
4	(steroid, C_8H_{17})	No reaction	NaOEt–DMF, 20 h, 20 °C (93%)

phosphonate and an alkyl halide (S4.18c) comprising the alkyl group of the trialkyl phosphite. The reaction is usually done at high temperature, and the alkyl halide by-product typically *may vaporize out of the reaction mixture, which should be trapped at low temperature outside the reaction if it is given off in large amounts.* As with phosphonium salts, there are other methods to prepare phosphonates, as illustrated in Scheme 4.18. Two involve the addition of the methyl phosphonate fragment, one *via* an organocopper reagent[75] (S4.18d) and another *via* a Rh catalyzed Negishi-like (organozinc) coupling[76] (S4.18e). There is also a method for generating phosphonates from alcohols[77] (S4.18f) involving the use of a trialkyl phosphite and I_2 to generate the benzylic iodide from the alcohol *in situ*, which rapidly reacts with trialkyl phosphite *via* an Arbuzov reaction to give the phosphonate.

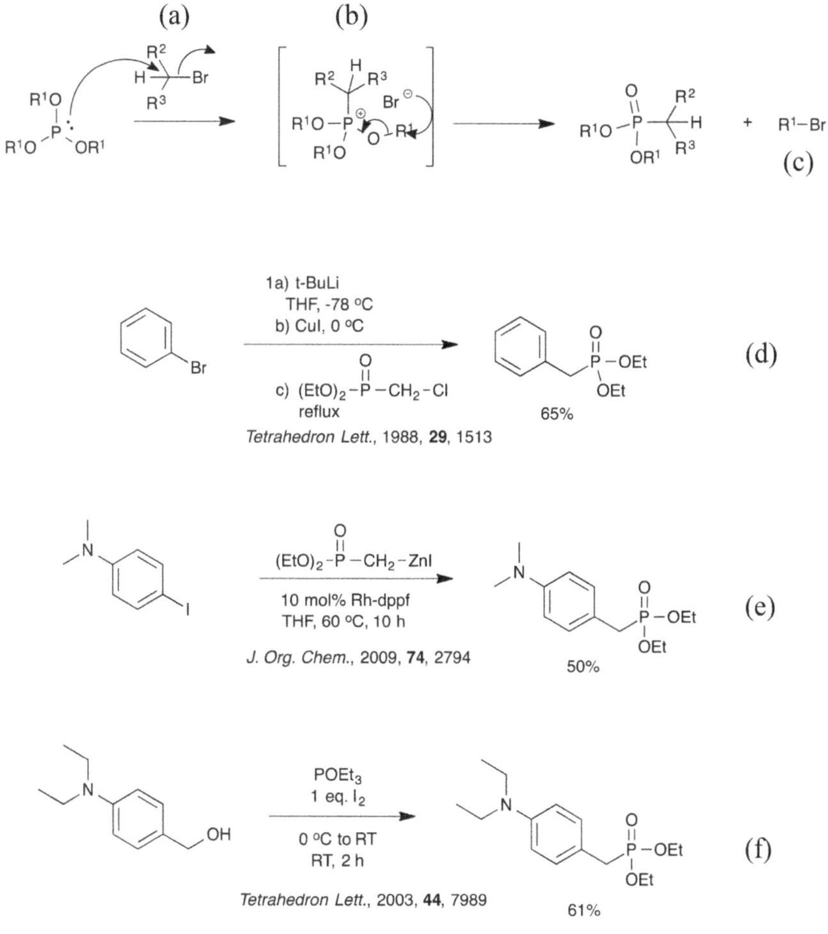

Scheme 4.18 Methods of preparing dialkylphosphonates–HWE reagents precursors.[75–77]

This method may be particularly useful if the benzylic bromide or iodide is too unstable to purify, which is often the case when a strong electron donating group is *para* to the benzylic position.

Some example HWE couplings are shown in Scheme 4.19. The first demonstrates common conditions where aldehyde **100** and phosphonate **101** are co-dissolved in a polar solvent and treated with KO*t*-Bu to give the alkene **102** in high yield.[78] The second example illustrates a

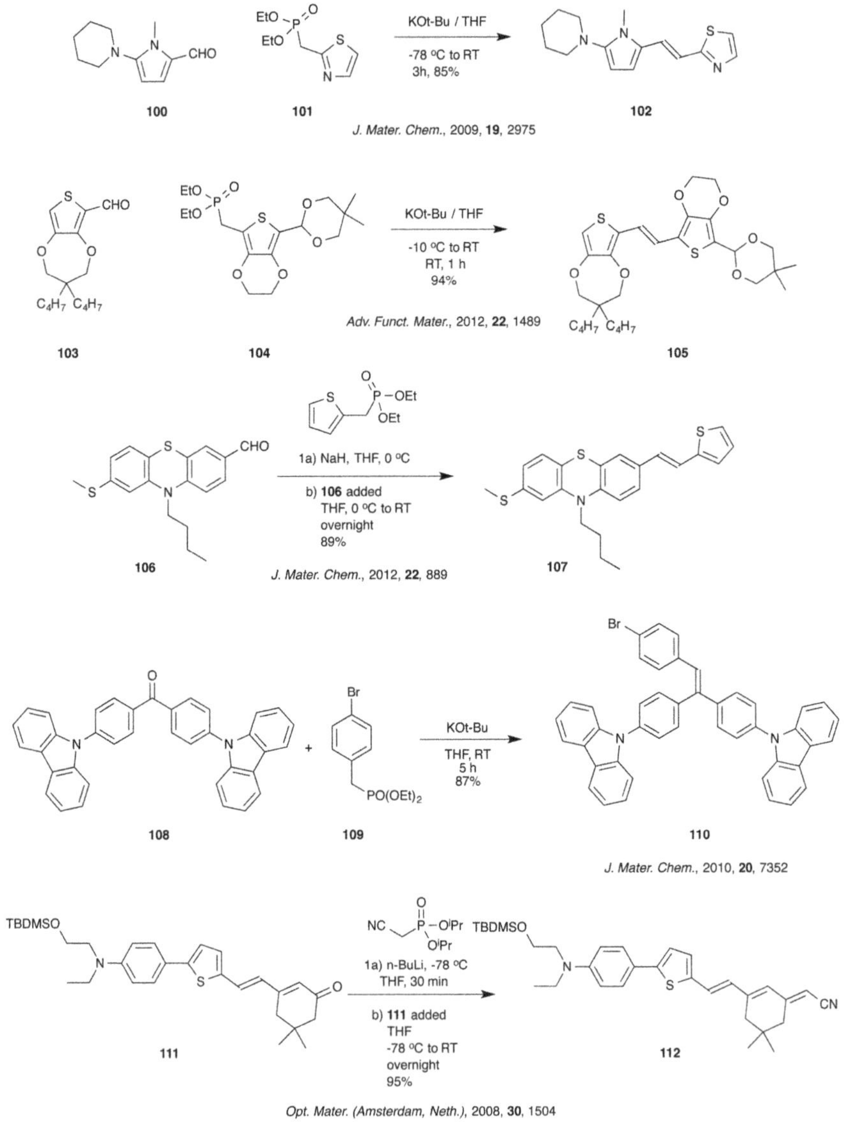

Scheme 4.19   HWE reactions on aldehyde and ketone substrates.[78–82]

HWE reaction with a deactivated aldehyde **103** being used with phosphonate **104** having an acetal (*i.e.*, a protected aldehyde) to give the corresponding alkene **105** in high yield.[79] Note that both the aldehydes in **100** and **103** are conjugated to a strong electron donor group, which typically reduces the electrophilicity of aldehyde, but the HWE reagents are still sufficiently reactive to provide high yield of the olefinated product. Another example shows the use of a non-alkoxide base (NaH) in the conversion of **106** to **107** in high yield.[80] Bases such as NaH may be useful if substrates are sensitive to nucleophilic attack by alkoxides. Two other examples show HWE conditions with ketones: one of diarylketone **108** with phosphonate **109** to give alkene **110** in high yield[81] and the second of more sterically hindered ketone **111** to give **112** in high yield.[82] Note that both these reactions of ketones would be difficult with Wittig reagents, but the synthesis of **112** would be *very* challenging with Wittig reagents due to the relatively sterically hindered ketone *and* stabilization of the attacking anion by the cyano group.

Multiple reactions on a substrate are also possible with the HWE coupling, as illustrated in Scheme 4.20, and they are typically higher yielding than similar Wittig couplings since the HWE reagents are more reactive. The first is coupling of a diphosphonate **113** with carbazole-aldehyde **114** under typical conditions to give the corresponding doubly coupled product **115** in moderate yield.[56] This has also been applied to a four-fold reaction of tetraphosphonate **116** with bithiophene-aldehyde **117** to give corresponding tetraalkene **118** in high yield.[83] Finally, diphosphonate **119** can be reacted with approximately one equivalent of aldehyde to give the monoalkene-phosphonate **120** that can then be reacted with a terephthalaldehyde to give **121** in high yield.[84] Note that the relatively high molecular weight **121** with extended π-conjugation was synthesized in only two steps from readily available intermediate **119** and *N,N*-(dibutylamino)benzaldehyde by this method.

4.3.3 Aldehyde π-Extension *via* Phosphorous Reagents

Section 4.2.4 discussed methods of carbonyl π-extension using a lithium reagent addition/carbonyl-transposition methodology. There are also reagents and methods to accomplish aldehyde π-extensions using phosphorous reagents that are noteworthy and summarized in Scheme 4.21. One such reagent and methodology utilizes the moderately shelf-stable (typically decomposes to other products after ~6 months) tri-*n*-butylphosphonium Wittig salt **122** having a dioxolane (a protected aldehyde) group that can then be deprotected in one

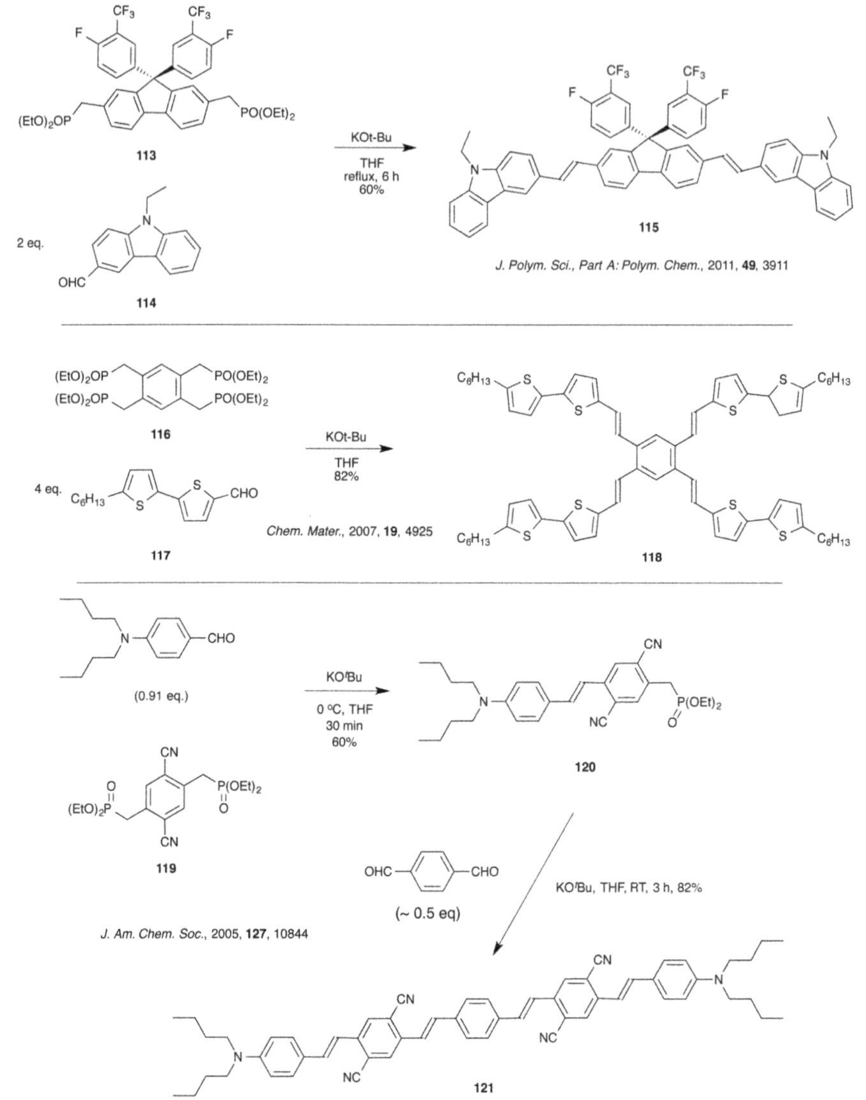

Scheme 4.20 Multiple HWE reactions on substrates with multiple aldehydes.[56,84,89]

pot by acid to give the alkene-aldehyde. The first example illustrated in Scheme 4.21 shows reaction of Wittig salt **122** with 3,4-dibutylthiophene carboxaldehyde and one pot deprotection to give the corresponding aldehyde **123** extended by one alkene in high yield.[85] Using the same conditions, aldehyde **123** can be extended to **124** in moderate yield and then to **125** in moderate-to-low yield in an iterative process. The reaction yield decrease may be due to sensitivity

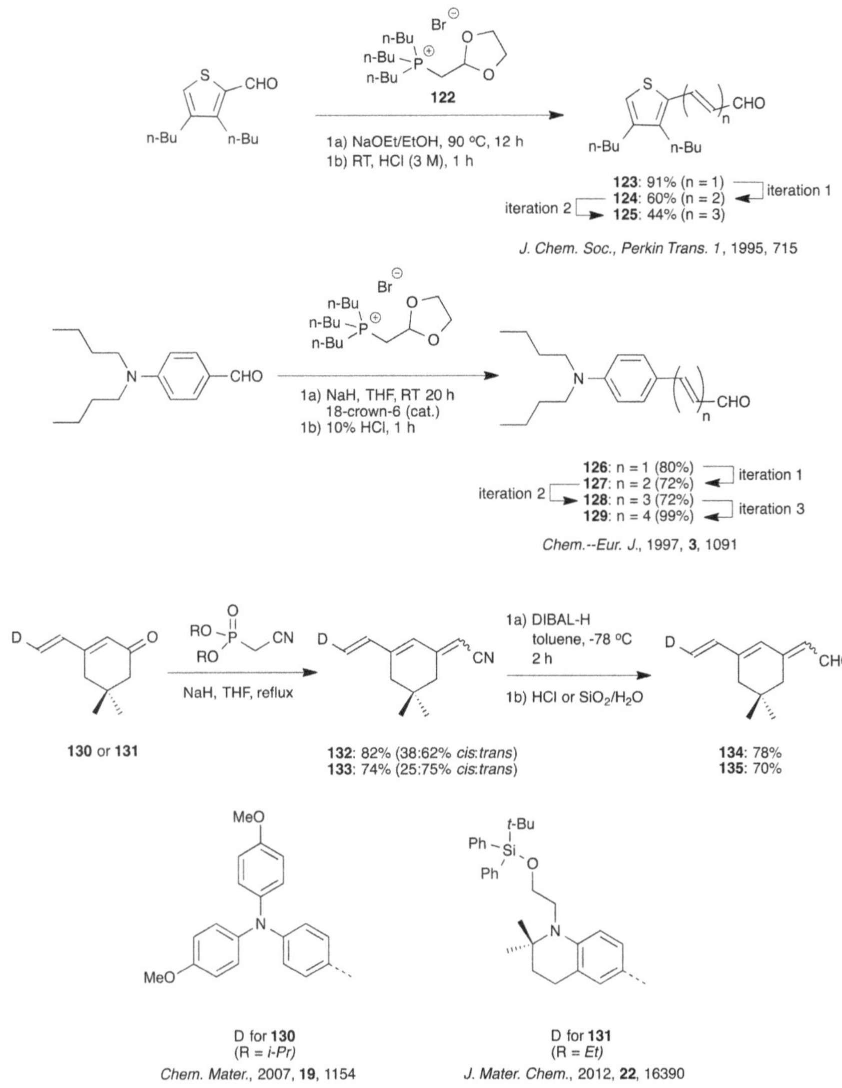

Scheme 4.21 Aldehyde π-extension of aldehydes by Wittig and HWE reagents.[85–88]

of the aldehydes to the basic/nucleophilic conditions (NaOEt/HOEt) with increasingly extended conjugation. Another example with multiple iterative extensions—using **122** employing NaH/THF as a base in less nucleophilic conditions than NaOEt/EtOH—results in extension of 4-(dibutylamino)benzaldehyde by one alkene (**125**) to four alkenes (**129**) with high yields for each step.[86] For aldehyde extension from ketones, a stepwise HWE coupling/hydride reduction methodology has been developed and is particularly useful for the synthesis of ring-locked π-bridges. Examples are shown in Scheme 4.21 for a polyene π-bridge having a diphenylamino donor (**130**)[87] and a ring-locked dialkylamino donor (**131**).[88] Reaction of the ketones with diisopropyl (cyanomethyl)-phosphonate gives the corresponding cyanoalkenes **132** and **133**, respectively, in high yields with a mixture of *cis* (*Z*) and *trans* (*E*) isomers. Reduction of the nitriles **132** and **133** gives the corresponding aldehydes **134** and **135**, respectively, in high yields. This methodology using HWE chemistry contrasts to the aldehyde extension using Wittig salt **122** in that the intermediate alkene is isolated; however, the low reactivity of Wittig reagents towards ketones in general favors the use of a phosphonate for extension of ketones. Also note that when **134** and **135** are coupled to a strong acceptor in a subsequent step (a Knoevenagel condensation, Chapter 6), the reaction gives only the *trans* isomer in moderate and high yields.

4.4 CONCLUSION

This chapter has covered some of the most widely utilized reactions in π-conjugated materials synthesis in aryl-halide and aryl-aldehyde formylation. In many reaction schemes for π-conjugated materials, both halogenation and formylation are usually present, and are critical steps in the overall transformations. Although the synthesis of aryl-halides were illustrated in this chapter to be useful intermediates in the synthesis of an aryl-aldehyde *via* lithiation–formylation, in Chapter 5 the aryl-halides themselves are very often the fundamentally important functional group for aryl–aryl organometallic coupling. In addition to the synthesis of alkenes *via* Wittig and HWE methods, aldehydes are often used as a functional group to attach acceptors, as will be illustrated in Chapter 6.

REFERENCES

1. R. Larock, in *Comprehensive Organic Transformations: A Guide to Functional Group Preparations*, John Wiley & Sons, New York, 2nd edn, 1999, pp. 620–626.
2. C.-Y. Kuo, W. Nie, H. Tsai, H.-J. Yen, A. D. Mohite, G. Gupta, A. M. Dattelbaum, D. J. William, K. C. Cha, Y. Yang, L. Wang and H.-L. Wang, *Macromolecules*, 2014, **47**, 1008.
3. I. A. Wright, P. J. Skabara, J. C. Forgie, A. L. Kanibolotsky, B. González, S. J. Coles, S. Gambino and I. D. W. Samuel, *J. Mater. Chem.*, 2011, **21**, 1462.

4. B. Burkhart, P. P. Khlyabich, T. Cakir Canak, T. W. LaJoie and B. C. Thompson, *Macromolecules*, 2011, **44**, 1242.
5. H. T. Nguyen, O. Coulembier, J. Winter, P. Gerbaux, X. Crispin and P. Dubois, *Polym. Bull.*, 2010, **66**, 51.
6. P. van Rijn, D. Janeliunas, A. M. Brizard, M. C. A. Stuart, G. J. M. Koper, R. Eelkema and J. H. van Esch, *New J. Chem.*, 2011, **35**, 558.
7. Y. Nie, B. Zhao, P. Tang, P. Jiang, Z. Tian, P. Shen and S. Tan, *J. Polym. Sci., Part A: Polym. Chem.*, 2011, **49**, 3604.
8. P. Paoprasert, J. W. Spalenka, D. L. Peterson, R. E. Ruther, R. J. Hamers, P. G. Evans and P. Gopalan, *J. Mater. Chem.*, 2010, **20**, 2651.
9. J.-C. Li, S.-J. Kim, S.-H. Lee, Y.-S. Lee, K. Zong and S.-C. Yu, *Macromol. Res.*, 2009, **17**, 356.
10. P. Blanchard, P. Verlhac, L. Michaux, P. Frere and J. Roncali, *Chem.–Eur. J.*, 2006, **12**, 1244.
11. Y. He, W. Wu, G. Zhao, Y. Liu and Y. Li, *Macromolecules*, 2008, **41**, 9760.
12. M. He and F. Zhang, *J. Org. Chem.*, 2007, **72**, 442.
13. R. Tkachov, V. Senkovskyy, H. Komber and A. Kiriy, *Macromolecules*, 2011, **44**, 2006.
14. J. Li, M. Yan, Y. Xie and Q. Qiao, *Energy Environ. Sci.*, 2011, **4**, 4276.
15. T. Yamamoto, K. Toyota and N. Morita, *Tetrahedron Lett.*, 2010, **51**, 1364.
16. R. S. Loewe, S. M. Khersonsky and R. D. McCullough, *Adv. Mater.*, 1999, **11**, 250.
17. A. Yokoyama, R. Miyakoshi and T. Yokozawa, *Macromolecules*, 2004, **37**, 1169.
18. H. T. Nguyen, B. C. Dong and N. H. Nguyen, *Macromol. Res.*, 2013, **22**, 85.
19. T. Beryozkina, V. Senkovskyy, E. Kaul and A. Kiriy, *Macromolecules*, 2008, **41**, 7817.
20. S. Wu, L. Huang, H. Tian, Y. Geng and F. Wang, *Macromolecules*, 2011, **44**, 7558.
21. S. D. Boyd, A. K. Y. Jen and C. K. Luscombe, *Macromolecules*, 2009, **42**, 9387.
22. P. M. Beaujuge, H. N. Tsao, M. R. Hansen, C. M. Amb, C. Risko, J. Subbiah, K. R. Choudhury, A. Mavrinskiy, W. Pisula, J. L. Bredas, F. So, K. Mullen and J. R. Reynolds, *J. Am. Chem. Soc.*, 2012, **134**, 8944.
23. A. C. Heinrich, B. Thiedemann, P. J. Gates and A. Staubitz, *Org. Lett.*, 2013, **15**, 4666.
24. J. I. Tietz, A. J. Seed and P. Sampson, *Org. Lett.*, 2012, **14**, 5058.

25. X. Guo, S. R. Puniredd, M. Baumgarten, W. Pisula and K. Mullen, *J. Am. Chem. Soc.*, 2012, **134**, 8404.
26. L. Y. Lin, C. H. Tsai, K. T. Wong, T. W. Huang, L. Hsieh, S. H. Liu, H. W. Lin, C. C. Wu, S. H. Chou, S. H. Chen and A. I. Tsai, *J. Org. Chem.*, 2010, **75**, 4778.
27. H. Ebata, E. Miyazaki, T. Yamamoto and K. Takimiya, *Org. Lett.*, 2007, **9**, 4499.
28. T. Dohi, N. Yamaoka, S. Nakamura, K. Sumida, K. Morimoto and Y. Kita, *Chem.–Eur. J.*, 2013, **19**, 2067.
29. C. Du, W. Li, C. Li and Z. Bo, *J. Polym. Sci., Part A: Polym. Chem.*, 2013, **51**, 383.
30. M. He, T. M. Leslie and J. A. Sinicropi, *Chem. Mater.*, 2002, **14**, 4662.
31. M. Yamamoto and S. Zheng, *US Pat. pub.*, 2010/0152338, 2010.
32. P. Wuts and T. Greene, in *Greene's Protective Groups in Organic Synthesis*, John Wiley & Sons, Inc., Hoboken, NJ, 2007, pp. 165–221.
33. L. Shi, C. He, D. Zhu, Q. He, Y. Li, Y. Chen, Y. Sun, Y. Fu, D. Wen, H. Cao and J. Cheng, *J. Mater. Chem.*, 2012, **22**, 11629.
34. H. Xiao, B. Leng and H. Tian, *Polymer*, 2005, **46**, 5707.
35. L. Guo, K. F. Li, M. S. Wong and K. W. Cheah, *Chem. Commun.*, 2013, **49**, 3597.
36. S. H. Kim, I. Cho, M. K. Sim, S. Park and S. Y. Park, *J. Mater. Chem.*, 2011, **21**, 9139.
37. N. Berton, I. Fabre-Francke, D. Bourrat, F. Chandezon and S. Sadki, *J. Phys. Chem. B*, 2009, **113**, 14087.
38. J.-H. Cho, Y.-S. Ryu, S.-H. Oh, J.-K. Kwon and E.-K. Yum, *Bull. Korean Chem. Soc.*, 2011, **32**, 2461.
39. X. Zong, M. Liang, T. Chen, J. Jia, L. Wang, Z. Sun and S. Xue, *Chem. Commun.*, 2012, **48**, 6645.
40. J. Youn, P.-Y. Huang, Y.-W. Huang, M.-C. Chen, Y.-J. Lin, H. Huang, R. P. Ortiz, C. Stern, M.-C. Chung, C.-Y. Feng, L.-H. Chen, A. Facchetti and T. J. Marks, *Adv. Funct. Mater.*, 2012, **22**, 48.
41. D. Mysliwiec and M. Stepien, *Angew. Chem., Int. Ed.*, 2013, **52**, 1713.
42. J. He, F. Guo, X. Li, W. Wu, J. Yang and J. Hua, *Chem.–Eur. J.*, 2012, **18**, 7903.
43. K. F. Chen, C. W. Chang, J. L. Lin, Y. C. Hsu, M. C. Yeh, C. P. Hsu and S. S. Sun, *Chem.–Eur. J.*, 2010, **16**, 12873.
44. U. Lawrentz, W. Grahn, K. Lukaszuk, C. Klein, R. Wortmann, A. Feldner and D. Scherer, *Chem.–Eur. J.*, 2002, **8**, 1573.
45. A. Miyake, M. Suzuki, M. Sumino, Y. Iizuka and T. Ogawa, *Org. Process Res. Dev.*, 2002, **6**, 922.

46. A. Rajput and P. Girase, *Int. J. Pharm., Chem. Biol. Sci.*, 2013, **3**, 25.

47. S. Subramaniyan, H. Xin, F. S. Kim and S. A. Jenekhe, *Macromolecules*, 2011, **44**, 6245.

48. S. Delgado Ledesma, R. Ponce Ortiz, M. C. Ruiz Delgado, Y. Vida, E. Perez-Inestrosa, J. Casado, V. Hernandez, O. K. Kim, J. M. Lehn and J. T. Lopez Navarrete, *Chem.–Eur. J.*, 2004, **10**, 3805.

49. M. Xu, M. Zhang, M. Pastore, R. Li, F. De Angelis and P. Wang, *Chem. Sci.*, 2012, **3**, 976.

50. J. Nakayama and T. Fujimori, *Heterocycles*, 1991, **32**, 991.

51. A. H. Younes, L. Zhang, R. J. Clark, M. W. Davidson and L. Zhu, *Org. Biomol. Chem.*, 2010, **8**, 5431.

52. A. Yassar, C. Videlot and A. Jaafari, *Sol. Energy Mater. Sol. Cells*, 2006, **90**, 916.

53. X. Cheng, M. Liang, S. Sun, Y. Shi, Z. Ma, Z. Sun and S. Xue, *Tetrahedron*, 2012, **68**, 5375.

54. G. Li, K.-J. Jiang, P. Bao, Y.-F. Li, S.-L. Li and L.-M. Yang, *New J. Chem.*, 2009, **33**, 868.

55. S.-L. Li, K.-J. Jiang, K.-F. Shao and L.-M. Yang, *Chem. Commun.*, 2006, 2792.

56. C. Bian, G. Jiang, H. Tong, Y. Cheng, Z. Xie, L. Wang, X. Jing and F. Wang, *J. Polym. Sci., Part A: Polym. Chem.*, 2011, **49**, 3911.

57. Y. Song, C. A. Di, Z. Wei, T. Zhao, W. Xu, Y. Liu, D. Zhang and D. Zhu, *Chem.–Eur. J.*, 2008, **14**, 4731.

58. R. R. Petrov, L. Knight, S. R. Chen, J. Wager-Miller, S. W. McDaniel, F. Diaz, F. Barth, H. L. Pan, K. Mackie, C. N. Cavasotto and P. Diaz, *Eur. J. Med. Chem.*, 2013, **69**, 881.

59. M. He, R. J. Twieg, U. Gubler, D. Wright and W. E. Moerner, *Chem. Mater.*, 2003, **15**, 1156.

60. W. Leng, H. Y. Woo, D. Vak, G. C. Bazan and A. Myers Kelley, *J. Raman Spectrosc.*, 2006, **37**, 132.

61. L. F. v. Staden, D. Gravestock and D. J. Ager, *Chem. Soc. Rev.*, 2002, **31**, 195.

62. I. Gosney and A. Powley, in *Organophosphorous Reagents in Organic Synthesis*, Academic Press, London, 1979, pp. 17–154.

63. B. E. Maryanoff and A. B. Reitz, *Chem. Rev.*, 1989, **89**, 863.

64. H. J. Bestmann, *Angew. Chem., Int. Ed. Engl.*, 1965, **4**, 583.

65. J.-X. Zhang, P. Dubois and R. Jérôme, *Synth. Commun.*, 1996, **26**, 3091.

66. M. Blanchard-Desce, L. Porrès and B. K. Bhatthula, *Synthesis*, 2003, 1541.

67. J. Jia, K. Cao, P. Xue, Y. Zhang, H. Zhou and R. Lu, *Tetrahedron*, 2012, **68**, 3626.

68. X. Piao, X. Zhang, Y. Mori, M. Koishi, A. Nakaya, S. Inoue, I. Aoki, A. Otomo and S. Yokoyama, *J. Polym. Sci., Part A: Polym. Chem.*, 2011, **49**, 47.

69. Z. Q. Liu, Q. Fang, D. Wang, D. X. Cao, G. Xue, W. T. Yu and H. Lei, *Chem.–Eur. J.*, 2003, **9**, 5074.

70. R. Chen, X. Yang, H. Tian, X. Wang, A. Hagfeldt and L. Sun, *Chem. Mater.*, 2007, **19**, 4007.

71. M. Rumi, J. E. Ehrlich, A. A. Heikal, J. W. Perry, S. Barlow, Z. Hu, D. McCord-Maughon, T. C. Parker, H. Röckel, S. Thayumanavan, S. R. Marder, D. Beljonne and J.-L. Brédas, *J. Am. Chem. Soc.*, 2000, **122**, 9500.

72. R. P. Ortiz, M. C. Ruiz Delgado, J. Casado, V. Hernandez, O. K. Kim, H. Y. Woo and J. T. Lopez Navarrete, *J. Am. Chem. Soc.*, 2004, **126**, 13363.

73. P. Leriche, J.-M. Raimundo, M. Turbiez, V. Monroche, M. Allain, F. o.-X. Sauvage, J. Roncali, P. Frère and P. J. Skabara, *J. Mater. Chem.*, 2003, **13**, 1324.

74. B. Walker, in *Organophosphorous Reagents in Organic Synthesis*, Academic Press, London, 1979, pp. 155–206.

75. M. K. Poindexter and T. J. Katz, *Tetrahedron Lett.*, 1988, **29**, 1513.

76. H. Takahashi, S. Inagaki, N. Yoshii, F. Gao, Y. Nishihara and K. Takagi, *J. Org. Chem.*, 2009, **74**, 2794.

77. S. Zheng, S. Barlow, T. C. Parker and S. R. Marder, *Tetrahedron Lett.*, 2003, **44**, 7989.

78. X. Ma, F. Ma, Z. Zhao, N. Song and J. Zhang, *J. Mater. Chem.*, 2009, **19**, 2975.

79. K. N. Shivananda, I. Cohen, E. Borzin, Y. Gerchikov, M. Firstenberg, O. Solomeshch, N. Tessler and Y. Eichen, *Adv. Funct. Mater.*, 2012, **22**, 1489.

80. M. Marszalek, S. Nagane, A. Ichake, R. Humphry-Baker, V. Paul, S. M. Zakeeruddin and M. Grätzel, *J. Mater. Chem.*, 2012, **22**, 889.

81. Z. Yang, Z. Chi, B. Xu, H. Li, X. Zhang, X. Li, S. Liu, Y. Zhang and J. Xu, *J. Mater. Chem.*, 2010, **20**, 7352.

82. A. J. P. Akelaitis, B. C. Olbricht, P. A. Sullivan, Y. Liao, S. K. Lee, D. H. Bale, D. B. Lao, W. Kaminsky, B. E. Eichinger, D. H. Choi, P. J. Reid and L. R. Dalton, *Opt. Mater.*, 2008, **30**, 1504.

83. J. Pei, J. Ni, X.-H. Zhou, X.-Y. Cao and Y.-H. Lai, *J. Org. Chem.*, 2002, **67**, 4924.

84. S. J. Chung, M. Rumi, V. Alain, S. Barlow, J. W. Perry and S. R. Marder, *J. Am. Chem. Soc.*, 2005, **127**, 10844.

85. C. W. Spangler and M. He, *J. Chem. Soc., Perkin Trans. 1*, 1995, 715.

86. M. Blanchard-Desce, V. Alain, P. V. Bedworth, S. R. Marder, A. Fort, C. Runser, M. Barzoukas, S. Lebus and R. Wortmann, *Chem.–Eur. J.*, 1997, **3**, 1091.
87. Y.-J. Cheng, J. Luo, S. Hau, D. H. Bale, T.-D. Kim, Z. Shi, D. B. Lao, N. M. Tucker, Y. Tian, L. R. Dalton, P. J. Reid and A. K. Y. Jen, *Chem. Mater.*, 2007, **19**, 1154.
88. X.-H. Zhou, J. Luo, J. A. Davies, S. Huang and A. K. Y. Jen, *J. Mater. Chem.*, 2012, **22**, 16390.
89. K. H. Kim, Z. G. Chi, M. J. Cho, J. I. Jin, M. Y. Cho, S. J. Kim, J. S. Joo and D. H. Choi, *Chem. Mater.*, 2007, **19**, 4925.

CHAPTER 5

Creating Conjugation: Organometallic Coupling

5.1 INTRODUCTION

Organometallic coupling methods have been used widely in the synthesis of π-conjugated materials to create π-conjugation between aromatic and heteroaromatic moieties. Table 5.1 attempts to summarize some of the main reaction chemistries and their commonly referred to names. First note that all the reactions use a catalyst except the C—C Ullmann coupling, which in Section 3.3.3 was first introduced as the *N*-aryl Ullmann coupling variant in the synthesis of donors. Also note that, in general, the Ullmann coupling works well mostly for homocoupling (Ar—Ar), and as such the structural variety of the final product is relatively limited compared to that available using cross coupling methods based on Pd catalysis (or Ni catalysis for the Kumada Coupling). Pd coupling reactions can be used to create aryl–aryl cross-coupled products (Stille, Suzuki, Kumada, and direct arylation), alkene-linked aryl cross-coupled products (Heck), and alkyne-linked aryl cross-coupled products (Sonogashira). Moreover, in addition to creating the fundamentally important π-conjugation link, they can be used to install pendant alkyl, aryl, and functional groups for solubility, processing, crosslinking, surface modification, and property modification (*e.g.*, IE and EA). *The enormous chemical and structural scope of organometallic coupling is the main reason they are*

Synthetic Methods in Organic Electronic and Photonic Materials: A Practical Guide
By Timothy C. Parker and Seth R. Marder
© Timothy C. Parker and Seth R. Marder, 2015
Published by the Royal Society of Chemistry, www.rsc.org

Table 5.1 Organometallic coupling reactions. X = halogen or pseudohalogen; M = metal, metalloid, or metal equivalent.

$$Ar-X \; + \; M\cdots R \;\; \xrightarrow{\text{Catalyst}} \;\; Ar-R \; + \; M-X$$

Entry	M	R	Catalyst	Name
1	Cu···	···Ar	None	Ullmann
2	R$_3$Sn···	···Ar'	Pd	Stille
3	RO$_2$B···	···Ar'	Pd	Suzuki
4	Mg···	···Ar' or Alkyl	Ni	Kumada
5	Zn---	···Ar' or Alkyl	Pd	Negishi
6	R$_3$Si----	···Ar' or Alkyl	Pd	Hiyama
7	Cu···	···≡—Ar' or Alkyl	Pd	Sonogashira
8	H--	Ar' or FG	Pd	Heck
9	H--	···Ar or Ar'	Pd or Ir	Direct arylation

a central synthetic method utilized in the synthesis of both small molecule and polymer π-conjugated materials. The purpose of this chapter is to provide a strong fundamental basis for each organometallic method by illustrating various conditions for the synthesis of small molecule π-conjugated materials. The use of organometallic coupling in the synthesis of polymers is a large topic itself and is covered in Sections 7.3.3–7.4.4.

5.2 BASIC ORGANOMETALLIC REACTIONS

The organometallic reactions presented in this chapter have overall mechanisms that include two or more basic organometallic reactions in combination. There are a number of good organometallic textbooks that provide many examples electron counting, oxidation state determination, and basic organometallic processes.[1] Herein, palladium will be used to illustrate the basic organometallic reactions due to its central importance in catalytic reactions used in the synthesis of π-conjugated materials as summarized in Table 5.1. Examples of Pd complexes are shown in Figure 5.1. Typically, Pd complexes have a square planar geometry with the Pd in an oxidation state of 0 or 2, *i.e.*, Pd(0) or Pd(II). Complex F5.1a is an example of an electronically *saturated* Pd(0) complex since the total electron count is 18. Complex F5.1b has a dissociated ligand (shown) and is an example of a Pd(0) complex that is electronically *unsaturated* since the total electron count is 16. Complex F5.1c is an example of a Pd(II) complex since the total charge

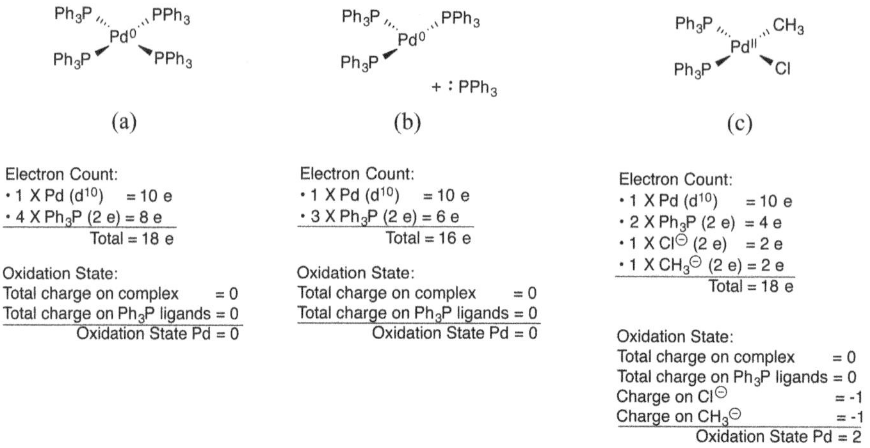

(a) (b) (c)

Electron Count:
• 1 X Pd (d^{10}) = 10 e
• 4 X Ph$_3$P (2 e) = 8 e
  Total = 18 e

Oxidation State:
Total charge on complex   = 0
Total charge on Ph$_3$P ligands = 0
  Oxidation State Pd = 0

Electron Count:
• 1 X Pd (d^{10}) = 10 e
• 3 X Ph$_3$P (2 e) = 6 e
  Total = 16 e

Oxidation State:
Total charge on complex   = 0
Total charge on Ph$_3$P ligands = 0
  Oxidation State Pd = 0

Electron Count:
• 1 X Pd (d^{10}) = 10 e
• 2 X Ph$_3$P (2 e) = 4 e
• 1 X Cl$^{\ominus}$ (2 e) = 2 e
• 1 X CH$_3^{\ominus}$ (2 e) = 2 e
  Total = 18 e

Oxidation State:
Total charge on complex   = 0
Total charge on Ph$_3$P ligands = 0
Charge on Cl$^{\ominus}$ = -1
Charge on CH$_3^{\ominus}$ = -1
  Oxidation State Pd = 2

Figure 5.1 Electron counting and oxidation state of: (a) a neutral palladium complex; (b) a dissociated palladium complex; and (c) a palladium complex with organic ligand.

on the complex is 0 the chloride and methyl ligands each have a *formal* charge of −1 (oxidation state = total complex charge − Σ ligand charges). The types of Pd(0) and Pd(II) complexes shown in Figure 5.1 play a central role in organometallic coupling reactions that are used in the synthesis of π-conjugated materials. In Sections 5.2.1–5.2.5 below, the basic organometallic reactions that include Pd(0) and Pd(II) complexes are discussed, including synthetically important reasons for varying ligands and reaction conditions. It should be noted, however, that Pd(IV) complexes (even those with and electron counts of 20) are known and have been demonstrated to be effective catalysts.[2]

5.2.1 Ligand Exchange – Transmetallation

Exchange of one ligand for another ligand is a common process in organometallic chemistry. Scheme 5.1 demonstrates the process in a Pd(II) complex (S5.1a) where the chloride ligand is exchanged for the more nucleophilic methyl group of a Grignard reagent (S5.1b) to give a Pd(II) complex with two carbanion ligands (S5.1c) and a magnesium salt (MgCl$_2$). In this example substitution occurs because the methyl anion is more nucleophilic than the chloride anion. In other cases, similarly nucleophilic ligands may exchange and one may precipitate from solution if the salt is insoluble. From the perspective of the methyl group, the Mg(II) metal has been exchanged for a Pd(II) metal, thus the process is also referred *transmetallation*.

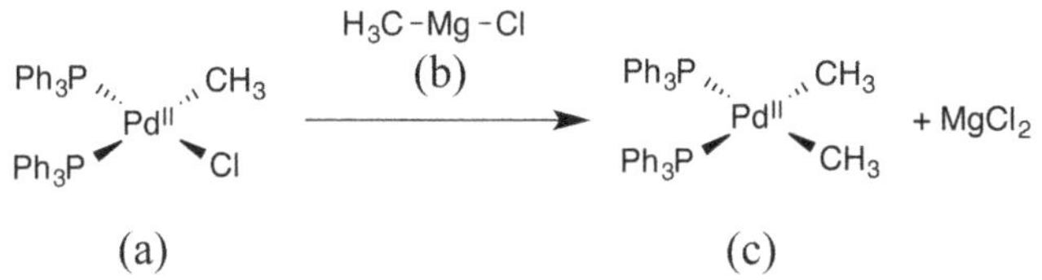

Scheme 5.1 Ligand exchange on palladium complexes.

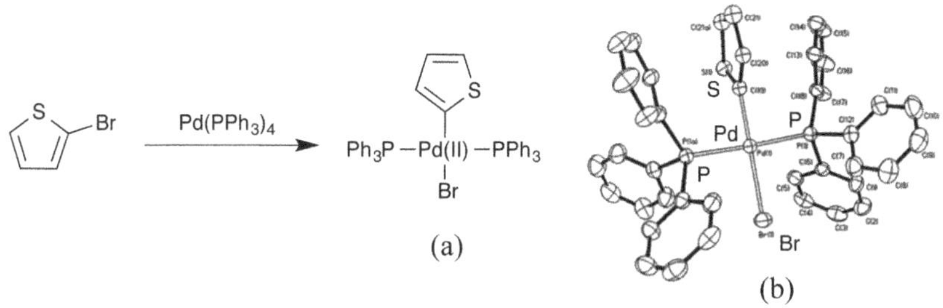

Scheme 5.2 Oxidative addition of palladium to aryl halides. Reprinted with permission from *Organometallics*, 1998, **17**, 3988. Copyright 1998 American Chemical Society.

5.2.2 Oxidative Addition

Reaction of a metal with an organic halide to give a complex where the metal has been inserted into a halogen bond is one of the most important steps in catalysis. An example is shown in Scheme 5.1 for the reaction of 2-bromothiophene with $Pd(PPh_3)_4$ to give the corresponding Pd(II) complex (S5.2a). The geometry of the halide and organic ligand around the palladium is generally *trans* as shown in the X-ray crystal structure (S5.2b) of S5.2a in Scheme 5.2. During the process, the ligands dissociate so that the metal becomes coordinatively unsaturated before inserting into the Ar—X bond (X is a halide). During the process, the oxidation state of Pd increases from 0 to 2 $(Pd(0) \rightarrow Pd(II))$. Although the mechanisms for oxidative addition may be complex, including radicals in some cases, the process typically behaves as if the metal were a nucleophile and the reaction is a nucleophilic aromatic substitution. As such, Ar—X substrates with higher electron affinity often undergo oxidative insertion more readily, and substrates with electron withdrawing groups are often referred to as "activated" towards oxidative addition.

5.2.3 Reductive Elimination

Reductive elimination is another fundamentally important process in organometallic chemistry and in most cases two ligands are removed from the metal and form a bond while the oxidation state

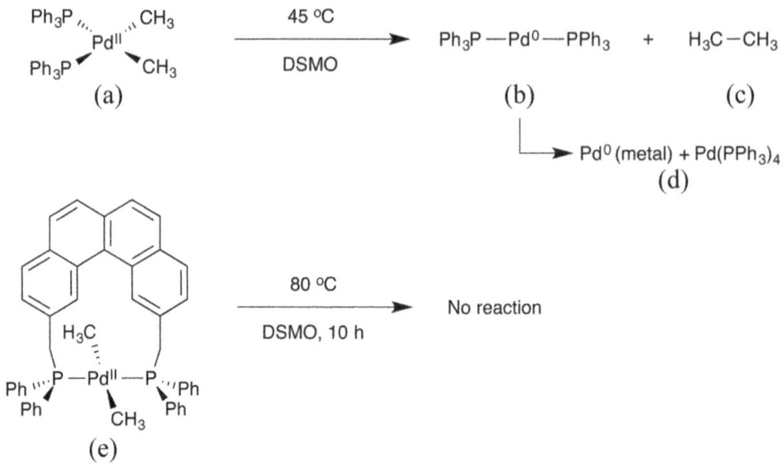

Scheme 5.3 Reductive elimination from palladium complexes.

of the metal is reduced by two. An example of reductive elimination with a Pd(II) complex is shown in Scheme 5.3. The *cis*-dimethyl Pd(II) complex (S5.3a) reductively eliminates the two methyl ligands to give a Pd(0) species (S5.3b) and ethane (S5.3c). It should be noted that the coordinately and electronically unsaturated complex S5.3b is typically unstable and disproportionates (S5.3d) to give Pd metal and Pd(PPh$_3$)$_4$, which is important in certain catalytic cycles with specific Pd catalysts, as discussed below in Section 5.4. The two ligands often need to be adjacent to each other around the metal for reductive elimination to occur, which for Pd(II) square planar complexes requires a *cis* geometry. For example, although S5.3a undergoes relatively facile reductive elimination, the Pd(II) complex S5.3e does not reductively eliminate at relatively high temperatures since the bridging ligand locks the two methyl groups in a *trans* geometry.[3] For Pd(II), more coordinating solvents such as THF and DMF tend to promote reductive elimination along with Pd ligands that can readily dissociate.

5.2.4 Migratory Insertion

Migratory insertion is another fundamental organometallic reaction important in the catalytic cycle of the Heck reaction discussed in Section 5.4.6 below. An example is shown for a Pd(II) complex (S5.4a) in Scheme 5.4. Note that the complexed alkene is a neutral ligand accounting for two electrons so that complex S5.4a is electronically saturated with 18 electrons. An arrow formalism representation

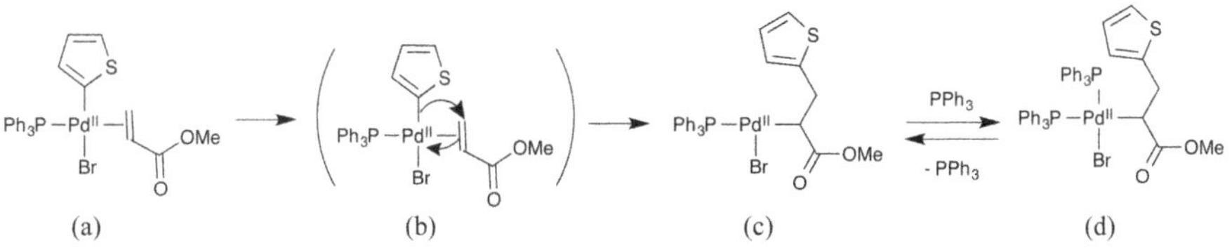

Scheme 5.4 Migratory insertion on alkene–palladium complexes.

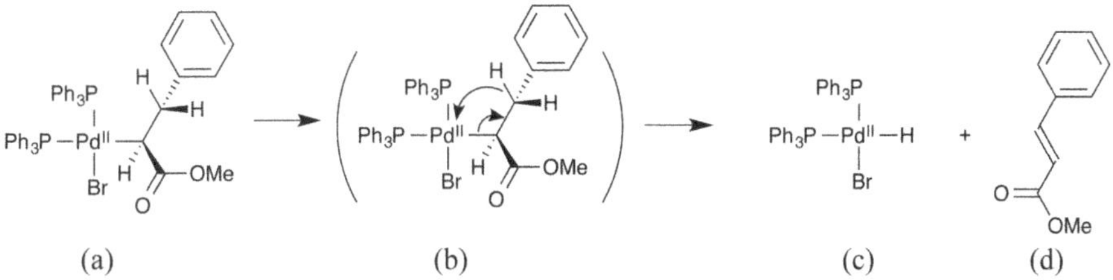

Scheme 5.5 β-Hydride elimination from palladium complexes.

of migratory insertion is shown in S5.4b to give the Pd(II) complex S5.4c, which is electronically unsaturated with 16 electrons. In this case, PPh$_3$ (or other ligands) can coordinate to the metal to stabilize the Pd(II) as complex S5.4d. Note that the arrow formalism in S5.4b treats the process as a nucleophilic attack of the thiophene ligand on the alkene ligand, and indeed alkenes that have electron withdrawing groups react almost exclusively in migratory insertions to give Michael addition.

5.2.5 β-Hydride Elimination

β-Hydride elimination is another fundamental organometallic reaction that is important in the mechanism of the Heck reaction discussed in Section 5.4.6. An example, again in terms of a Pd(II) complex, is shown in Scheme 5.5 for the complex S5.5a. The arrow formalism representation is shown in S5.5b. In this case, a hydride at the β position on the ligand displaces the ligand resulting in a Pd(II) complex with a hydride ligand (S5.5c) and an alkene (S5.5d), although in some cases the alkene may remain η^2-bonded to the metal.

5.3 C—C ULLMANN COUPLING

The Ullmann coupling has been known since the beginning of the 20th century and a substantial review of aryl–aryl Ullmann coupling appeared in 1974.[4] As illustrated in Scheme 5.6, the reaction involves

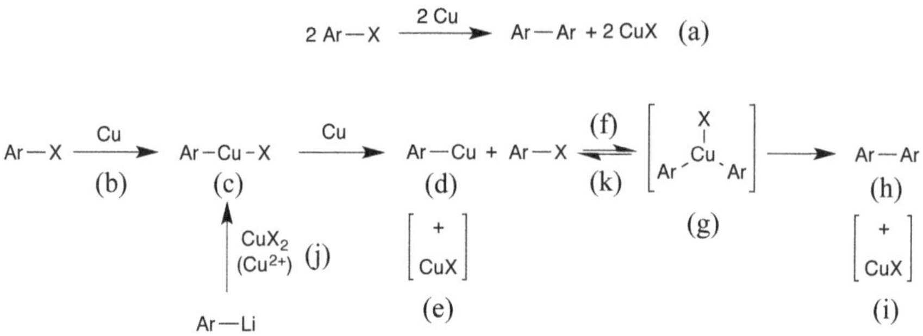

$$2\,Ar{-}X \xrightarrow{\;2\,Cu\;} Ar{-}Ar + 2\,CuX \quad (a)$$

$$Ar{-}X \xrightarrow{\;Cu\;}_{(b)} Ar{-}Cu{-}X \xrightarrow{\;Cu\;}_{(c)} Ar{-}Cu + Ar{-}X \underset{(k)}{\overset{(f)}{\rightleftharpoons}} \begin{bmatrix} & X & \\ & | & \\ Ar^{\diagup} & Cu & \diagdown Ar \end{bmatrix} \longrightarrow \begin{matrix} Ar{-}Ar \\ (h) \end{matrix}$$

$$\Big\uparrow\; \begin{matrix} CuX_2 \\ (Cu^{2+}) \end{matrix}\;(j) \qquad \begin{bmatrix} + \\ CuX \end{bmatrix} \qquad\qquad (g) \qquad \begin{bmatrix} + \\ CuX \end{bmatrix}$$

$$Ar{-}Li \qquad\qquad\qquad (e) \qquad\qquad\qquad\qquad\qquad (i)$$

Scheme 5.6 General illustration of the C–C Ullmann coupling.

the coupling of an aryl halide with a stoichiometric amount of copper (S5.6a). Mechanistically, the first step of the reaction is thought to be insertion (S5.6b) of the copper metal into the aryl halide bond to give an organocopper species (S5.6c). With this insertion, the copper metal goes from oxidation state 0 to oxidation state 2 (Cu[0] → Cu[II]) and thus is another example of oxidative addition discussed above in Section 5.2.2 and throughout Chapter 5. The organocopper(II) species is reduced to an organocopper(I) species by another equivalent of copper metal (S5.6d), which gives an equivalent of copper(I) halide as a by-product (S5.6e). Organocopper(I) may insert into another aryl halide (S5.6f) to give an organocopper(III) species (S5.6g), which can eliminate to give the coupled biaryl product (S5.6h) and another equivalent of copper(I) halide (S5.6i). This elimination takes copper(III) to copper(I) and is another example reductive elimination discussed above in Section 5.2.3 and throughout Chapter 5. It should be noted that the organocopper(II) species (S5.6c) can often be generated by reaction with an aryl lithium and copper(II) halide salt (S5.6j) at generally lower temperatures than oxidative addition of copper from the aryl halide (S5.6b), although the remainder of the reaction mechanism may be slightly different.

As with the C–N Ullmann coupling, the classic C–C Ullmann coupling is often limited to aryl iodides and generally requires high temperatures.[5] Additionally, functional groups in the substrate such as amines, alcohols, and carboxylic acids can be reactive during the process and provide C–N or C–O[6] coupled products instead of the desired C–C coupling product. However, electron-withdrawing groups on the aryl halide substrate, in particular on the carbon adjacent to the aryl halide carbon, are good substrates for this reaction and the reactions often proceed in high yield. Most importantly, *cross-coupling Ullmann C–C reactions are rare*, mainly because the

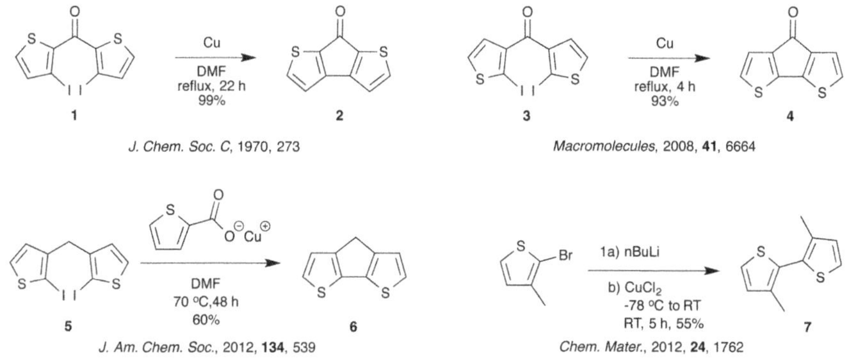

Scheme 5.7 C—C Ullmann coupling on several substrates.[7–10]

organocopper(III) intermediate (S5.6g) can revert (S5.6k) to the aryl halide and the organocopper(I) species (S5.6d). Such insertion and reversion effectively means that the aryl groups are "scrambled," which practically limits the reaction to "homocoupling" of identical aryl groups. Scheme 5.7 outlines some reactions where C—C Ullmann coupling has been used in the synthesis of π-conjugated materials. One reaction that works particularly well *via* Ullmann conditions is the intramolecular coupling of thiophene substituted ketones, such as the reaction of dithiophene ketone **1** to give **2** in high yield[7] and isomeric dithiophene ketone **3** to give **4** in high yield.[8] However, note that even with a favorable reaction substrate (intramolecular coupling and the electron withdrawing ketone), the reaction conditions require relatively high temperature and long reaction times. An activated copper(I) salt of thiophene carboxylic acid may be used to reduce the temperature of the reaction, as is shown for the intramolecular coupling of **5** to **6** in moderate yield.[9] Finally, the reaction temperature and time can be reduced further by avoiding the oxidative addition of copper metal through forming the organocopper(II) halide from the organolithium and a copper(II) halide salt such as with the reaction of 3-methyl-2-bromothiophene to give the homocoupled bithiophene **7** in moderate yield.[10]

5.4 PD-CATALYZED COUPLINGS

5.4.1 Introduction

A Nobel Prize was awarded for the development of palladium catalyzed cross-coupling reactions in 2010 to Professors Richard F. Heck (University of Delaware), Ei-ichi Negishi (Purdue University), and

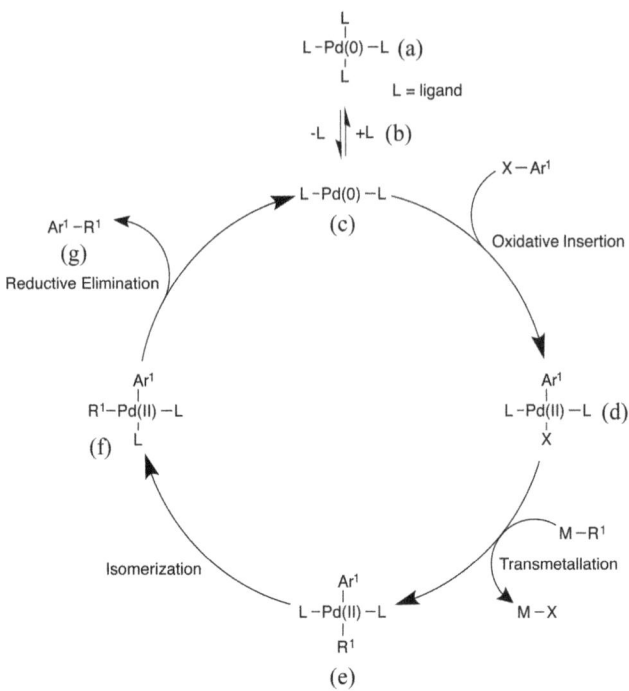

Scheme 5.8   Catalytic cycle for palladium(0) catalyzed coupling.

Akira Suzuki (Hokkaido University). These reactions have enabled chemists to achieve important transformations in a simple fashion that might be synthetically difficult to achieve otherwise. In the synthesis of π-conjugated small molecules and polymers, Pd-catalyzed coupling reactions are nearly ubiquitous. A general mechanism for Pd-catalyzed coupling is illustrated in Scheme 5.8. Most active Pd-catalysts have Pd in the 0 oxidation state (S5.8a), which can undergo ligand dissociation (S5.8b) to give a Pd(0) complex with open coordination site/s (S5.8c). Typically, the Pd(0) complex undergoes oxidative addition into an active Ar—X bond (where X is usually a halide) to give a Pd(II) complex (S5.8d). This complex can undergo a substitution of the halide with a nucleophilic group on an organometallic compound, referred to as a transmetallation step, to give a Pd(II) species having the two organic coupling partners (*i.e.*, Ar^1 and R^1) as ligands on Pd, which are shown in a *trans* orientation around the metal in S5.8e. As discussed in Section 5.2.3 above, typically, such a *trans* complex cannot undergo reductive elimination and would have to isomerize to the *cis* isomer (S5.8f) to favor reductive elimination and form the coupled product S5.8g and the Pd(0) complex that can reenter the catalytic cycle (S5.8c). It should be noted that in some cases, the Pd(II) complex S5.8e can undergo oxidative insertion with another aryl halide to give a Pd(IV) species, which may give mixtures in cross coupling reactions,[3] although it should be noted that other Pd(IV) species are highly active participants in more recently developed Pd coupling reactions.[2]

One of the main general features of Pd-catalyzed couplings is that the rate of oxidative addition is often the rate-limiting step. Typically, the aromatic substrate (Ar—X) is an aryl halide, but X can also be a "pseudohalide" such as a sulfonate (*e.g.*, triflate, tosylate, or mesylate)[11] and in some cases a diazonium salt.[12] Such pseudohalides can be useful when aryl halide substrates are not stable or are difficult to synthesize. In the synthesis of π-conjugated materials, aryl halides have been used to for a vast majority of reactions. Among the halides and triflate (OTf, the most commonly used pseudohalide), the rate of the oxidative addition usually follows the trend I > OTf > Br >> Cl. These differences in reactivity are often sufficiently large that when different halides and pseudohalides are both present in the substrate, it is possible to produce selectively a single cross-coupled product derived from reaction of the more reactive halide or triflate. In general, electron-donating groups on the substrate tend to decrease the rate of oxidative addition, and electron-withdrawing groups tend to increase the rate of oxidative addition. As will be illustrated in this chapter, iodides and bromides are the aryl halides that are most widely employed, with iodide being the prevalent choice when the substrate is an electron donor and more resistant to oxidative addition.

The main point of differentiation in the various Pd catalyzed couplings is the organometallic compound used in the transmetallation step. Different metals or metalloids (B, Si) often have slight variations in the catalytic cycle mechanism presented in Scheme 5.8, which are important to know in the context of substrate compatibility, and these differences will be presented in each section below. The organometal can be generated in a separate step and in some cases purified (Stille, Suzuki, Hiyama), can be generated in one pot as the first step or *in situ* with stoichiometric use of the metal (Kumada, Negishi), or can be generated *in situ* catalytically (Sonogashira). Suzuki and Stille reagents are particularly useful since they can often be purified under normal laboratory conditions and are often shelf-stable reagents (this ability to obtain highly pure reagents is often critically important for the synthesis of polymers as will be discussed in Section 7.2). As will be illustrated, the Heck and direct arylation mechanisms, after common oxidative addition steps, are substantially different from the reactions that use an organometallic compound for transmetallation, and their mechanisms will be briefly presented as well. In the synthesis of π-conjugated materials, the vast majority of coupling reactions employed have been the Stille, Suzuki, Kumada, or Sonogashira methods, and as such, the Negishi[13] and Hiyama[14] couplings will not be broadly discussed.

In some cases, particularly in π-conjugated polymers, the products may be contaminated with Pd metal nanoparticles from the catalyst, which can affect the electrical or photophysical properties significantly.[15] Additionally, these Pd metal nanoparticles may be present in small amounts and may be difficult to detect by normal characterization techniques such as ^{1}H NMR, HPLC, and UV/vis spectroscopy.[16] In these cases, addition of a Pd scavenger that strongly complexes the Pd metal such as *N,N*-diethylphenylazothioformamide,[17] diethylammonium diethyldithiocarbamate,[18] or silica supported scavengers[19] may be used as part of the reaction workup or as a separate purification step to remove traces of Pd. The use of these or similar scavengers can reduce the concentration of palladium in the material to less than 0.1 ppm.[16]

Given the enormous significance and usage in every branch of synthetic chemistry, there are numerous reviews articles and books written on Pd-catalyzed cross coupling and the various specific cross couplings. Many of these articles provide historical context, and as such are valuable resources not only for providing numerous chemical examples but also for learning how the scientific process evolved. Here are a few references that are particularly useful both for chemical instruction and historical perspective:

- C. C. C. Johansson-Seechurn, M. O. Kitching, T. J. Colacot and V. Snieckus, Palladium Catalyzed Cross-Coupling: A Historical Contextual Perspective to the 2010 Nobel Prize, *Angew. Chem., Int. Ed.*, 2012, **51**, 5062–5085;[20]
- X.-F. Wu, P. Anbarasan, H. Neumann and M. Beller, From Nobel Metal to Nobel Prize: Palladium-Catalyzed Coupling Reactions as Key Methods in Organic Synthesis, *Angew. Chem., Int. Ed.*, 2010, **49**, 9047–9050;[21]
- Themed issue: "Cross coupling reactions in organic synthesis", *Chem. Soc. Rev.*, 2011, **40**(10), 4877–5203 (Whole issue with numerous articles devoted to recent aspects of C—C cross coupling, including microwave assistance and nanoparticle catalysis);[22] and
- B. H. Lipshutz, A. R. Abela, Z. V. Boskovic, T. Nishikata, C. Duplais and A. Krasovskiy, "Greening up" Cross-Coupling Chemistry, *Top. Catal.*, 2010, **53**(15–18), 985–990.[23]

5.4.2 Stille Coupling

The general reactivity of the Stille coupling is outlined in Scheme 5.9. The coupling was developed largely by John Stille and coworkers[24] and it has been widely used in many aspects in organic synthesis. As an indication of the breadth and usefulness of the method, it is generally acknowledged that Stille would have been an extremely likely candidate for a Nobel Prize if not for his premature death in the crash of United Airlines Flight 232 at Sioux City, Iowa, USA on July 19, 1989. The Stille coupling has many general features that underpin its synthetic utility, which are noted in Scheme 5.9. The substrate can include aryl, heteroaryl, or alkene halides or sulfonates and the organometal can be an aryl, heteroaryl, alkene, alkyne, allyl, or benzyl group, which enables a broad range of cross coupled products. Another useful feature of the reaction is that the Sn reagent can be purified and is typically shelf stable. The Sn groups typically used are either tributyltin (n-Bu$_3$Sn−) or trimethyltin (Me$_3$Sn−), with the trimethyltin reagents being more reactive; however, *trimethyl tin reagents are significantly more toxic than tributyl tin reagents and should be treated with due caution* (see below). A third feature of the reaction that is highly useful is that Sn reagents tend to be reasonably tolerant towards functional groups such as aldehydes, ketones, esters, *etc.* that are not stable to organometals such as Grignard or lithium reagents, which allows many Stille couplings to be employed without the use of protecting groups.

The mechanism of the Stille reaction is largely a prototypical Pd-catalyzed coupling as illustrated in Scheme 5.8. As such, one potential drawback is that when the Stille reagent has a relatively high electron affinity, such as with several acceptors used in the synthesis of π-conjugated materials, the transmetallation step can be very sluggish or non-operative and the Stille reagent itself may be unstable. Additionally, when substrates with relatively low IP are used, such as with

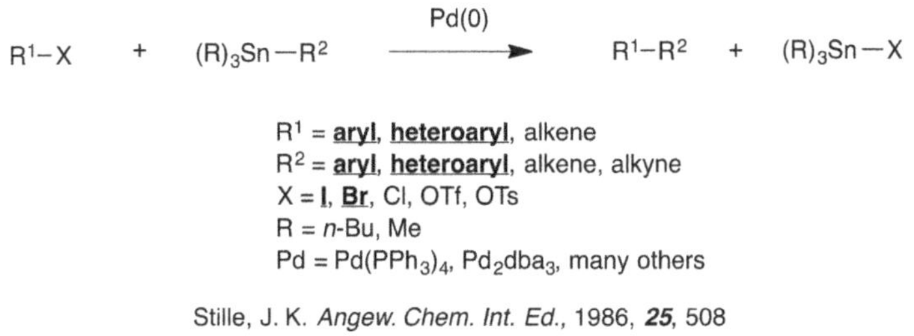

$$R^1\text{−}X \quad + \quad (R)_3Sn\text{−}R^2 \quad \xrightarrow{\text{Pd(0)}} \quad R^1\text{−}R^2 \quad + \quad (R)_3Sn\text{−}X$$

R^1 = **aryl**, **heteroaryl**, alkene
R^2 = **aryl**, **heteroaryl**, alkene, alkyne
X = **I**, **Br**, Cl, OTf, OTs
R = n-Bu, Me
Pd = Pd(PPh$_3$)$_4$, Pd$_2$dba$_3$, many others

Stille, J. K. *Angew. Chem. Int. Ed.*, 1986, **25**, 508

Scheme 5.9 General illustration of the Stille coupling.[24]

many strong donors in π-conjugated materials, the rate of oxidative addition may be reduced and the reaction may be less high yielding in general, especially if long reaction times and high temperatures lead to significant decomposition of the starting materials, products, or catalysts. In the synthesis of π-conjugated materials, both the substrates and the Stille reagents tend to be aryl or heteroaryl compounds, and iodide and bromide substrates are typically employed; however, efficient Stille couplings are known for chloride[25] and sulfonate[26] substrates that utilize highly active Pd catalysts and/or Sn reagent activators such as CsF, even with synthetically difficult bulky substrates and Sn reagents.[27] These chloride and sulfonate reactions may be preferable when, for example, the iodides or bromides are too unstable or are too synthetically challenging. Copper(I) salts and fluoride salts may be used to enhance the reaction yield, in some cases improving the yield from 0% to nearly quantitative.[28] This is thought to occur from a combination of (1) the "Copper Effect," where Cu(I) coordinates the free ligand/s of the Pd catalyst so that more coordination sites and less steric hindrance are on the R–Pd(II)–X (S5.8d) complex for transmetallation; and (2) the fluoride ion acting to coordinate Sn, increasing nucleophilicity of the Sn reagent, and improving the transmetallation step overall.[29]

Stille reagents, by-products, and side-products of the cross-coupling reactions are generally considered toxic. Thus, care should be taken that all waste is properly segregated and disposed of starting with the preparation of Stille reagents through the final steps of purification of the cross coupled products. In general, Sn by- and side-products can be treated with aqueous KF or CsF in organic solvents (where $R_3Sn–F$ may be insoluble in the organic solvent and filtered off), or may be scavenged with silica supported cysteine, so-called "tin scavengers". Some methods for the preparation of Stille reagents are illustrated in Scheme 5.10. Most commonly, Sn reagents are prepared

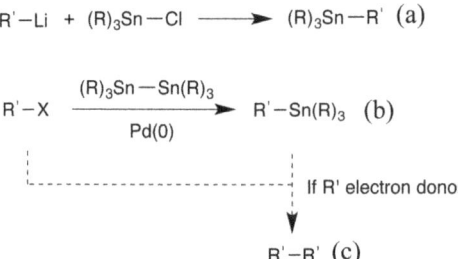

Scheme 5.10   General preparation of Stille reagents (organostannanes).

from organolithiums or Grignard reagents and a trialkyltin chloride (S5.10a). Often, the reaction is clean and sufficiently efficient to allow use of the resulting tin reagent without further purification; however, when purification is necessary—as may be the case in polymerization reactions (discussed in Section 7.2)—the Sn reagents are often sufficiently stable to be purified by distillation, crystallization, or column chromatography. Distillation or chromatography is typically used for purification of tributyltin reagents since they tend to be liquids or oils. When the Sn reagent is sensitive to even slightly acidic normal-phase silica gel, neutralizing the silica gel with 2–5% triethylamine in the column packing solvent, and maybe the eluent solvent, may retard decomposition during chromatography. If the Sn reagent is too sensitive to neutralized silica, then reverse-phase chromatography may be an effective alternative. To avoid chromatography, trimethyltin chloride may be used instead of tributyltin chloride as the stannylating agent to potentially provide a crystalline product that may be purified by precipitation or recrystallization; however, *trimethyl tin chloride is significantly more toxic than tributyl tin chloride* (Me$_3$Sn—Cl: NFPA Health = 4, LD50 (rats) = 12.6 mg kg^{-1}; ^{n}Bu$_3$Sn—Cl: NFPA Health = 2, LD50 (rats) = 129 mg kg^{-1}), so considerable caution should be used when preparing Me$_3$Sn reagents, but both Me$_3$Sn and Bu$_3$Sn are considered toxic. One drawback to the preparation of Sn reagents from alkyllithium reagents is that the functional group tolerance is limited since the nucleophilic lithium or Grignard reagent will add to many carbonyl compounds. A protecting group may be used, but the protection/deprotection cycle and respective product isolation may be undesirable, in which case an "*in situ* protecting group" may be used.[30] In some cases, another method that is more tolerant of many functional groups (including carbonyls and nitriles) may be used to prepare the Stille reagent, which is the reaction of an aryl halide with a (hexalkyl)ditin (R$_3$Sn—SnR$_3$) under Pd catalysis (S5.10b), although if the aryl halide substrate is electron donating, the resultant Sn reagent may be more reactive than R$_3$Sn—SnR$_3$ and will likely couple with the remaining aryl halide (S5.10c).

Some specific examples of Stille reagent synthesis are shown in Scheme 5.11. Both lithiation of thienothiophene and 3-bromo-*N*-ethylcarbazole followed by reaction with tributyltin chloride at −78 °C to room temperature provide the corresponding thienothiophene stannane **8**[31] and carbazole stannane **9**[32] in high yields. Both mono- and disubstituted stannanes may be prepared from thiophene-containing compounds with more than one active α-position, such as with the mono deprotonation/lithiation of **10** to give **11**[33] after reaction

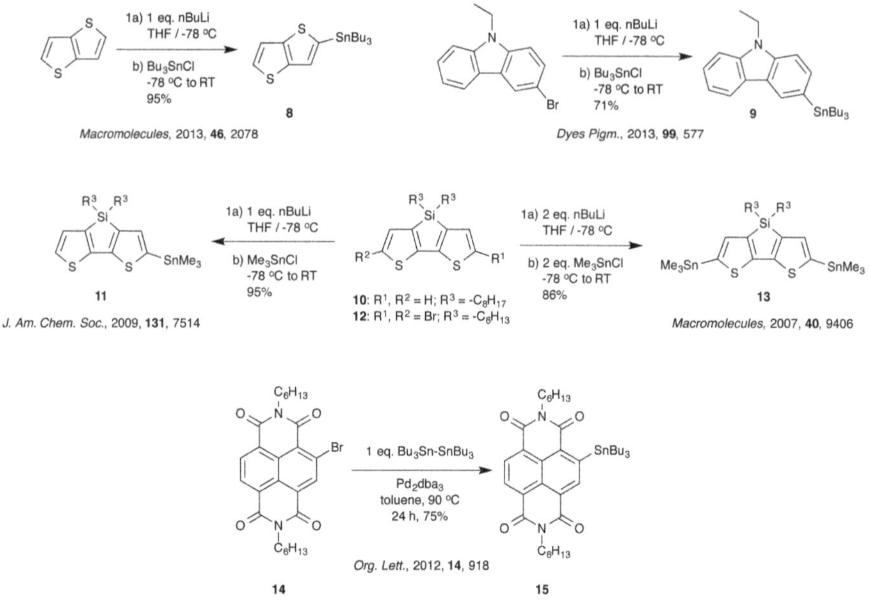

Macromolecules, 2013, **46**, 2078

Dyes Pigm., 2013, **99**, 577

J. Am. Chem. Soc., 2009, **131**, 7514

10: R^1, R^2 = H; R^3 = -C_8H_{17}
12: R^1, R^2 = Br; R^3 = -C_6H_{13}

Macromolecules, 2007, **40**, 9406

Org. Lett., 2012, **14**, 918

Scheme 5.11   Examples or Stille reagent preparation on several substrates.[31–35]

with trimethyltin chloride and a double bromine–lithium exchange with **12** to give bis(trimethylstannane) **13** [34] in high yields. However, it should be noted that controlling the lithiation step to produce just a mono lithiated thiophene can sometimes be difficult and a mixture of monostannane, distannane, and starting material may result. Finally, one example of the reaction of hexabutylditin under Pd catalysis is illustrated by the reaction of **14** to give **15** [35] in high yield, which would be difficult to achieve by other means since the parent naphthalene diimide (NDI) **14** is typically unstable to lithiation.

Examples of the Stille cross coupling with thiophenes are shown in Scheme 5.12. In general, thiophene cross-couplings are good reactions to judge the efficiency of catalysts and coupling conditions since halothiophene substrates are moderate to strong donors and are therefore at least somewhat deactivated. The first example shows the coupling of thiophene stannane **16** to bromoterthiophene **17** to give the corresponding quarterthiophene **18** in moderate yield.[36] One example where the yield of the coupling can be improved by using activators such as CuI and CsF is shown by the double addition of 2-(tributylstannyl)thiophene to bithiophene **19** to give the corresponding quarterthiophene **20** in high yield.[37] Another double Stille coupling employed more active catalyst Pd$_2$(dba)$_3$ and the bulky ligand P(o-tol)$_3$ to couple 2,5-bis(trimethylstannyl)thiophene **21** and

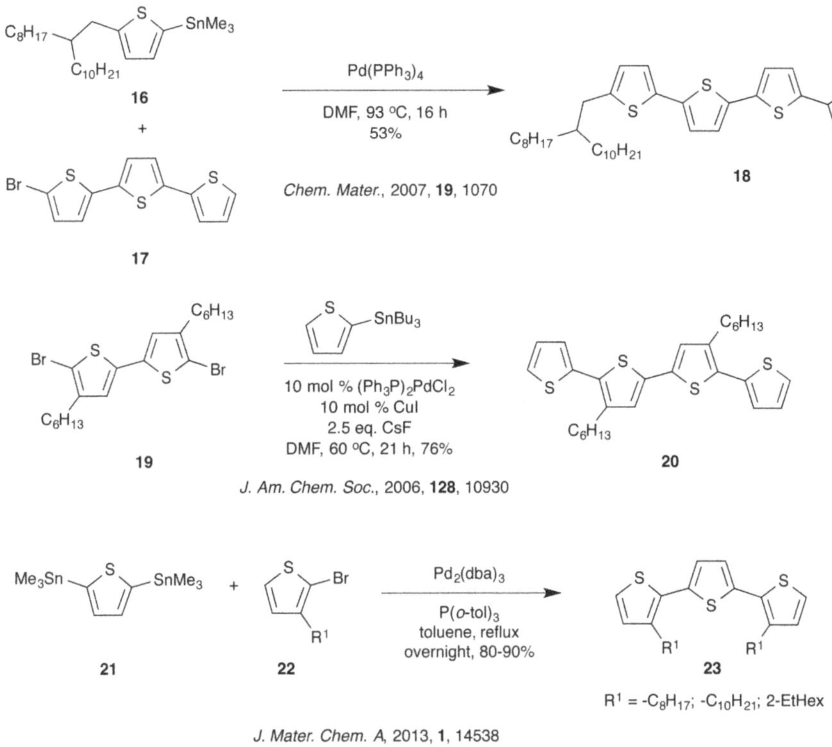

Scheme 5.12 Stille coupling of thiophene substrates.[36–38]

3-alkyl-2-bromothiophenes **22** to give the corresponding terthiophenes **23** in high yield.[38] The significantly higher yields for the double couplings of **19** and **22** compared to mono coupling of **17** is likely a result of the more active catalysts and CuI/CsF conditions used in **19** and **22**. These and similar activated conditions and catalysts in general may be good to use in the coupling of bromothiophenes, which are somewhat deactivated by the thiophene being an electron donor, particularly when they are more deactivated by the presence of bulky alkyl groups adjacent to the bromine such as in **19** and **22**.

Stille couplings that employ more favorable electron-donor stannanes and electron-acceptor substrates are illustrated in Scheme 5.13. The first example illustrates the functional group tolerance of the reaction with the coupling of stannane **24** to the bromo substrate **25**, having both a ketone and an aldehyde, to give the corresponding cross coupled product **26** in moderate yield.[39] Other examples include the cross coupling of electron-acceptor substrate **27** with thiophene stannanes **28** using (Ph$_3$P)$_2$PdCl$_2$ (a Pd(II) catalyst)[40] and the classic Pd(0) catalyst Pd(PPh$_3$)$_4$ under normal reflux conditions[41] and with

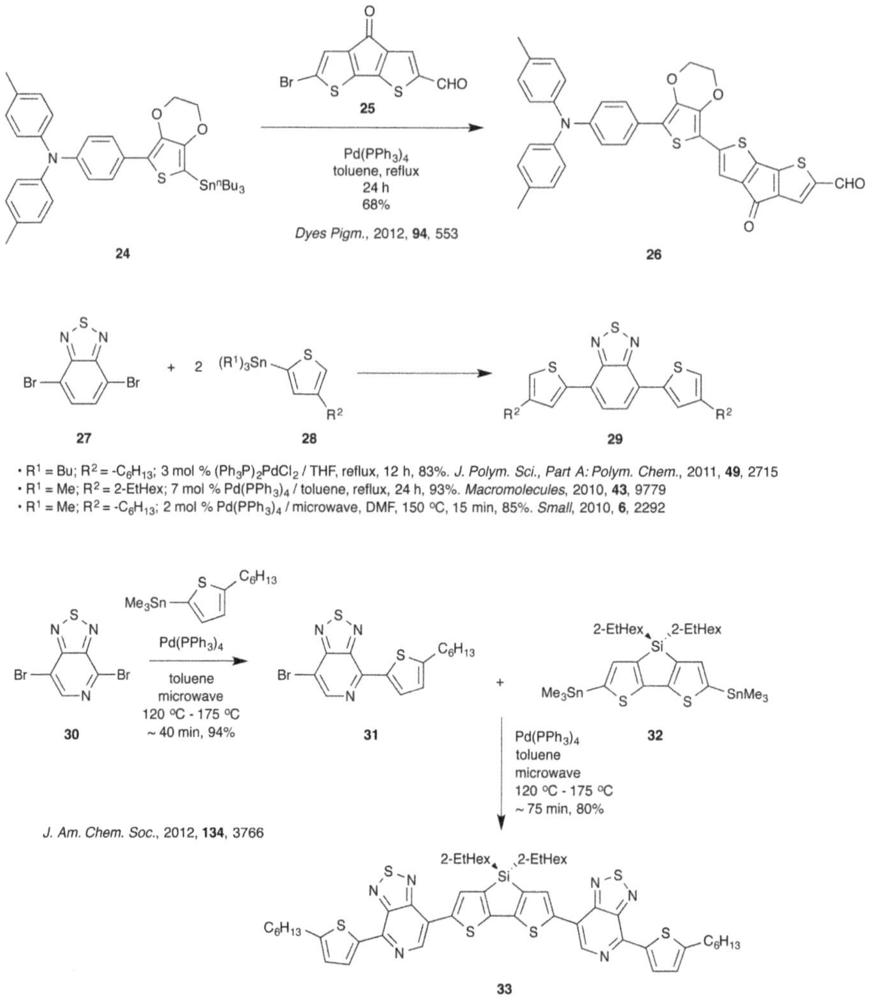

Scheme 5.13 Stille coupling substrates with electron accepting aryl halides.[39–43]

microwave assistance[42] to give corresponding doubly coupled **29** in high yields. Note that microwave assistance *substantially* reduces the reaction time from 24 h to 15 min, even with lower catalyst loading of 2% compared to 7%. This substantial improvement of reaction yields and reduction of reaction time is underscored in the reaction of pyridolothiadiazole **30** with 5-hexyl-2-(trimethylstannyl)thiophene to give the corresponding cross coupled product **31** in high yield followed by a double coupling of **31** to distannane **32** to give **33** in high yield.[43] The reaction of **30** also underscores the preference of Pd catalysts to insert into bonds with relatively high electron affinity since the first

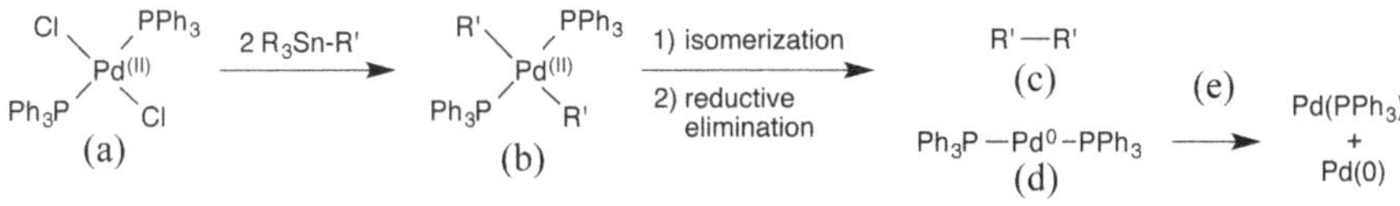

Scheme 5.14 Formation of palladium(0) catalyst from palladium(II) precursor complex *in situ*.

reaction on **30** occurs selectively at the C—Br bond adjacent to the electron withdrawing N of the pyridolo ring.

Note that $Pd(PPh_3)_2Cl_2$ may sometimes be used as a substitute for $Pd(PPh_3)_4$ in Pd coupling reactions, such as in the first example for the coupling of **27** and **28** in Scheme 5.13. This generally is because $Pd(PPh_3)_2Cl_2$ is much more air stable and moisture stable than $Pd(PPh_3)_4$; however, one caveat in the use of $Pd(PPh_3)_2Cl_2$ is that Pd(II) may be reduced to Pd(0) in the reaction mixture with certain consequences that are outlined in Scheme 5.14. The $Pd(PPh_3)_2Cl_2$ (S5.14a) may undergo a double transmetallation by the Stille reagent (or other organometal depending on the coupling being used) to give intermediate S5.14b that can isomerize and reductively eliminate to give the homocoupled product S5.14c and the Pd(0) species S5.14d. As noted in Scheme 5.3 above, one potential complication is disproportionation (S5.14e) of the two-ligand Pd(0) species to potentially inactive Pd(0) metal and $Pd(PPh_3)_4$. Although the $Pd(PPh_3)_4$ catalyst can then enter a normal Pd(0)–Pd(II) coupling cycle, precipitation of Pd(0) may practically reduce the amount of catalyst by half unless extra equivalents of PPh_3 are added. Another potential drawback is the presence of an unwanted homocoupled side-product in the same molar percentage as the original Pd(II) catalyst amount, which may be difficult to separate. If such Pd(II) to Pd(0) conversion and side-products are a problem, then $Pd(PPh_3)_4$ may be generated from $Pd(PPh_3)_2Cl_2$ and DIBAL-H and 2 equivalents of PPh_3 before adding the other reactants, although this may introduce DIBAL side-products as well.[44]

Two final examples of the Stille coupling shown in Scheme 5.15 demonstrate special cases that may be useful in the synthesis of certain π-conjugated materials. One is the use of half an equivalent of (hexabutyl)ditin to give a homocoupled product such as with the conversion of NDI **34** to **35** in one pot, albeit in low yield.[35] In this case, the intermediate tributyltin–NDI Stille reagent may be particularly deactivated since NDI is a relatively strong electron acceptor. The second example is the use of an aryl triflate **36** as a the substrate coupled with stannane **37** using the accelerant LiCl to give the corresponding

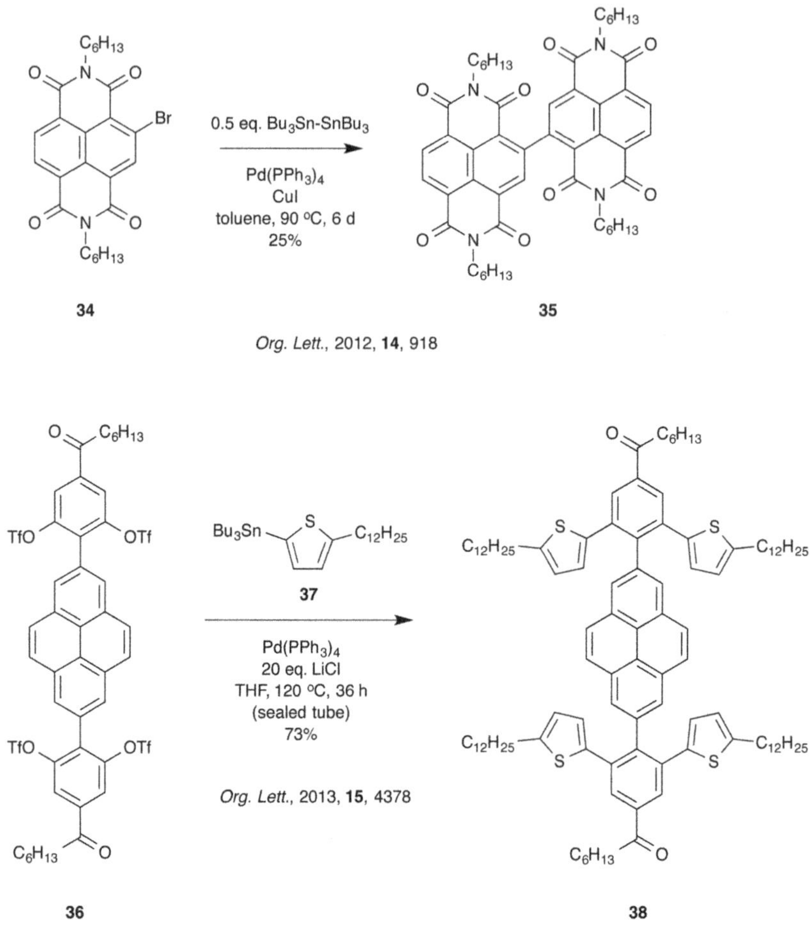

Scheme 5.15   Specialized Stille coupling.[35,45]

tetracoupled **38** in high yield.[45] The accelerating effect of LiCl on the Stille coupling has been known for some time.[24]

5.4.3   Suzuki Coupling

The general reactivity of the Suzuki coupling is outlined in Scheme 5.16. Akira Suzuki and coworkers accomplished much of the initial development, and other organometallic research groups around the world have developed many variations for the reaction.[46,47] The reaction cross couples a halide or pseudohalide substrate (S5.16a) with a boron containing organometal S5.16b to give the cross coupled product S5.16c and boronate acids/ester salts S5.16d. Like the Stille coupling, the organoboron reagent and the reaction conditions are

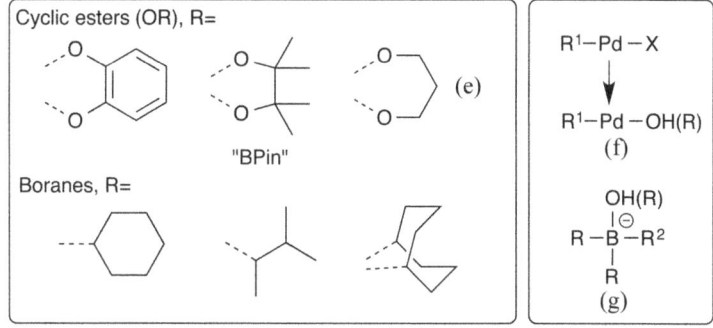

R[1] = **aryl**, **heteroaryl**, alkene, alkane
R[2] = **aryl**, **heteroaryl**, alkene, alkyne, alkane
X = **I**, **Br**, Cl, OTf, OTs
R = OH, OR (esters), alkyl (boranes), also trifluoroborates (K[+] salts)
base = NaOEt, NaOH, Na_2CO_3, K_2CO_3, K_3PO_4, $CsCO_3$, $BaOH_2$, others
Pd = $Pd(PPh_3)_4$, Pd_2dba_3, $PdCl_2dppf$, others

A. Suzuki, *Angew. Chem., Int. Ed.*, 2011, **50**, 6723 (Nobel Lecture/Review)
N. Miyaura and A. Suzuki, *Chemical reviews*, 1995, **95**, 2457

Scheme 5.16 General illustration of the Suzuki coupling.[46,47]

reasonably tolerant of nucleophile-sensitive functional groups such as aldehydes, esters, and nitriles; however, the necessity of base (usually in the form of hydroxide or alkoxide) may be problematic for base sensitive substrates. Also like the Stille reaction, the organoboron reagent may be shelf-stable and capable of being purified by conventional methods such as distillation, recrystallization, and/or column chromatography, especially when the organoboron is a boronic acid diester. In many cases, column chromatography may be accompanied by streaking and material loss, which has been speculated to occur through interaction of the vacant orbital of boron with nucleophilic sites on silica gel, and may be avoided by impregnating the silica gel with boric acid.[48] One *significant* advantage of the Suzuki coupling over the Stille coupling is that the boron reagents and by-products are generally less toxic or nontoxic and the reaction may not be as sensitive to electron-acceptor nucleophilic R[2] groups on the organoboron species.

The substrates and organoboron reagents can contain almost any variation for C−C bond forming reactions. Like most palladium couplings, the preferred halides on the substrate are typically I and Br, but catalysts and conditions have been developed for Cl and sulfonates as well.[49,50] There is a large variety organoboron reagents that have been used, including boronic acids, boronic acid esters, boranes, and trifluoroborate salts $(R^2-BF_3^-K^+)$. In particular, boronic acids and esters have been used in the synthesis of π-conjugated materials, with esters such as diEt, di(*i*-Pr) and cyclic esters (S5.16e, inset) such as from catechol and particularly from pinacol ("Bpin") being used widely. One potential issue is that boronic acids can be unstable, with a proton–boron exchange ("protodeborylation") occurring sometimes readily, especially with heteroaryl and electron acceptor boronic acids. Since the reaction solvents are usually polar and are often a mixture of organic solvent and water, protodeborylation can necessitate a large excess of boronic acid. In contrast, trifluoroborate salts $(R-BF_3^-K^+)$ are more stable, especially for heteroaryls that appear in many acceptors, and can often be purified by recrystallization, precipitation, and/or Soxhlet extraction.[51] Trifluoroborate salts are known to be tolerant to other reactions on the substrate, such as Wittig reactions, HWE reactions, nucleophilic substitutions, lithiations, and others.[52] One important difference between the Suzuki coupling and the Stille coupling is that the Suzuki coupling requires base typically to activate both the intermediate of the oxidative addition to a Pd(II) hydroxide or alkoxide (S5.16f, inset) and to activate the organoboron reagent (S5.16g, inset) for transmetallation. Largely in part because of the inorganic base component, reactions are generally run in polar protic, polar aprotic, or aqueous/organic solvent mixtures (*e.g.*, H_2O/toluene). Note that, since the reaction may be biphasic with water and an organic solvent, vigorous stirring may be important and often times reaction yields may benefit from strong mechanical stirring instead of the more convenient magnetic stirring. Because of the importance of the base, as well as the general low solubility of inorganic bases and organic substrates in common solvents, the use of a number of bases have been developed, including the use of NaOEt, NaOH, Na_2CO_3, K_2CO_3, K_3PO_4, $CsCO_3$, $BaOH_2$, and phase transfer bases, among others.

Most Suzuki reactions used in the synthesis of π-conjugated materials have historically employed either boronic acids or esters. Scheme 5.17 summarizes the number of methods that have been developed for the preparation and interconversion or boronic acid and boronate ester reagents. One widely used method is the reaction of lithium and Grignard reagents with trialkoxy borates (S5.17a), such as tri(isopropyl)borate or particularly 2-isopropoxy-4,4,5,5-tetramethyl-1,3,2-dioxaborolane

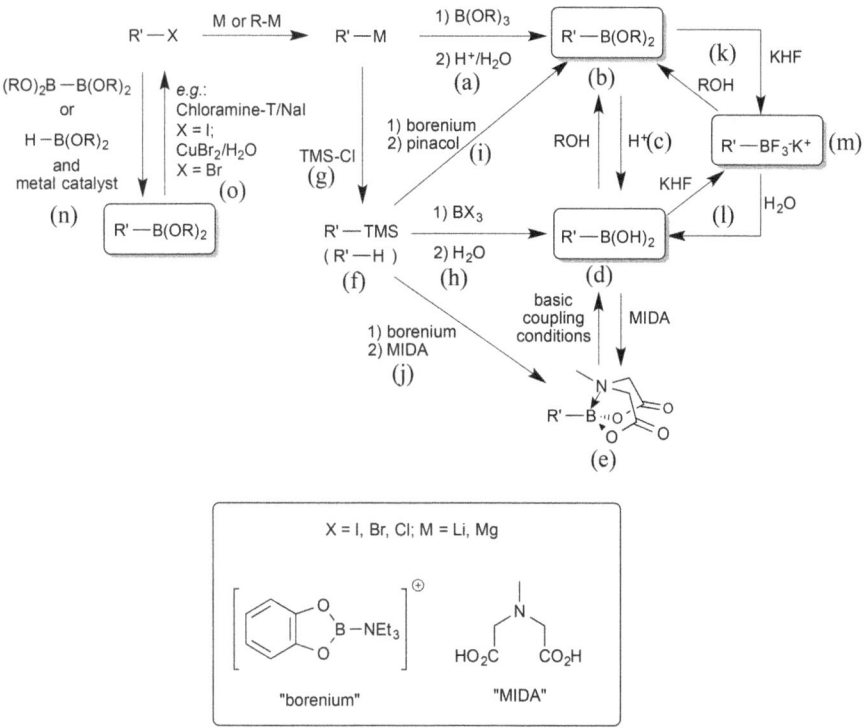

Scheme 5.17 Synthesis of and interconversion between Suzuki reagents and aryl halides.

(isopropoxy Bpin), to give the corresponding borate (S5.17b), usually in high yield. Acyclic boronates can be transesterified with other alcohols under acid catalysis, and transesterification using pinacol to give cyclic Bpin esters has been widely used. Acid hydrolysis (S5.17c) can convert the boronates to the corresponding boronic acid (S5.17d). One important modification of boronic acids is the conversion to B-MIDA reagents (S5.17e) by reaction with commercially available *N*-methyliminodiacetic acid (MIDA, Scheme 5.17 inset). B-MIDA reagents can be thought of as "slow release" boronic acids since they slowly hydrolyze to boronic acids in the basic Suzuki coupling medium and therefore can effectively act as protecting groups for unstable boronic acids.[53,54] A number of methods that use electrophilic substitution on either an arene or TMS-arene substrates (S5.17f) have been developed as well. The TMS-arene substrates are generally available from reaction of the organolithium with TMS-Cl (S5.17g). Electrophilic methods include reaction of the arene with a borontrihalide and either water hydrolysis (S5.17h) to provide the corresponding acid,[55] or esterification with pinacol to provide the Bpin ester.[56] Another electrophilic method is the reaction of "borenium" reagents to give either the Bpin (S5.17i)[57] or the corresponding B-MIDA reagents (S5.17j)[58] with the appropriate choice of workup conditions. Boronic acids (S5.17k) and esters (S5.17l) can be converted into potassium organotrifluoroborates (S5.17m) with potassium hydrogen fluoride, which can be used directly in Suzuki couplings[52] or can be converted back to the corresponding acid (S5.17k) or ester (S5.17l) with water or an alcohol, respectively, with silica gel catalysis.[59] Because of this mild and generally efficient interconversion between $-B(OH)_2/-B(OR)_2$ and $-BF_3^-K^+$, the $-BF_3^-K^+$ group can be thought of as a boronic acid/ester protecting group, given the stability of $-BF_3^-K^+$ to a broad array of reaction chemistry. One disadvantage to the above reactions is that the starting lithium and Grignard reagents are not tolerant of other functional groups such as aldehydes and ketones, among others. However, boron reagents can be prepared *via* Pd catalysis from the corresponding organohalide and diboronate such as $(Bpin)_2$[60] (S5.17n, the "Miyaura borylation"), which tolerates a variety of functional groups. Variations include: (1) $CuI/Bu_3P/KOt$-Bu;[61] (2) dialkylzinc;[62] (3) and Ir by C$-$H activation,[63] and even from aryl carbamates instead of aryl halides with Ni catalysis.[64] Finally, aryl boronates may be converted to iodides or bromides by a variety of methods.[65,66] As summarized in Scheme 5.17, the number of methods and substrates that have been developed for the preparation of boronic acids and esters is a good indication of the utility and popularity of the Suzuki coupling in synthetic chemistry.

Many examples of the synthesis of boronic acids and boronates exist in the literature of π-conjugated materials, but the examples given here are for the preparation of cyclic boronates since they can be purified more readily than boronic acids and have been used extensively, especially in polymer synthesis (Section 7.4.2). Examples of synthesis of cyclic boronates are shown in Scheme 5.18. These include the conversion of **39** to diBpin **40** through double deprotonation[67] and double lithium–halogen exchange[68] from the respective starting **39** in moderate to high yields using Bpin-O^{i}Pr as a borylating agent. Bpin-O^{i}Pr is a common reagent to make Bpin esters directly from lithium reagents, and the isopropoxide effectively acts as a leaving group. Electrophilic borylation with the "borenium" reagent is demonstrated for the conversion of **41** to Bpin **42** in high yield.[69] Another example with bithiophene illustrates conditions using transesterification with the lithiation and reaction with tributoxyborate and then one-pot transesterification with 2,2-dimethyl-1,3-propanediol to give the corresponding cyclic boronate **43** in high yield.[70] Other substrates that are of interest to the synthesis of π-conjugated materials have

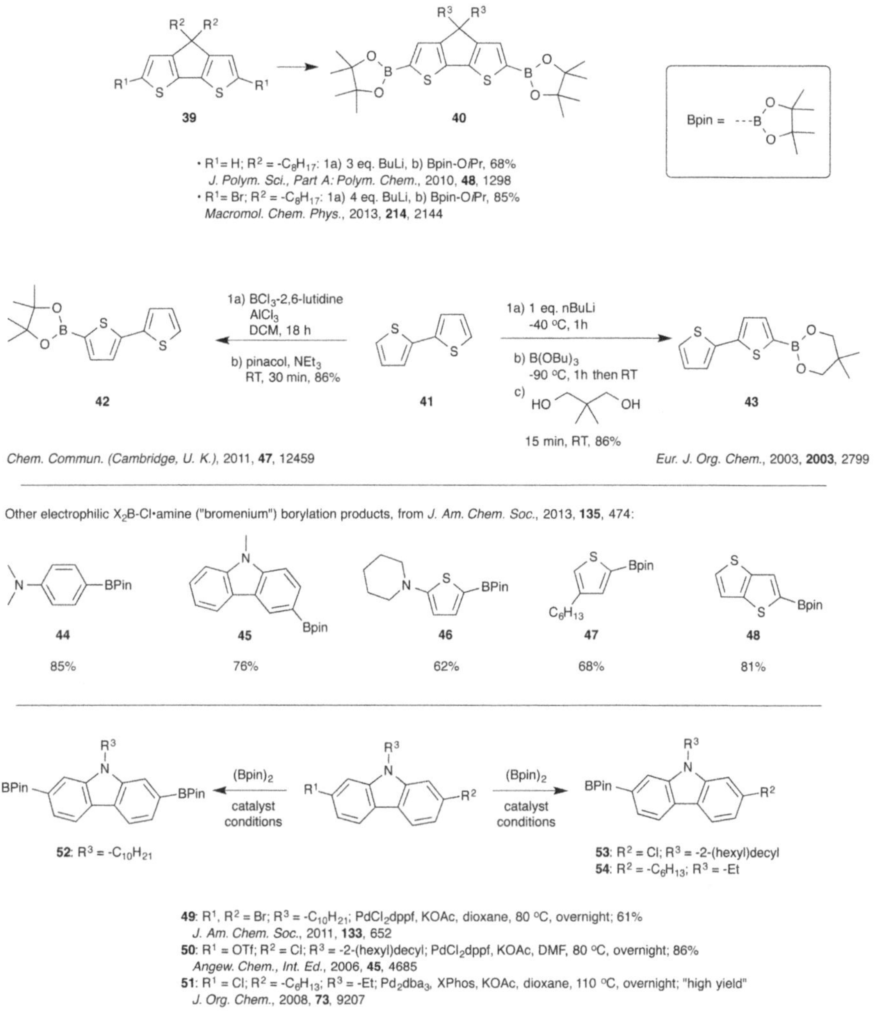

39
40
Bpin =
• R1= H; R2 = -C8H17: 1a) 3 eq. BuLi, b) Bpin-OiPr, 68%
J. Polym. Sci., Part A: Polym. Chem., 2010, 48, 1298
• R1= Br; R2 = -C8H17: 1a) 4 eq. BuLi, b) Bpin-OiPr, 85%
Macromol. Chem. Phys., 2013, 214, 2144
1a) BCl3-2,6-lutidine AlCl3 DCM, 18 h
b) pinacol, NEt3 RT, 30 min, 86%
1a) 1 eq. nBuLi -40 °C, 1h
b) B(OBu)3 -90 °C, 1h then RT
c)
15 min, RT, 86%
42
41
43
Chem. Commun. (Cambridge, U. K.), 2011, 47, 12459
Eur. J. Org. Chem., 2003, 2003, 2799
Other electrophilic X2B-Cl•amine ("bromenium") borylation products, from J. Am. Chem. Soc., 2013, 135, 474:
44
85%
45
76%
46
62%
47
68%
48
81%
(Bpin)2
catalyst conditions
52: R3 = -C10H21
53: R2 = Cl; R3 = -2-(hexyl)decyl
54: R2 = -C6H13; R3 = -Et
49: R1, R2 = Br; R3 = -C10H21; PdCl2dppf, KOAc, dioxane, 80 °C, overnight; 61%
J. Am. Chem. Soc., 2011, 133, 652
50: R1 = OTf; R2 = Cl; R3 = -2-(hexyl)decyl; PdCl2dppf, KOAc, DMF, 80 °C, overnight; 86%
Angew. Chem., Int. Ed., 2006, 45, 4685
51: R1 = Cl; R2 = -C6H13; R3 = -Et; Pd2dba3, XPhos, KOAc, dioxane, 110 °C, overnight; "high yield"
J. Org. Chem., 2008, 73, 9207

Scheme 5.18 Formation of Suzuki reagents from several substrates.[58,67–73]

been reacted with the bromenium borylating reagent to give **44–48** in moderate to high yields.[58] The synthesis of Bpin esters from various carbazoles **49–51** using organometallic catalysts are also illustrated. The first is a double borylation of carbazole **49** to the diBpin **52** with $(BPin)_2$ and $PdCl_2$dppf catalysis in moderate yield.[71] A second example shows selectivity of triflate (–OTf) over chloride with the reaction of **50** with palladium catalysis to give the monoBpin **53** in high yield.[72] Finally, chloride **51** may also be converted to a Bpin ester using a more activated Pd_2dba_3–XPhos catalyst with higher temperature to give **54** in high yield.[73] Given the interconvertibility of boronates and boronic acids illustrated in Scheme 5.17, a variety of other Suzuki reagents in principle should be available from any of these cyclic boronate synthesis reaction, so the reader is encouraged to explore other Suzuki reagents if Bpin is inefficient for any particular coupling.

Examples of Suzuki cross-coupling reactions are illustrated in Scheme 5.19. The first example demonstrates the functional group tolerance with the coupling of boronic acid **55** with bromoaldehyde **56** to give the corresponding aldehyde **57** in moderate yield.[74] Another example shows the tolerance to sulfone groups in the coupling of **58** with carbazole **59** to give the corresponding dicoupled **60** in high yield.[75] Both the reactions of the boronic acid **55** and the Bpin boronate **59** are common conditions that employ aqueous K_2CO_3 as a base. Note that, as mentioned earlier, these types of biphasic reactions may benefit from using vigorous mechanical stirring instead of the more convenient magnetic stirring. Another example uses a more active catalyst and base to couple moderately strong electron-donor selenophene dibromide **61** to sterically hindered Bpin ester **62** in high yield.[76] Another example uses an active system that includes KF as a fluoride source and microwave irradiation in the coupling of diiodide **64** with Bpin ester **65** to give the doubly coupled **66** in high yield in only 40 min,[77] which is a contrast to the 12–24 h that are often required. Finally, one interesting aspect of the Suzuki coupling is that since base activation is required, mixed Stille–Suzuki reagents can react selectively in non basic conditions at the organostannane to give extended Bpin esters, such as with the reactions of **67** and Stille–Suzuki reagent **68** to give the corresponding extended Bpin esters in high yield, and in the case of microwave irradiation, in only twenty minutes.[78] Also note the selectivity of the iodine over the bromine in **67**. As mentioned earlier, the general stability and ability to purify Suzuki reagents has resulted in many reactions involving the Suzuki coupling in the synthesis of π-conjugated materials, especially in the area of polymerization where high purity and high reaction yields are necessary to achieve high molecular weights. Thus, the Suzuki coupling will be discussed in Chapter 7 concerning polymerization.

5.4.4   Kumada and Negishi Coupling

The Kumada and Negishi reactions are summarized in Scheme 5.20. Both the Kumada (Mg) and Negishi (Zn) couplings are similar to the Stille coupling in mechanism, but the organometal Grignard and zinc reagents are not air/water stable and cannot be purified like Stille and Suzuki reagents. Historically, both couplings have been used relatively

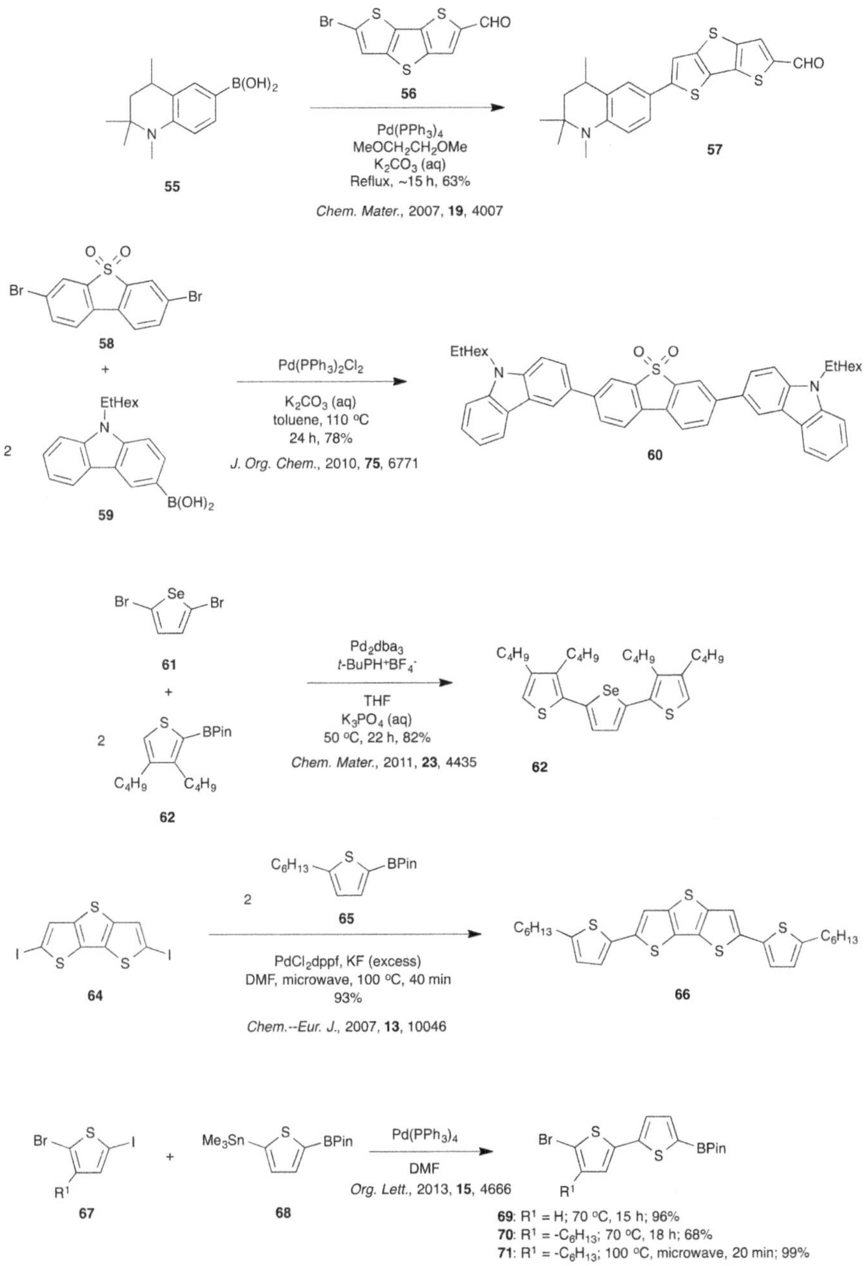

Scheme 5.19 Suzuki coupling on several substrates.[74–76,78,91]

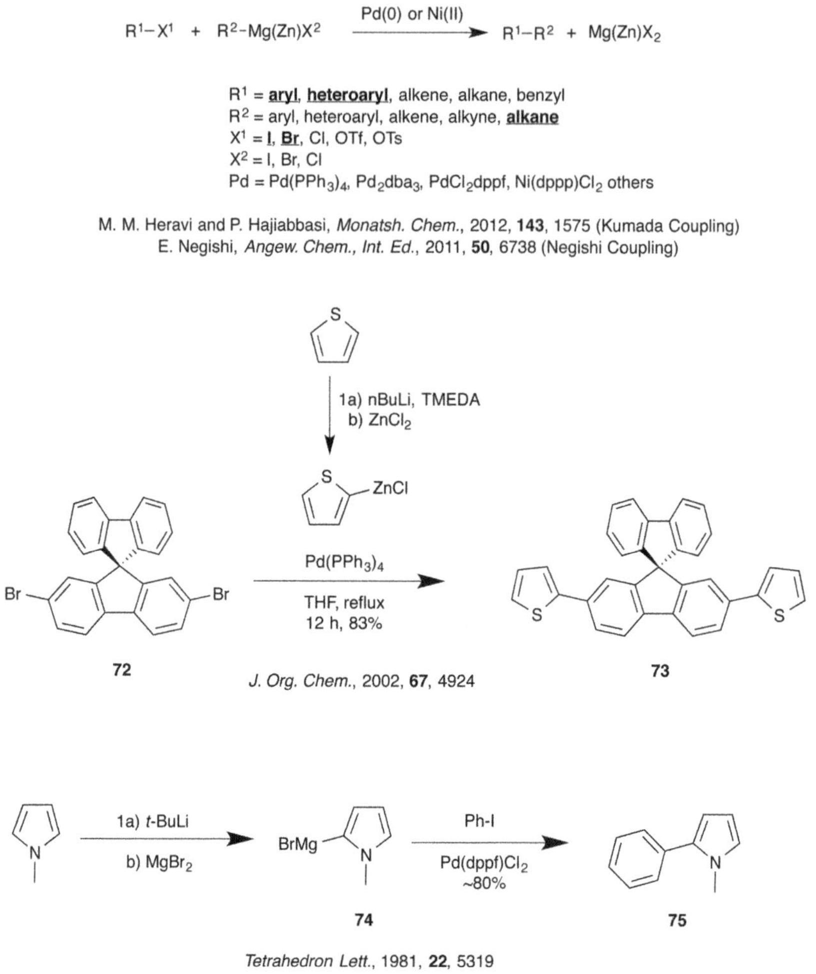
R1–X1 + R2–Mg(Zn)X2
Pd(0) or Ni(II)
R1–R2 + Mg(Zn)X2
R1 = aryl, heteroaryl, alkene, alkane, benzyl
R2 = aryl, heteroaryl, alkene, alkyne, alkane
X1 = I, Br, Cl, OTf, OTs
X2 = I, Br, Cl
Pd = Pd(PPh3)4, Pd2dba3, PdCl2dppf, Ni(dppp)Cl2 others
M. M. Heravi and P. Hajiabbasi, Monatsh. Chem., 2012, 143, 1575 (Kumada Coupling)
E. Negishi, Angew. Chem., Int. Ed., 2011, 50, 6738 (Negishi Coupling)
S
1a) nBuLi, TMEDA
b) ZnCl2
ZnCl
Pd(PPh3)4
THF, reflux
12 h, 83%
Br
72
73
J. Org. Chem., 2002, 67, 4924
1a) t-BuLi
b) MgBr2
BrMg
N
Ph-I
Pd(dppf)Cl2
~80%
74
75
Tetrahedron Lett., 1981, 22, 5319

Scheme 5.20 Kumada and Negishi coupling reactions.[13,40–43]

rarely in the formation of small molecule aryl–aryl bonds in the synthesis of π-conjugated materials; however, Kumada coupling has been used widely for the synthesis of alkyl thiophenes, which was discussed in Section 3.2.8. However, the scope of the Kumada and Negishi couplings is broad and a variant has been developed for almost every non-heteroatom carbon functional group, including alkenes, alkanes, benzyls, and aryl groups. When used for aryl cross couplings, the preferred substrates are aryl halides and bromides, but reactions using aryl chlorides with both Pd[79] and Ni[80] catalysis have been developed along with sulfonates.[81] Scheme 5.15 illustrates an example of the Negishi cross coupling that couples dibromo **72** and 2-thienylzinc chloride to give the corresponding dicoupled product **73** in high yield. Although several examples of the alkyl–aryl Kumada coupling are illustrated in Section 3.2.8, one example of aryl–aryl Kumada cross coupling is the reaction of Grignard **74** with iodobenzene to give phenyl pyrrole **75** in high yield. Note that both examples generate the Zn and Mg reagents by reaction of a lithiated heteroarenes with the respective metal(II)halides, which is a common way to generate these reagents from arenes. Although not used extensively in the synthesis of aryl–aryl bonds in small molecule π-conjugated materials, Kumada coupling *has* been used quite often for the synthesis of π-conjugated polymers, particularly polythiophenes, where it has the advantage of being a living or living-like polymerization.[82,83] Kumada coupling in polymerization is discussed in Section 7.4.3.

5.4.5 Sonogashira Coupling

General features of the Sonogashira coupling are summarized in Scheme 5.21. The reaction generally involves the Pd catalyzed coupling of an aryl halide or pseudohalide substrate (S5.21a) with a terminal alkyne (*i.e.*, with hydrogen on one acetylene carbon) (S5.21b) with the assistance of a catalytic amount of Cu(I) halide and a stoichiometric amount of an amine to give a disubstituted acetylene (S5.21c), with ammonium salts as the main by-product. The reaction mechanism is similar to that of the Stille reaction, except a copper acetylide transmetallating agent is formed catalytically *in situ*, which is also illustrated in Scheme 5.21.[84] The Cu(I) salt (S5.21d) is believed to form an η^2 complex (S5.21e) that increases the acidity of the acetylene (uncomplexed phenylacetylene $pK_a \sim 30$) and allows deprotonation by an amine base (S5.21f) to form a copper acetylide (S5.21g) and ammonium salt. The copper acetylide can then transmetallate (S5.21h) the Pd complex from the Pd catalysis cycle (S5.8d, Scheme 5.8) to regenerate the Cu(I) salt for another cycle of copper acetylide formation (S5.21d). One problematic side reaction is the oxidative coupling of the copper acetylide to give the homocoupled diyne (S5.21i), so degassing the solution with a stream of inert gas (Ar or N_2), or vacuum pump/fill cycles with an inert gas, and then strict exclusion of oxygen is necessary (a pale green color in the reaction is evidence of oxygen). Other oxidants may also affect the alkyne homocoupling. If the reaction is slow but the aryl halide substrate is otherwise active, copper acetylide formation may be slow. If copper acetylide formation is slow under the mild reaction conditions, the copper acetylide can be generated separately or in a first step by deprotonation of the

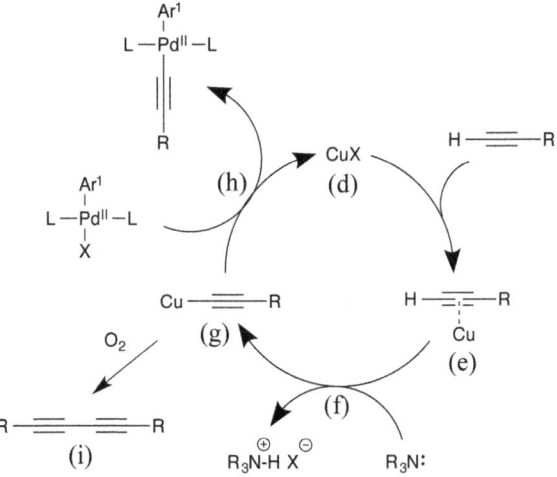

R¹ = **aryl, heteroaryl**, alkene
R² = **aryl, heteroaryl**, alkene, alkyne, alkane, **silane**
X¹ = **I, Br**, Cl, OTf, OTs
X² = **I**, Br
amine base = triethylamine, di(isopropyl)amine, others
Pd = Pd(PPh₃)₄, (PPh₃)₂PdCl₂, Pd₂dba₃, others

N. M. Jenny, M. Mayor and T. R. Eaton, *Eur. J. Org. Chem.*, 2011, **2011**, 4965

Scheme 5.21 General illustration of Sonogashira coupling and catalytic copper acetylide formation.

alkyne with *n*-BuLi, LDA, or a Grignard reagent followed by addition of stoichiometric CuI. In general, the reaction is mild and tolerant of functional groups including carbonyls. Many amine bases have been used, including triethylamine (TEA), which is often used as a solvent; however, diisopropylamine (DIPA) is an alternative choice in which many organic compounds are more soluble. In some cases, other solvents such as DMF or benzene can be used with stoichiometric amounts of amine bases such as TEA, DIPA, piperidine, di(isopropyl) ethyl amine (Hünig's base), or *n*-butyl amine. In particular, the cyclic amine bases piperidine and pyrrolidine are very effective, although they may react with some functional groups such as aldehydes.

There are numerous ways to prepare alkynes, and a good book has been written on the subject: L. Brandsma, *Preparative Acetylene*

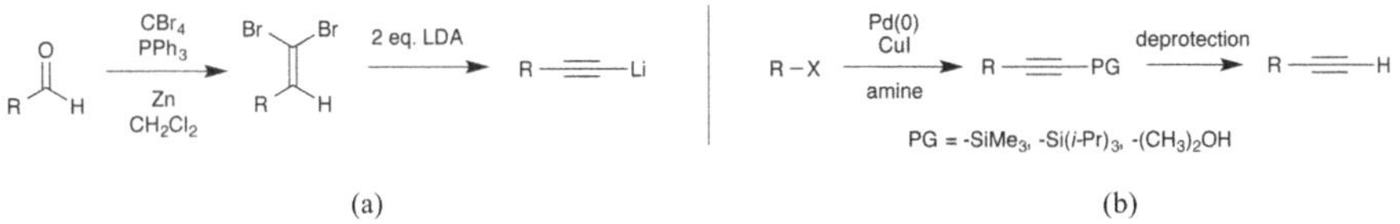

(a)   (b)

Scheme 5.22 Preparation of alkynes by: (a) dibromoolefination and elimination; and (b) Sonogashira coupling with a protecting group and deprotection.

Chemistry, No. 34 in the Series: Studies in Organic Chemistry, Elsevier, Amsterdam, 1988.[85] A few methods that are convenient on the preparative scale are summarized in Scheme 5.22. One method is a two-step process (S5.22a) from the reaction of aldehydes and $CBr_4/PPh_3/Zn$ to give a dibromoolefin followed by elimination with LDA or *n*-BuLi. The CBr_4 and PPh_3 form a Wittig reagent and bromine, and the bromine is reduced by Zn, so the reagent may be prepared in advance of adding the aldehyde to avoid any side reaction with bromine (in this case, the reaction mixture is usually a mauve color when formation of the Wittig reagent completes over 12 hours). One equivalent of LDA or *n*-BuLi eliminates the dibromoolefin to a bromoalkyne, and then the second equivalent of LDA or *n*-BuLi exchanges with the bromine on the alkyne to give the lithium acetylide, which can be protonated by water (or reacted with an electrophile). Another method (S5.22b) is the Sonogashira coupling of an appropriate substrate with a commercially available (or easily prepared) alkyne with a protecting group (PG) to give the protected acetylene, which can then be deprotected with appropriate conditions to give a terminal alkyne that can then be used in a Sonogashira coupling. Commercially available alkynes used widely for this preparation are (trimethyl)silylacetylene (TMS-acetylene), (triisopropyl)silylacetylene (TIPS-acetylene), and 2-methyl-3-butyn-2-ol. The TMS group can be removed with $K_2CO_3/MeOH$ or tetrabutylammonium fluoride (TBAF) in THF, the TIPS group can be removed with TBAF/THF but is generally stable to $K_2CO_3/MeOH$, and the dimethylmethanol group of 2-methyl-3-butyn-2-ol can be removed with a KOH, but is generally stable to both $K_2CO_3/MeOH$ or TBAF/THF. The differences in these deprotection conditions allow two or more protected alkynes on a substrate to be deprotected chemoselectively.

Examples of the preparation of arylalkynes are illustrated in Scheme 5.23. The first example is a dibromoolefination of aldehyde **76** followed by elimination with *n*-BuLi to give the corresponding alkyne **77** in high yield.[86] Another example reacts aldehyde **78** that contains two bromines with the dibromoolefination reagent followed by elimination and trapping of the lithium acetylide with (triethyl)silylchloride

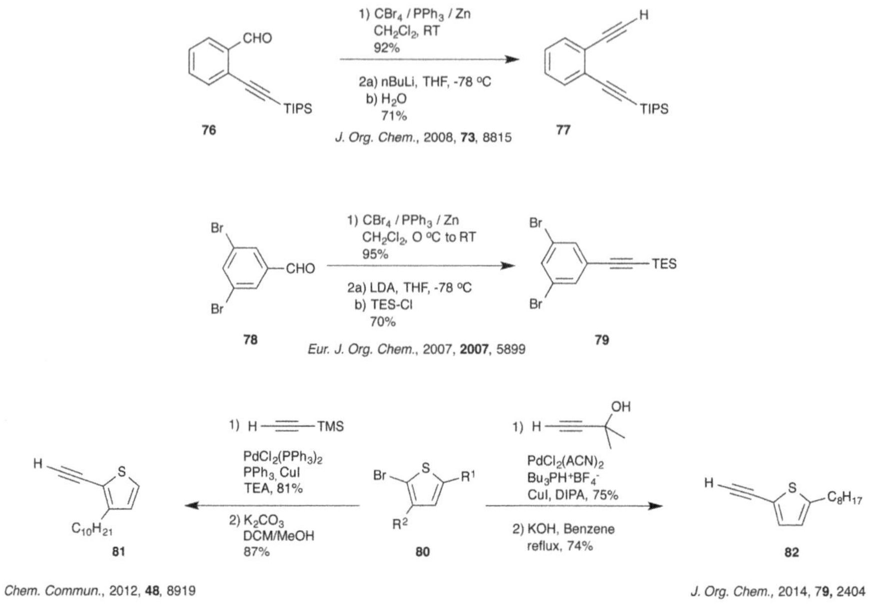
1) CBr4 / PPh3 / Zn
CH2Cl2, RT
92%
2a) nBuLi, THF, -78 °C
b) H2O
71%
J. Org. Chem., 2008, 73, 8815
CHO
TIPS
H
76
77
1) CBr4 / PPh3 / Zn
CH2Cl2, 0 °C to RT
95%
2a) LDA, THF, -78 °C
b) TES-Cl
70%
Eur. J. Org. Chem., 2007, 2007, 5899
Br
CHO
TES
78
79
1) H—≡—TMS
PdCl2(PPh3)2
PPh3, CuI
TEA, 81%
2) K2CO3
DCM/MeOH
87%
1) H
OH
PdCl2(ACN)2
Bu3PH+BF4-
CuI, DIPA, 75%
2) KOH, Benzene
reflux, 74%
S
R1
R2
C10H21
C8H17
81
80
82
Chem. Commun., 2012, 48, 8919
J. Org. Chem., 2014, 79, 2404

Scheme 5.23 Preparation of terminal alkynes for Sonogashira coupling from aldehyde and aryl halide substrates.[86–89]

to give alkyne **79** in high yield.[87] Note that LDA was used for the elimination of the dibromoolefin since *n*-BuLi would likely exchange the bromines on the benzene ring. Other examples in Scheme 5.23 include the Sonogashira coupling to thiophene substrates (**80**), including the coupling of TMS-acetylene followed by deprotection with K_2CO_3/MeOH to give the corresponding alkyne **81** in high yield.[88] In the last example 2-methyl-3-butyn-2-ol is coupled with a thiophene substrate and deprotected with KOH to give alkyne **82** in high yield.[89] There are many examples of the use of both these methods to prepare aryl and heteroaryl terminal alkynes in the scientific literature.

Examples of Sonogashira coupling on various substrates and using several different alkynes are shown in Scheme 5.24. The first example uses THF as a solvent and DIPA as a base to couple dibromoanthracene **83** to TMS-acetylene to give the dialkyne **84** in high yield.[90] Substrate **83** can also be coupled to alkyne **85** with TEA as the solvent to give the corresponding dialkyne **86** in high yield.[91] Note that, in both of these reactions, $Pd(PPh_3)_2Cl_2$ is used as the catalyst, yet the active catalyst is Pd(0). When $Pd(PPh_3)_2Cl_2$ is used, Pd(0) is generated first by coupling two of the alkynes in a directly analogous process, as outlined in Scheme 5.14 in Section 5.4.2. Typically, with the use of $Pd(PPh_3)_2Cl_2$, diynes from homocoupling of the acetylene should

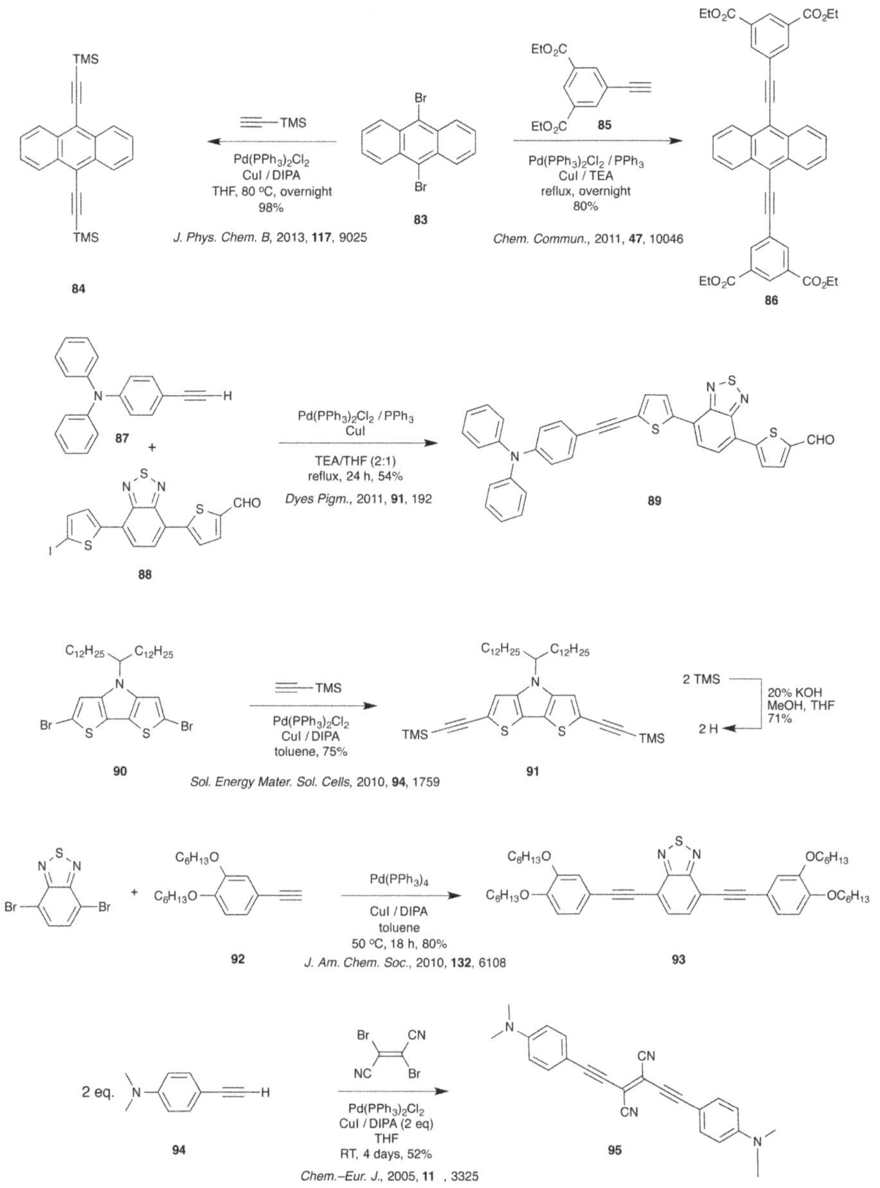

Scheme 5.24 Sonogashira coupling on several aryl halide substrates.[77,90,92–95]

be present in the reaction mixture in approximately the same molar percentage as the $Pd(PPh_3)_2Cl_2$. Deactivation of a portion of the Pd of $Pd(PPh_3)_2Cl_2$ (S5.14e) by precipitation can be avoided by adding an extra two equivalents of ligand per mole of $Pd(PPh_3)_2Cl_2$, which is illustrated with the coupling of alkyne **87** to aldehyde **88** to give the poorly soluble **89** in moderate yield.[92] Another example uses a strong electron donor substrate **90** to couple to TMS-acetylene using toluene as the solvent to give the corresponding protected dialkyne **91** in high yield.[93] Toluene or benzene is a good solvent for the Sonogashira coupling when extended π-conjugated intermediates may have limited solubility in TEA or DIPA as a solvent. Note that the protected dialkyne **91** was deprotected with KOH/MeOH in high yield. Another example also uses toluene except with the Pd(0) catalyst Pd(PPh$_3$)$_4$ to couple 4,7-dibromobenzothiadiazole to alkyne **92** to give doubly coupled **93** in high yield.[94] Toluene is a particularly good solvent for this reaction since 4,7-dibromobenzothiadiazole has limited solubility in amines or THF. Another example illustrates the Sonogashira coupling of **94** to the alkene 1,2-dibromo-1,2-dicyanoethylene to give the corresponding ene-dialkyene **95** in moderate yield.[95] The Sonogashira coupling to haloalkenes is generally more facile than to aryl halides and can often be accomplished from chloroalkenes at room temperature.[96]

Another example of the Sonogashira coupling is illustrated in Scheme 5.25 and demonstrates the utility of the (dimethyl)methanol

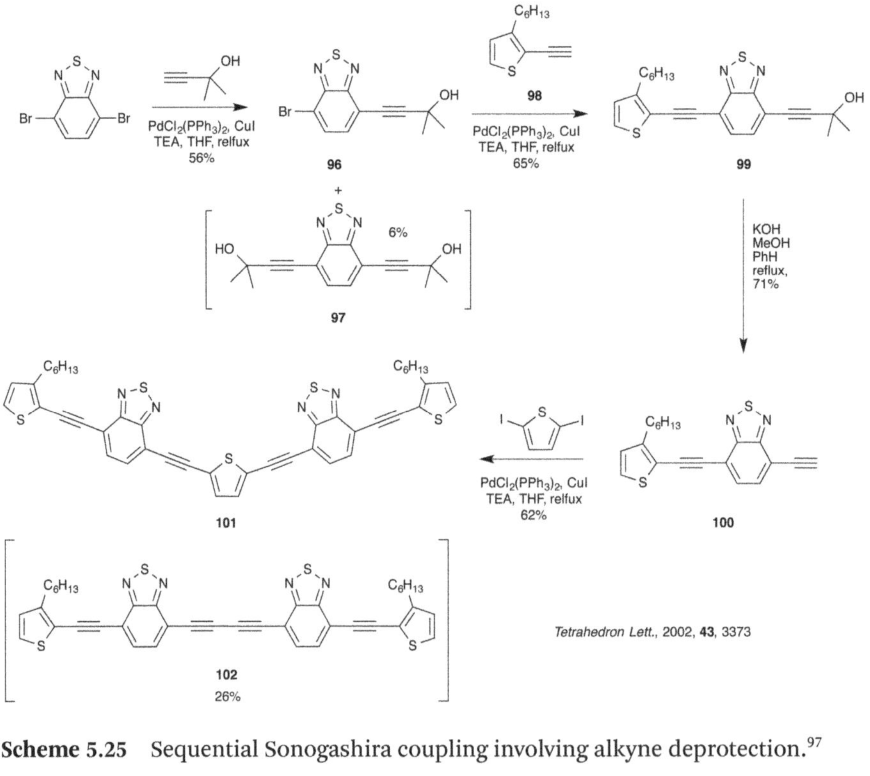

Scheme 5.25 Sequential Sonogashira coupling involving alkyne deprotection.[97]

alkyne protecting group. The first reaction couples 4,7-dibromoben-zothiadiazole to one equivalent of 2-methyl-3-butyn-2-ol to give **96** in moderate yield, which can be easily separated chromatographically from the doubly coupled side-product **97**, since **97** has two alcohols and **96** has only one.[97] Bromo **96** can be coupled with alkyne **98** to give the corresponding product **99** in moderate yield followed by basic deprotection to **100** and coupling to 2,5-diiodothiophene to give the extended π-conjugated **101** in moderate yield along with a significant amount of the homocoupled side-product/by-product alkyne **102**. Thus, in only four steps, three of which are Sonogashira couplings, compound **101** with extended π-conjugation was synthesized from readily available starting materials.

5.4.6 Heck Coupling

The Heck reaction has been widely used in the synthesis of many π-conjugated materials as well as natural products. For the importance and applicability of the methods, Heck shared the 2010 Nobel Prize with Negishi and Suzuki, and the reaction has been reviewed extensively.[98–100] General features of the Heck coupling are shown in Scheme 5.26. The reaction typically involves the coupling of an aryl or heteroaryl halide substrate (S5.26a) with an alkene (S5.26b) to substitute the aryl or heteroaryl group of the substrate for a hydrogen in the 2 position of the alkene to provide a new 1,2-disubstituted alkene (S5.26c). The geometry of the substitution around the alkene is typically *trans*, but *cis* isomers may sometimes result in small percentages. For most of the substrates used in the synthesis of π-conjugated materials, the 1,1 side-product (S5.26d) is not formed to an appreciable extent. One significant difference between the Heck and other Pd-catalyzed reactions such as the Stille, Suzuki, Negishi, and other cross coupling reactions is that the proposed mechanism does not involve a transmetallation step. Rather, as shown in Scheme 5.26, after oxidative addition (S5.26e), the alkene π-coordinates to the Pd (S5.26f, after loss of a ligand, L) and then a migratory insertion occurs (S5.26h) to give an alkylated Pd intermediate (S5.26i), as described generally in Section 5.2.4 above. The Pd-alkylated intermediate undergoes a β-hydride elimination (S5.26j) to give the 1,2-alkene and a protonated Pd intermediate (S5.26k) that reductively eliminates with an equivalent of amine (S5.26l) to regenerate the Pd catalyst. β-Hydride elimination generally was described above in Section 5.2.5. The main substrates are aryl and heteroaryl halide and the alkenes typically are substituted with an aryl, heteroaryl, or a functional group. As

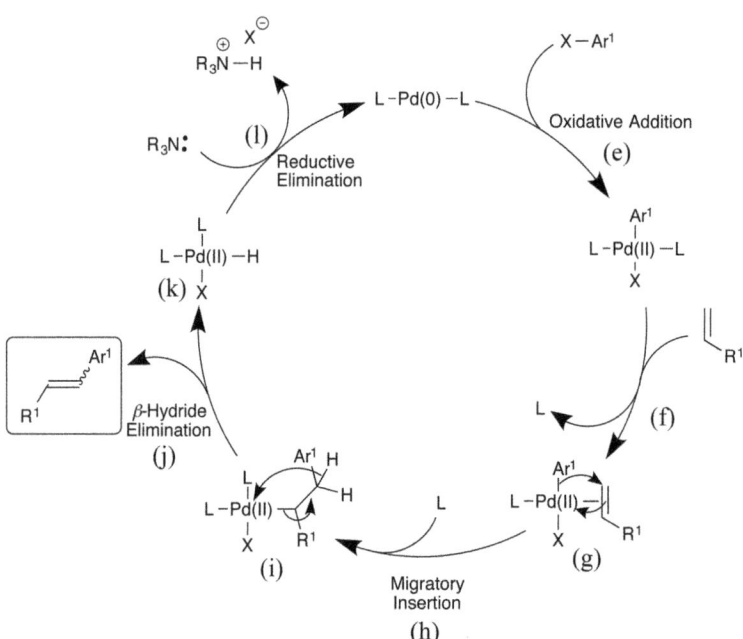

R¹ = **aryl**, **heteroaryl** (alkenes)
R² = **aryl**, **heteroaryl**, **functional group**, alkene, alkyne
X¹ = **I**, **Br**, Cl, -N₂BF₄ (OTf, OTs)
amine base = **triethylamine**, **di(isopropyl)amine**, inorganics
Pd = **PdOAc₂**, Pd(PPh₃)₄, others

G. T. Crisp, *Chem. Soc. Rev.*, 1998, **27**, 427
I. P. Beletskaya and A. V. Cheprakov, *Chem. Rev.*, 2000, **100**, 3009
N. J. Whitcombe, K. K. Hii and S. E. Gibson, *Tetrahedron*, 2001, **57**, 7449

Scheme 5.26 General illustration and mechanism of the Heck coupling.[98,99]

the migratory insertion attack of the aryl group on the alkene is nuc-leophilic, alkenes with *electron withdrawing* functional groups such as esters, ketones, nitriles, and electron acceptor aryls and heteroaryls react relatively quickly and favor the 1,2-product. When the alkene is substituted with electron donor groups (such as methoxy, amino, and *N*-amide), 1,1-substitution is favored almost exclusively and in fact has been developed as a method to synthesize 1,1-disubstituted alkenes.[101] The base used for the reductive elimination/regeneration is typically an amine, and often the solvent, although the use of inorganic bases has been developed for substrates that may be sensitive to nucleophilic amines. Diazonium salts have been used in some cases as a pseudohalide on aryl substrates, but triflates[102] and tosylates[103] have been used mostly for alkene substrates. Various different substrates, catalysts, ligands, bases, and solvent conditions have been developed to address reactivity, regiochemical, and stereochemical issues, so care should be taken to match conditions with desired aryl and heteroaryl substrates and alkenes.

Scheme 5.27 illustrates some examples of Heck coupling of interest to π-conjugated materials synthesis. The first example employs triethylamine (TEA) as the solvent with the bulky ligand P(o-tol)$_3$ at high temperature to couple the somewhat deactivated arylamine substrate **103** to alkene **104** to give alkene **105** in moderate yield over 18 hours.[104] The catalyst Pd(OAc)$_2$, which contains Pd(II), is often used as an air stable source of Pd, and generation of Pd(0) for the catalytic process has been proposed to occur through reduction of the Pd(II) with the phosphine ligand to give Pd(0) and phosphine oxide.[100] The classic Pd(OAc)$_2$/P(o-tol)$_3$/TEA conditions may also be used with a solvent such as DMF to improve solubility, such as with the coupling of **106** to a reactive ester-substituted alkene to give **107** in moderate yield.[105] Similar conditions were used to couple 5-bromothiophene carboxaldehyde, which is less deactivated than just thiophene due to the electron-withdrawing aldehyde, to dialkene **108** to give **109** in high yield.[106] Carbazole can also be used as a substrate such as with the reaction of **110** with vinyl pyridine (**111**) to give the corresponding coupled product **112** in moderate yield.[107] The final two examples illustrate the use of inorganic bases with iodides such as KOAc for the coupling of **113** to **114** in moderate yield[108] and the coupling of **116** to **117** in moderate yield.[109] Both reactions demonstrate the compatibility of the reaction with often-sensitive functional groups such as aldehydes.

5.5 DIRECT ARYLATION

The application of direct arylation in the synthesis of π-conjugated materials is still in a relatively early stage but is rapidly growing.[110,111] General direct arylation reactions that have been used in π-conjugated material synthesis are illustrated in Scheme 5.28 and include: (1) oxidative direct aryl–aryl coupling (S5.28a) to give homocoupled products; (2) direct aryl–aryl cross-coupling (S5.28b) to an aryl or heteroaryl halide; (3) and aryl–heteroatom cross coupling (S5.28c).

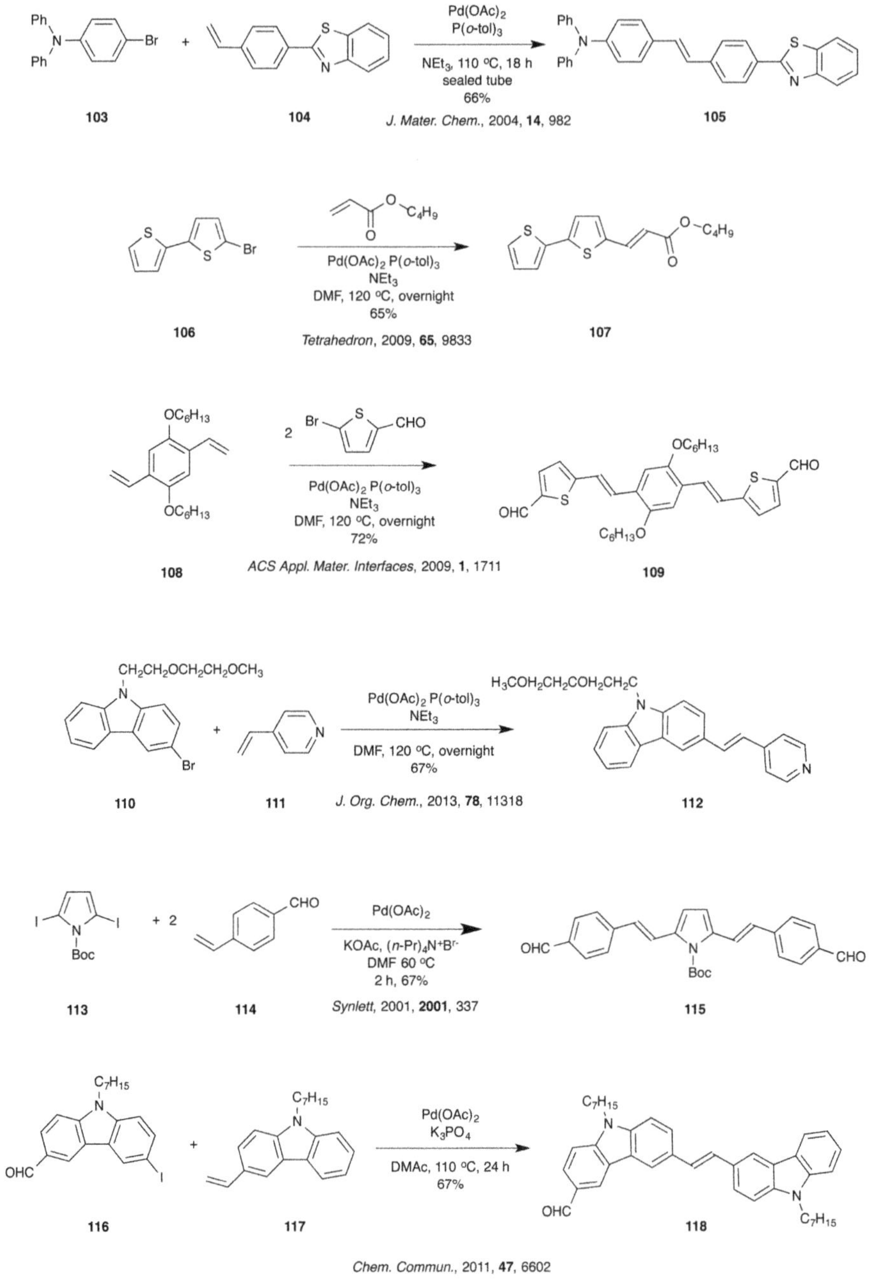

Scheme 5.27   Examples of Heck coupling on several substrates.[104–109]

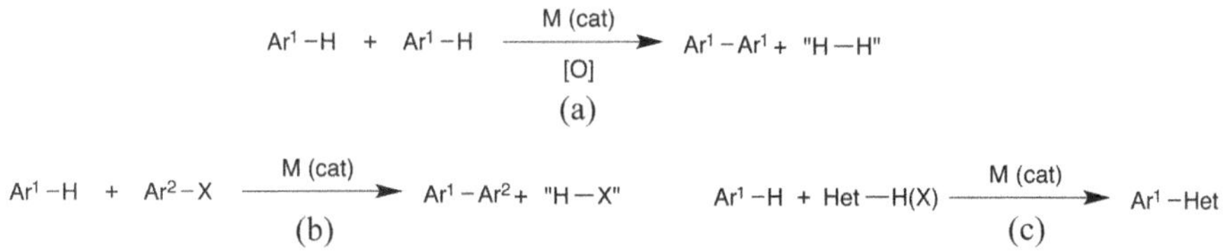

$$Ar^1-H \quad + \quad Ar^1-H \quad \xrightarrow[\text{[O]}]{\text{M (cat)}} \quad Ar^1-Ar^1 + \text{"H}-\text{H"}$$

(a)

$$Ar^1-H \quad + \quad Ar^2-X \quad \xrightarrow{\text{M (cat)}} \quad Ar^1-Ar^2 + \text{"H}-\text{X"} \qquad Ar^1-H \ + \ Het-H(X) \quad \xrightarrow{\text{M (cat)}} \quad Ar^1-Het$$

(b) (c)

D. Alberico, M. E. Scott and M. Lautens, *Chem. Rev.*, 2007, **107**, 174
J. Roger, A. L. Gottumukkala and H. Doucet, *ChemCatChem*, 2010, **2**, 20
S. Messaoudi, J.-D. Brion and M. Alami, *Eur. J. Org. Chem.*, 2010, **2010**, 6495

Scheme 5.28 General illustration of C—H activated coupling methods.

Note that in these reactions there is no metallating agent and in some cases no aryl halide; *this is one of the distinct advantages* of direct aryl coupling in that the need for making metallating reagents and aryl halides separately is wholly or partially avoided, which reduces the number of reactions steps, the synthesis time, the use of sometimes toxic and hazardous reagents (*n*-BuLi, Br$_2$, tin reagents, *etc.*), and the generation of waste products. The US NFPA hazards of many reagents used in the synthesis of π-conjugated materials are summarized in Figure 5.2. Such "green" chemistry is attractive for general safety reasons as well as both to industrial and research laboratory hygiene and cost reduction. Typically, the metal catalyst is Pd, but at this point Ir has also been used in some cases. As the application of direct arylation to the synthesis of π-conjugated materials is still relatively new, there are relatively few representative examples of each of the chemistries illustrated in Scheme 5.28; however, the rapid growth of the field will continue to provide useful reactions for the synthesis of π-conjugated materials.

Examples of oxidative direct aryl homocoupling are illustrated in Scheme 5.29. These include the coupling of bromothiophene **119** with Pd catalysis and AgNO$_3$ as oxidant to give the corresponding dibromobithiophene **120** in high yield.[112] Similar conditions, except with the use of AgF as oxidant, homocouples **121** to give **122** in moderate yield after 24 hours.[105] A final example homocouples difluoro dithienocyclopentadiene **123** to give **124** in moderate yield.[113] One disadvantage of these reactions is that they require more than stoichiometric amounts of Ag oxidant and relatively high temperatures, both of which drawbacks will likely be addressed in the near future.

Direct aryl–aryl cross coupling is under development busily as a replacement, in some cases, for the Stille or Suzuki cross-coupling reactions. Examples of direct aryl–aryl cross coupling are illustrated in Scheme 5.30. Unlike the direct oxidative coupling, an aryl halide is needed in at least one of the coupling partners in the direct aryl–aryl

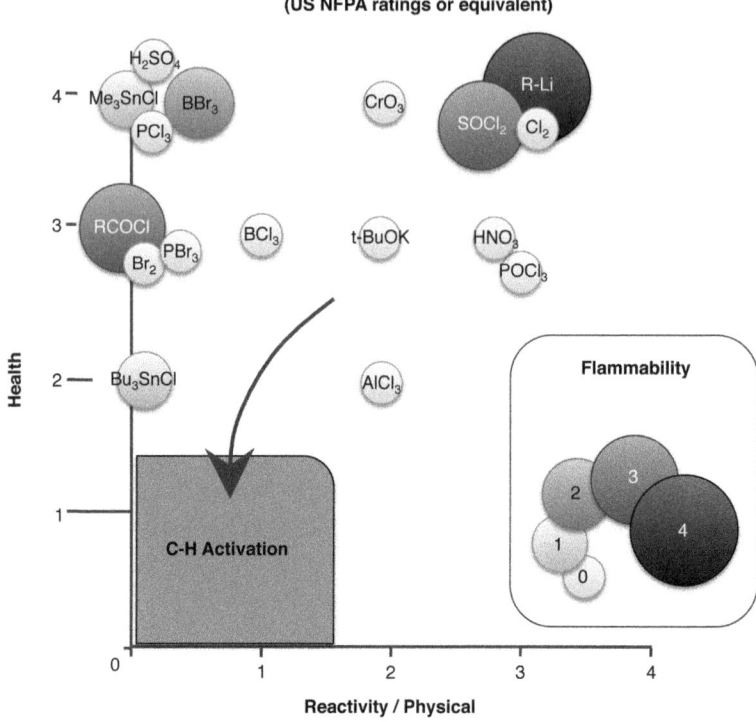

Figure 5.2   Bubble graph of common reagent hazards in organic synthesis.

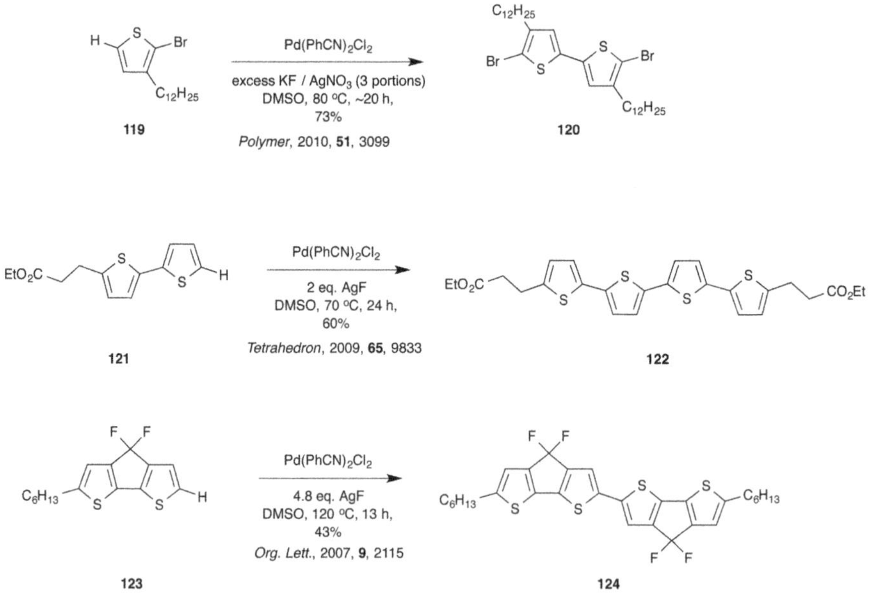

Scheme 5.29   Oxidative C—H homocoupling.[105,112,113]

cross couplings developed thus far; however, the need for synthesizing and purifying organostannanes and boronic acid esters is still avoided. The first example illustrates the direct arylation of the electron donor EDOT with electron donor halide **125** to give dicoupled compound **126** in high yield after 24 hours, with Cs_2CO_3 used stoichiometrically as a base.[114] Another example is the coupling of iodocarbazole **127** to bromothiophene **128** to give bromothiophene product **129.1** in high yield.[115] Of interest to oligomer synthesis is the iterative elongation of **129.1** outlined in the Scheme 5.30 inset. This strategy involves lithiating the bromothiophene with *n*-BuLi and converting to a new iodothiophene (with I_2) and then another reaction with bromothiophene **128** to give an oligothiophene extended by one thiophene unit and end-capped with bromine atom, which can enter the cycle again. Iodination gives high yields for oligomers **129.1** and **129.2** and

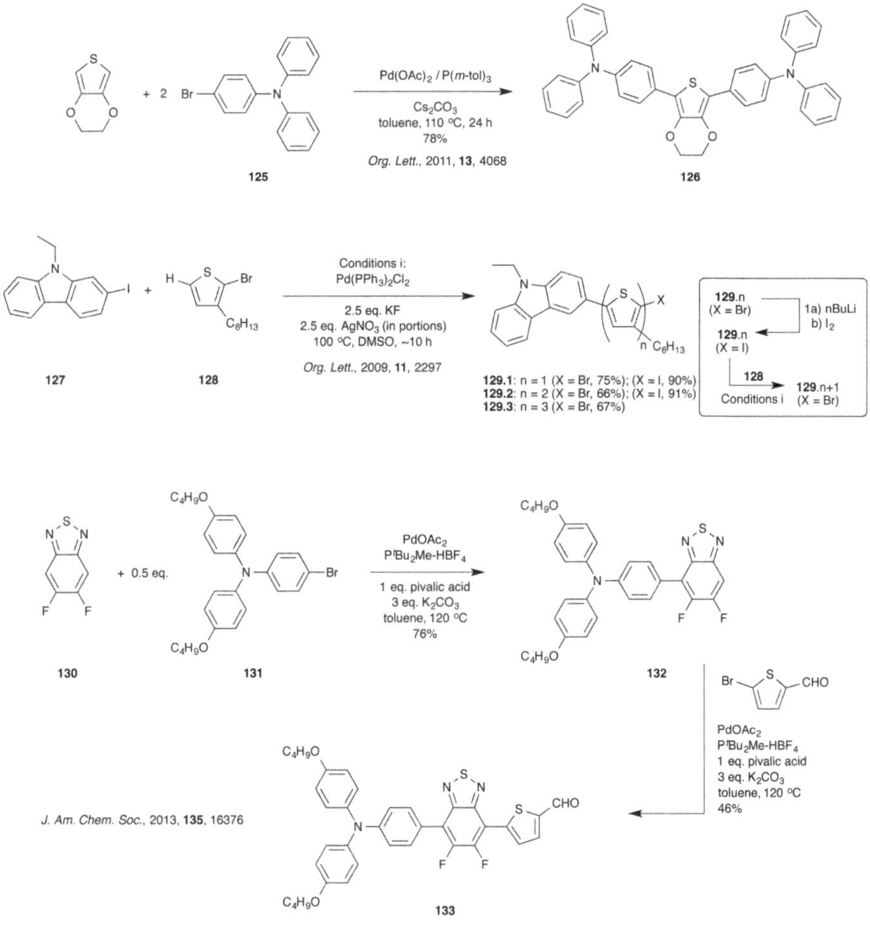

Scheme 5.30 Direct arylation C–H cross coupling with aryl halides.[114–116]

the direct aryl–aryl coupling of the iodothiophene oligomers with bromothiophene **128** give high to moderate yields for **129.1–129.3**. Conceptually, the oligomers could be extended with further iterations. A final example shows the coupling of an excess of 5,6-difluorobenzothiadiazole with bromoarylamine **131** to give a monocoupled **132** in high yield that in turn can be directly arylated with 5-bromothiophene carboxaldehyde to give the differentially substituted benzothiadiazole **133** in moderate yield overall, without the need to synthesize Stille or Suzuki reagents.[116] It should be pointed out that the direct arylation of 5,6-difluorobenzothiadiazole and similar electron acceptors appears to be highly efficient, whereas Stille or Suzuki reagents from such electron acceptor intermediates would likely react sluggishly. Thus, for the synthesis of π-conjugated materials having an electron acceptor moiety, direct arylation seems to be a synthetic strength, whereas for Stille and Suzuki coupling, electron acceptor moieties typically are a synthetic weakness.

Scheme 5.31 illustrates some examples of direct aryl–heteroatom coupling with Ir catalysts that result in borylation to an aryl Bpin,

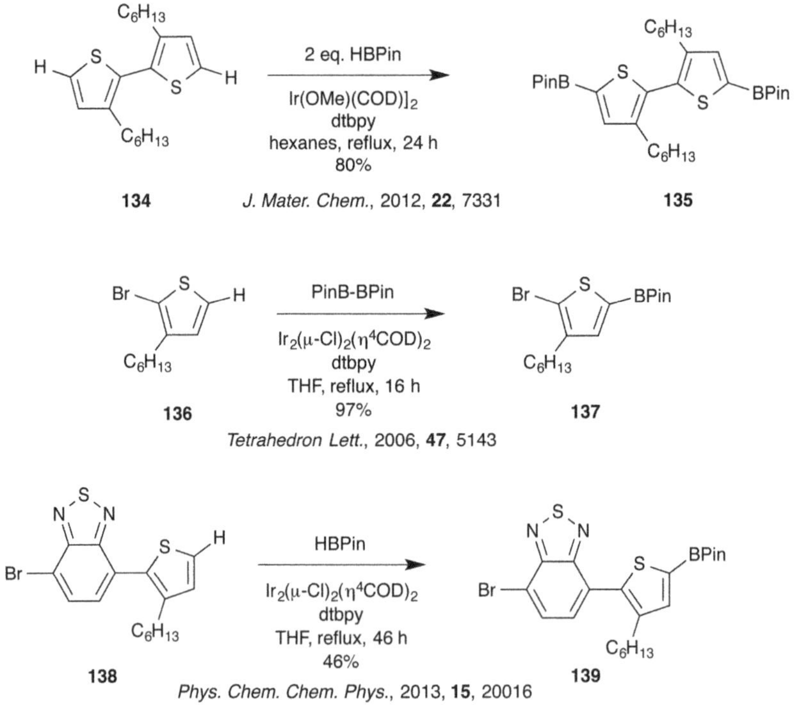

Scheme 5.31 Formation of aryl boronate Suzuki reagents *via* C–H activated coupling.[117–119]

which is of important synthetic utility in the synthesis of both small molecule and polymer π-conjugated materials as discussed in Sections 5.4.3 and 7.4.2, respectively. These examples include the double coupling of H-Bpin to bithiophene **134** to give diBpin **135** in high yield after 24 hours.[117] Another example uses (Bpin)$_2$ and the bromothiophene **136** to give the bromo Bpin **137** in high yield.[118] A final example illustrates that the reaction can also be used on a thiophene substituted with an electron acceptor moiety such as the reaction of **138** to **139** in moderate yield.[119]

Another important feature of C–H activation in the synthesis of π-conjugated materials is the possibility of regioselectivity that is different from that available with traditional functionalization such as electrophilic bromination, Vilsmeier reactions, and deprotonation. Examples of this are shown in Scheme 5.32. In the first case, pyrene (**140**, shown with the numbering scheme) can be reacted with 1.1 equivalents of (Bpin)$_2$ to give the 2-Bpin-pyrene **141** in good yield after 16 h.[120] With 2.2 equivalents of (Bpin)$_2$, the 2,7-diBpin-pyrene **142** is produced in high yield. Bpin groups can then be replaced

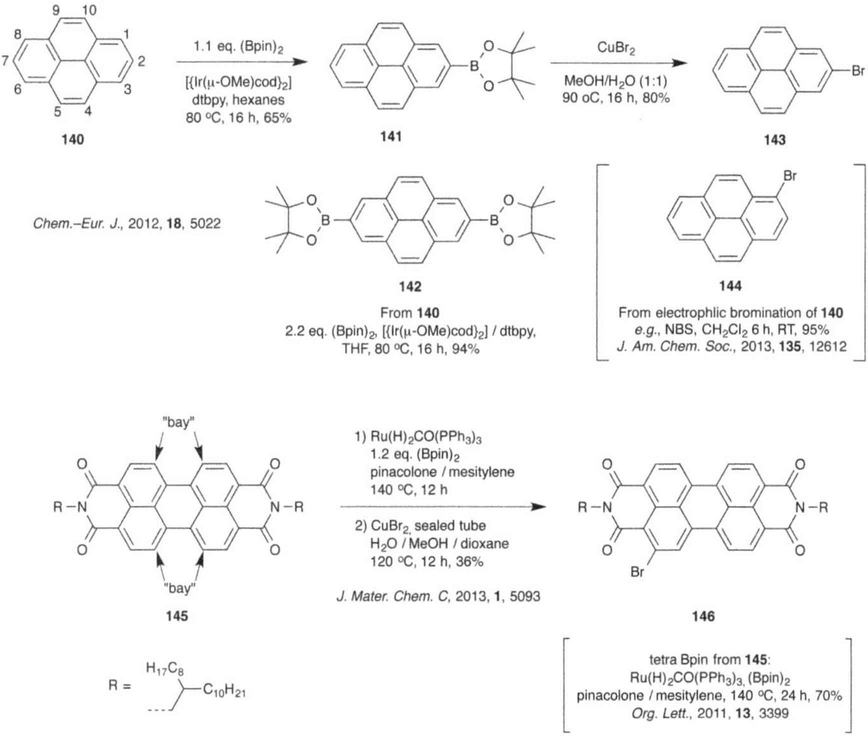

Scheme 5.32 Directed C–H coupling to form aryl boronates.[120–123]

with bromines by reaction with CuBr$_2$ in H$_2$O/MeOH to give the corresponding bromopyrene, as in the case of **141** reacting to give **143**. This strategy allows different regiochemical control than straight electrophilic bromination of pyrene, which gives 1-bromopyrene **144** from **140** under a variety of brominating conditions including NBS in CH$_2$Cl$_2$.[121] The difference in regiochemistry is thought to mainly arise from deleterious steric interactions between the hydrogen on the 10-position of pyrene with the organometallic catalyst, which makes the more sterically accessible 2-position more reactive. Another example of different regiochemistry is the transformation of PDI **145** to **146** with 1.2 equivalents of (Bpin)$_2$ and a ruthenium catalyst followed by boron–bromine exchange with CuBr$_2$.[122] It should be noted that the tetra Bpin is also available by this method.[123] The electrophilically active positions on PDI are the "bay" positions, indicated on **145** in Scheme 5.32. In this case of C−H activation of PDI by the Ru catalyst, the main factor influencing the regiochemical outcome is believed to be coordination of the Ru to the imide oxygen, which makes reaction at the position proximal to the imide rather than at the more distal bay position.

5.6 CONCLUSION

This chapter has covered the broad topic of organometallic coupling to extend the conjugation of π-conjugated systems. The power of these methods lies in the fact that related reactions and processes extend conjugation through aryl–aryl, aryl–alkene, and aryl–alkyne linkages with differing electron acceptor and electron donor substitutions on the coupling partners, mostly with good tolerance of functional groups and with high yields. These features make organometallic coupling reactions attractive for the synthesis of a range of π-conjugated materials, and the high reaction yields of the processes make them attractive for polymerizations, as is presented in Chapter 7. One potential disadvantage of these methods is that synthesizing the transmetallating agent often requires bromination or lithiation, which is acceptable for electron donor intermediates, but poses a synthetic challenge for electron acceptor intermediates. Alternatively, direct arylation through C−H has provided a viable method to synthesize π-conjugated materials and intermediates with the use of less hazardous or non-hazardous reagents. Almost as significantly, C−H activation potentially allows functionalization and reaction of electron acceptor substrates that are resistant to electrophilic reaction or unstable to nucleophilic lithiation, while in some cases providing a path for different regiochemistry.

Given the sensitivity of acceptors to nucleophilic lithiation and their resistance to electrophilic substitution, installing acceptors in π-conjugated materials often is planned for the last step, or among the last steps, in the synthesis. As such, after discussing the synthesis and functionalization of donors (Chapters 3 and 4) and extending π-conjugation (Chapters 4 and 5), Chapter 6 is devoted to the synthesis and installation of acceptors.

REFERENCES

1. D. Steinborn, *Fundamentals of Organometallic Catalysis*, Wiley-VCH Verlag & Co. FGaA, Weinheim, DE, 2012.
2. K. Muniz, *Angew. Chem., Int. Ed.*, 2009, **48**, 9412.
3. A. Gillie and J. K. Stille, *J. Am. Chem. Soc.*, 1980, **102**, 4933.
4. P. E. Fanta, *Synthesis*, 1974, **1974**, 9.
5. J. Hassan, M. Sévignon, C. Gozzi, E. Schulz and M. Lemaire, *Chem. Rev.*, 2002, **102**, 1359.
6. K. Kunz, U. Scholz and D. Ganzer, *Synlett*, 2003, 2428.
7. P. Jordens, G. Rawson and H. Wynberg, *J. Chem. Soc. C*, 1970, 273.
8. B. Pal, W.-C. Yen, J.-S. Yang, C.-Y. Chao, Y.-C. Hung, S.-T. Lin, C.-H. Chuang, C.-W. Chen and W.-F. Su, *Macromolecules*, 2008, **41**, 6664.
9. G. L. Gibson, T. M. McCormick and D. S. Seferos, *J. Am. Chem. Soc.*, 2012, **134**, 539.
10. T. Lei, Y. Cao, X. Zhou, Y. Peng, J. Bian and J. Pei, *Chem. Mater.*, 2012, **24**, 1762.
11. C. M. So and F. Y. Kwong, *Chem. Soc. Rev.*, 2011, **40**, 4963.
12. A. Roglans, A. Pla-Quintana and M. Moreno-Manas, *Chem. Rev.*, 2006, **106**, 4622.
13. E. Negishi, *Angew. Chem., Int. Ed.*, 2011, **50**, 6738.
14. Y. Nakao and T. Hiyama, *Chem. Soc. Rev.*, 2011, **40**, 4893.
15. F. C. Krebs, R. B. Nyberg and M. Jørgensen, *Chem. Mater.*, 2004, **16**, 1313.
16. K. T. Nielsen, K. Bechgaard and F. C. Krebs, *Macromolecules*, 2005, **38**, 658.
17. Z. R. Owczarczyk, W. A. Braunecker, A. Garcia, R. Larsen, A. M. Nardes, N. Kopidakis, D. S. Ginley and D. C. Olson, *Macromolecules*, 2013, **46**, 1350.
18. D. G. Patel, K. R. Graham and J. R. Reynolds, *J. Mater. Chem.*, 2012, **22**, 3004.
19. N. Galaffu, S. P. Man, R. D. Wilkes and J. R. H. Wilson, *Org. Process Res. Dev.*, 2007, **11**, 406.

20. C. C. Johansson Seechurn, M. O. Kitching, T. J. Colacot and V. Snieckus, *Angew. Chem., Int. Ed.*, 2012, **51**, 5062.
21. X. F. Wu, P. Anbarasan, H. Neumann and M. Beller, *Angew. Chem., Int. Ed.*, 2010, **49**, 9047.
22. M. Beller, *Chem. Soc. Rev.*, 2011, **40**, 4891.
23. B. H. Lipshutz, A. R. Abela, Ž. V. Bošković, T. Nishikata, C. Duplais and A. Krasovskiy, *Top. Catal.*, 2010, **53**, 985.
24. J. K. Stille, *Angew. Chem., Int. Ed. Engl.*, 1986, **25**, 508.
25. A. F. Littke and G. C. Fu, *Angew. Chem., Int. Ed.*, 1999, **38**, 2411.
26. S. L. Buchwald, J. R. Naber, B. P. Fors, X. Wu and J. T. Gunn, *Heterocycles*, 2010, **80**, 1215.
27. W. Su, S. Urgaonkar, P. A. McLaughlin and J. G. Verkade, *J. Am. Chem. Soc.*, 2004, **126**, 16433.
28. S. P. Mee, V. Lee and J. E. Baldwin, *Angew. Chem., Int. Ed.*, 2004, **43**, 1132.
29. P. Espinet and A. M. Echavarren, *Angew. Chem., Int. Ed.*, 2004, **43**, 4704.
30. F. Roschangar, J. C. Brown, B. E. Cooley, M. J. Sharp and R. T. Matsuoka, *Tetrahedron*, 2002, **58**, 1657.
31. K. Kawabata, M. Takeguchi and H. Goto, *Macromolecules*, 2013, **46**, 2078.
32. Y.-C. Chang, S.-C. Yeh, Y.-H. Chen, C.-T. Chen, R.-H. Lee and R.-J. Jeng, *Dyes Pigm.*, 2013, **99**, 577.
33. P. M. Beaujuge, W. Pisula, H. N. Tsao, S. Ellinger, K. Mullen and J. R. Reynolds, *J. Am. Chem. Soc.*, 2009, **131**, 7514.
34. L. Liao, L. Dai, A. Smith, M. Durstock, J. Lu, J. Ding and Y. Tao, *Macromolecules*, 2007, **40**, 9406.
35. L. E. Polander, A. S. Romanov, S. Barlow, K. Hwang do, B. Kippelen, T. V. Timofeeva and S. R. Marder, *Org. Lett.*, 2012, **14**, 918.
36. S. Ellinger, U. Ziener, U. Thewalt, K. Landfester and M. Möller, *Chem. Mater.*, 2007, **19**, 1070.
37. M. Takahashi, K. Masui, H. Sekiguchi, N. Kobayashi, A. Mori, M. Funahashi and N. Tamaoki, *J. Am. Chem. Soc.*, 2006, **128**, 10930.
38. K.-H. Kim, H. Yu, H. Kang, D. J. Kang, C.-H. Cho, H.-H. Cho, J. H. Oh and B. J. Kim, *J. Mater. Chem. A*, 2013, **1**, 14538.
39. C. Qin, A. Islam and L. Han, *Dyes Pigm.*, 2012, **94**, 553.
40. S. Wen, J. Pei, P. Li, Y. Zhou, W. Cheng, Q. Dong, Z. Li and W. Tian, *J. Polym. Sci., Part A: Polym. Chem.*, 2011, **49**, 2715.
41. L. Biniek, S. Fall, C. L. Chochos, D. V. Anokhin, D. A. Ivanov, N. Leclerc, P. Lévêque and T. Heiser, *Macromolecules*, 2010, **43**, 9779.

42. A. Midya, V. Mamidala, J. X. Yang, P. K. Ang, Z. K. Chen, W. Ji and K. P. Loh, *Small*, 2010, **6**, 2292.

43. Z. B. Henson, G. C. Welch, T. van der Poll and G. C. Bazan, *J. Am. Chem. Soc.*, 2012, **134**, 3766.

44. R. Arnold, S. A. Matchett and M. Rosenblum, *Organometallics*, 1988, 7, 2261.

45. L. Zou, X. Y. Wang, K. Shi, J. Y. Wang and J. Pei, *Org. Lett.*, 2013, **15**, 4378.

46. A. Suzuki, *Angew. Chem., Int. Ed.*, 2011, **50**, 6723.

47. N. Miyaura and A. Suzuki, *Chem. Rev.*, 1995, **95**, 2457.

48. S. Hitosugi, D. Tanimoto, W. Nakanishi and H. Isobe, *Chem. Lett.*, 2012, **41**, 972.

49. J. Yang, S. Liu, J.-F. Zheng and J. S. Zhou, *Eur. J. Org. Chem.*, 2012, **2012**, 6248.

50. T. E. Barder, S. D. Walker, J. R. Martinelli and S. L. Buchwald, *J. Am. Chem. Soc.*, 2005, **127**, 4685.

51. G. A. Molander, B. Canturk and L. E. Kennedy, *J. Org. Chem.*, 2009, **74**, 973.

52. S. Darses and J. P. Genet, *Chem. Rev.*, 2008, **108**, 288.

53. D. M. Knapp, E. P. Gillis and M. D. Burke, *J. Am. Chem. Soc.*, 2009, **131**, 6961.

54. E. P. Gillis and M. D. Burke, *J. Am. Chem. Soc.*, 2007, **129**, 6716.

55. Y. Sato, R. Yamasaki and S. Saito, *Angew. Chem., Int. Ed.*, 2009, **48**, 504.

56. T. T. Nguyen, M. Baumgarten, A. Rouhanipour, H. J. Rader, I. Lieberwirth and K. Mullen, *J. Am. Chem. Soc.*, 2013, **135**, 4183.

57. A. Del Grosso, P. J. Singleton, C. A. Muryn and M. J. Ingleson, *Angew. Chem., Int. Ed.*, 2011, **50**, 2102.

58. V. Bagutski, A. Del Grosso, J. A. Carrillo, I. A. Cade, M. D. Helm, J. R. Lawson, P. J. Singleton, S. A. Solomon, T. Marcelli and M. J. Ingleson, *J. Am. Chem. Soc.*, 2013, **135**, 474.

59. G. A. Molander, L. N. Cavalcanti, B. Canturk, P. S. Pan and L. E. Kennedy, *J. Org. Chem.*, 2009, **74**, 7364.

60. T. Ishiyama, M. Murata and N. Miyaura, *J. Org. Chem.*, 1995, **60**, 7508.

61. C. Kleeberg, L. Dang, Z. Lin and T. B. Marder, *Angew. Chem., Int. Ed.*, 2009, **48**, 5350.

62. Y. Nagashima, R. Takita, K. Yoshida, K. Hirano and M. Uchiyama, *J. Am. Chem. Soc.*, 2013, **135**, 18730.

63. T. Ishiyama, J. Takagi, Y. Yonekawa, J. F. Hartwig and N. Miyaura, *Adv. Synth. Catal.*, 2003, **345**, 1103.

64. K. Huang, D. G. Yu, S. F. Zheng, Z. H. Wu and Z. J. Shi, *Chem.–Eur. J.*, 2011, **17**, 786.
65. Y. Salma, S. Ballereau, C. Maaliki, S. Ladeira, N. Andrieu-Abadie and Y. Genisson, *Org. Biomol. Chem.*, 2010, **8**, 3227.
66. J. I. Perlmutter, L. T. Forbes, D. J. Krysan, K. Ebsworth-Mojica, J. M. Colquhoun, J. L. Wang, P. M. Dunman and D. P. Flaherty, *J. Med. Chem.*, 2014, **57**, 8540.
67. M.-C. Yuan, M.-H. Su, M.-Y. Chiu and K.-H. Wei, *J. Polym. Sci., Part A: Polym. Chem.*, 2010, **48**, 1298.
68. F. V. Drozdov, E. N. Myshkovskaya, D. K. Susarova, P. A. Troshin, O. D. Fominykh, M. Y. Balakina, A. V. Bakirov, M. A. Shcherbina, J. Choi, D. Tondelier, M. I. Buzin, S. N. Chvalun, A. Yassar and S. A. Ponomarenko, *Macromol. Chem. Phys.*, 2013, **214**, 2144.
69. A. Del Grosso, M. D. Helm, S. A. Solomon, D. Caras-Quintero and M. J. Ingleson, *Chem. Commun.*, 2011, **47**, 12459.
70. M. Frigoli, C. Moustrou, A. Samat and R. Guglielmetti, *Eur. J. Org. Chem.*, 2003, **2003**, 2799.
71. F. A. Lemasson, T. Strunk, P. Gerstel, F. Hennrich, S. Lebedkin, C. Barner-Kowollik, W. Wenzel, M. M. Kappes and M. Mayor, *J. Am. Chem. Soc.*, 2011, **133**, 652.
72. S. H. Jung, W. Pisula, A. Rouhanipour, H. J. Rader, J. Jacob and K. Mullen, *Angew. Chem., Int. Ed.*, 2006, **45**, 4685.
73. P. Gao, X. Feng, X. Yang, V. Enkelmann, M. Baumgarten and K. Mullen, *J. Org. Chem.*, 2008, **73**, 9207.
74. R. Chen, X. Yang, H. Tian, X. Wang, A. Hagfeldt and L. Sun, *Chem. Mater.*, 2007, **19**, 4007.
75. K. C. Moss, K. N. Bourdakos, V. Bhalla, K. T. Kamtekar, M. R. Bryce, M. A. Fox, H. L. Vaughan, F. B. Dias and A. P. Monkman, *J. Org. Chem.*, 2010, **75**, 6771.
76. S. Haid, A. Mishra, C. Uhrich, M. Pfeiffer and P. Bäuerle, *Chem. Mater.*, 2011, **23**, 4435.
77. M. Melucci, L. Favaretto, C. Bettini, M. Gazzano, N. Camaioni, P. Maccagnani, P. Ostoja, M. Monari and G. Barbarella, *Chem.–Eur. J.*, 2007, **13**, 10046.
78. A. C. Heinrich, B. Thiedemann, P. J. Gates and A. Staubitz, *Org. Lett.*, 2013, **15**, 4666.
79. J. Huang and S. P. Nolan, *J. Am. Chem. Soc.*, 1999, **121**, 9889.
80. N. Liu and Z. X. Wang, *J. Org. Chem.*, 2011, **76**, 10031.
81. M. E. Limmert, A. H. Roy and J. F. Hartwig, *J. Org. Chem.*, 2005, **70**, 9364.
82. R. Miyakoshi, A. Yokoyama and T. Yokozawa, *J. Am. Chem. Soc.*, 2005, **127**, 17542.

83. I. Osaka and R. D. McCullough, *Acc. Chem. Res.*, 2008, **41**, 1202.
84. P. Bertus, F. Fécourt, C. Bauder and P. Pale, *New J. Chem.*, 2004, **28**, 12.
85. L. Brandsma, *Preparative Acetylene Chemistry*, Elsevier, Amsterdam, 1988.
86. G. Bucher, A. A. Mahajan and M. Schmittel, *J. Org. Chem.*, 2008, **73**, 8815.
87. J. N. Clifford, A. Gégout, S. Zhang, R. Pereira de Freitas, M. Urbani, M. Holler, P. Ceroni, J.-F. Nierengarten and N. Armaroli, *Eur. J. Org. Chem.*, 2007, **2007**, 5899.
88. B. M. Kobilka, A. V. Dubrovskiy, M. D. Ewan, A. L. Tomlinson, R. C. Larock, S. Chaudhary and E. M. Jeffries, *Chem. Commun.*, 2012, **48**, 8919.
89. J. B. Giguere, J. Boismenu-Lavoie and J. F. Morin, *J. Org. Chem.*, 2014, **79**, 2404.
90. Y. Li, M. E. Kose and K. S. Schanze, *J. Phys. Chem. B*, 2013, **117**, 9025.
91. B. Gole, S. Shanmugaraju, A. K. Bar and P. S. Mukherjee, *Chem. Commun.*, 2011, **47**, 10046.
92. D. H. Lee, M. J. Lee, H. M. Song, B. J. Song, K. D. Seo, M. Pastore, C. Anselmi, S. Fantacci, F. De Angelis, M. K. Nazeeruddin, M. Gräetzel and H. K. Kim, *Dyes Pigm.*, 2011, **91**, 192.
93. R. S. Ashraf, J. Gilot and R. A. J. Janssen, *Sol. Energy Mater. Sol. Cells*, 2010, **94**, 1759.
94. F. Silvestri, A. Marrocchi, M. Seri, C. Kim, T. J. Marks, A. Facchetti and A. Taticchi, *J. Am. Chem. Soc.*, 2010, **132**, 6108.
95. N. N. Moonen, W. C. Pomerantz, R. Gist, C. Boudon, J. P. Gisselbrecht, T. Kawai, A. Kishioka, M. Gross, M. Irie and F. Diederich, *Chem.–Eur. J.*, 2005, **11**, 3325.
96. D. Chemin and G. Linstrumelle, *Tetrahedron*, 1994, **50**, 5335.
97. C. Kitamura, K. Saito, M. Nakagawa, M. Ouchi, A. Yoneda and Y. Yamashita, *Tetrahedron Lett.*, 2002, **43**, 3373.
98. G. T. Crisp, *Chem. Soc. Rev.*, 1998, **27**, 427.
99. I. P. Beletskaya and A. V. Cheprakov, *Chem. Rev.*, 2000, **100**, 3009.
100. N. J. Whitcombe, K. K. Hii and S. E. Gibson, *Tetrahedron*, 2001, **57**, 7449.
101. J. Ruan and J. Xiao, *Acc. Chem. Res.*, 2011, **44**, 614.
102. S. Cacchi, *Pure Appl. Chem.*, 1996, **68**, 45.
103. X. Fu, S. Zhang, J. Yin, T. L. McAllister, S. A. Jiang, C.-H. Tann, T. K. Thiruvengadam and F. Zhang, *Tetrahedron Lett.*, 2002, **43**, 573.
104. T.-C. Lin, G. S. He, P. N. Prasad and L.-S. Tan, *J. Mater. Chem.*, 2004, **14**, 982.

105. G. Romanazzi, F. Marinelli, P. Mastrorilli, L. Torsi, A. Sibaouih, M. Räisänen, T. Repo, P. Cosma, G. P. Suranna and C. F. Nobile, *Tetrahedron*, 2009, **65**, 9833.
106. J. A. Mikroyannidis, M. M. Stylianakis, P. Balraju, P. Suresh and G. D. Sharma, *ACS Appl. Mater. Interfaces*, 2009, **1**, 1711.
107. X. J. Feng, P. Z. Tian, Z. Xu, S. F. Chen and M. S. Wong, *J. Org. Chem.*, 2013, **78**, 11318.
108. L. Tietze and G. Nordmann, *Synlett*, 2001, **2001**, 337.
109. B. Xu, J. He, Y. Dong, F. Chen, W. Yu and W. Tian, *Chem. Commun.*, 2011, **47**, 6602.
110. K. Okamoto, J. Zhang, J. B. Housekeeper, S. R. Marder and C. K. Luscombe, *Macromolecules*, 2013, **46**, 8059.
111. L. G. Mercier and M. Leclerc, *Acc. Chem. Res.*, 2013, **46**, 1597.
112. M. A. M. Leenen, F. Cucinotta, W. Pisula, J. Steiger, R. Anselmann, H. Thiem and L. De Cola, *Polymer*, 2010, **51**, 3099.
113. Y. Ie, M. Nitani, M. Ishikawa, K. Nakayama, H. Tada, T. Kaneda and Y. Aso, *Org. Lett.*, 2007, **9**, 2115.
114. C. Y. Liu, H. Zhao and H. H. Yu, *Org. Lett.*, 2011, **13**, 4068.
115. N. Masuda, S. Tanba, A. Sugie, D. Monguchi, N. Koumura, K. Hara and A. Mori, *Org. Lett.*, 2009, **11**, 2297.
116. J. Zhang, W. Chen, A. J. Rojas, E. V. Jucov, T. V. Timofeeva, T. C. Parker, S. Barlow and S. R. Marder, *J. Am. Chem. Soc.*, 2013, **135**, 16376.
117. C.-Y. Chao, C.-H. Chao, L.-P. Chen, Y.-C. Hung, S.-T. Lin, W.-F. Su and C.-F. Lin, *J. Mater. Chem.*, 2012, **22**, 7331.
118. I. A. Liversedge, S. J. Higgins, M. Giles, M. Heeney and I. McCulloch, *Tetrahedron Lett.*, 2006, **47**, 5143.
119. A. Karolewski, A. Neubig, M. Thelakkat and S. Kummel, *Phys. Chem. Chem. Phys.*, 2013, **15**, 20016.
120. A. G. Crawford, Z. Liu, I. A. Mkhalid, M. H. Thibault, N. Schwarz, G. Alcaraz, A. Steffen, J. C. Collings, A. S. Batsanov, J. A. Howard and T. B. Marder, *Chem.–Eur. J.*, 2012, **18**, 5022.
121. R. S. Kathayat and N. S. Finney, *J. Am. Chem. Soc.*, 2013, **135**, 12612.
122. J. Zhang, S. Singh, D. K. Hwang, S. Barlow, B. Kippelen and S. R. Marder, *J. Mater. Chem. C*, 2013, **1**, 5093.
123. G. Battagliarin, Y. Zhao, C. Li and K. Mullen, *Org. Lett.*, 2011, **13**, 3399.

CHAPTER 6

Acceptors

6.1 INTRODUCTION

In the synthesis of π-conjugated materials, the installation of acceptors in is often a key step in modifying, for example, a material's
electron affinity (EA), ionization energy (IE), absorption wavelength
(λ_{max}), and nonlinear optical properties. However, the typically high
EA of electron acceptors often renders them sensitive to nucleophilic
attack, particularly by lithium reagents, and reduction by low IP materials. Additionally, the high IE also generally renders them difficult
to functionalize by electrophilic substitution reactions such as halogenation and Vilsmeier reactions. Because of these synthetic challenges, installation of acceptors has traditionally occurred as one of
the last steps in the synthesis of π-conjugated materials or as the very
last step, especially with strong acceptors, which may be thermally
unstable in addition to being sensitive to nucleophiles and resistant
to electrophilic substitution. Another consequence of this sensitivity
and low reactivity is that reactions to diversify the chemical structure
of acceptors is relatively limited compared to donors and π-bridges,
and as such the structural variation of acceptors is somewhat more
limited compared to that of typical donors and π-bridges. However,
given the central importance of acceptors to modification of many
important properties of π-conjugated materials, general reactions
and synthetic methodologies have been developed around certain key

Synthetic Methods in Organic Electronic and Photonic Materials: A Practical Guide
By Timothy C. Parker and Seth R. Marder
© Timothy C. Parker and Seth R. Marder, 2015
Published by the Royal Society of Chemistry, www.rsc.org

classes of acceptors to impart structural modifications that change acceptor strength, solubility/processability, and functionality.

This chapter covers one of the main methods of installing a wide variety of acceptors and introduces the synthesis and some key reactions of acceptor classes that have proven useful in π-conjugated materials and their synthesis.

6.2 KNOEVENAGEL CONDENSATION

The Knoevenagel Condensation was first reported in 1894 by Emil Knoevenagel and has been used extensively since. In an extensive review article by Jones published in 1967, there are over 1150 references and tables of substrates, conditions, and products that span nearly 300(!) pages.[1] The general reaction is illustrated in Scheme 6.1 and is between a ketone or aldehyde (S6.1a) and an activated methylene, *i.e.*, a methylene with at least one electron-withdrawing group (S6.1b) to give the corresponding alkene (S6.1c) and water (S6.1d). Addition (S6.1e) of the active methylene compound gives the intermediate adduct (S6.1f) that can undergo dehydration (S6.1g) to give the final product. One key aspect of the reaction is that there is typically an equilibrium between the starting materials, intermediate, and final products, where addition of water (S6.1h) and elimination (S6.1i) of the active methylene can regenerate the starting materials from the product. This process may limit the conversion and/or yield of the reaction or may cause a decomposition of the product if it is exposed to moisture or acidic media in subsequent steps (*e.g.*, during column chromatography on silica gel). Thus, removing water from the reaction may increase reaction yields, and protecting the product from water may be necessary, especially for very strong acceptors.

Knoevenagel condensations are generally catalyzed by bases, and especially by amine or nitrogen heterocycle bases. The mechanism can occur through more of an aldol-like condensation with a tertiary amine such as Et$_3$N or Hünig's base, strong hydroxide/alkoxide bases, or a nitrogen heterocyclic base like pyridine. The mechanism may occur through formation of imines or iminium salts from the aldehyde or ketones when primary or secondary amines (*e.g.*, butyl amine or piperidine, respectively) catalyze the reaction, such as illustrated in Scheme 6.1, S6.1g). The addition of acid, or use of an ammonium salt, may speed up the reaction much like an aldol condensation can be either acid or base catalyzed. Often, when an acid and amine or an ammonium salt is used as a catalyst, the effect of acid has been attributed to increasing the rate of formation of an imine intermediate

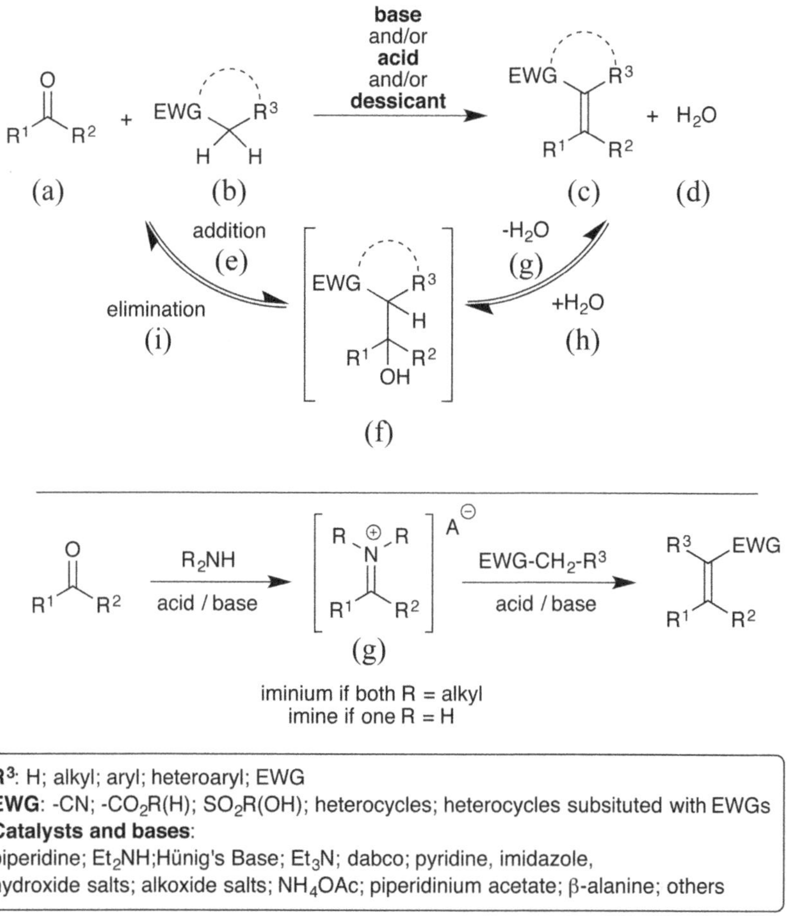

Scheme 6.1   General illustration of Knoevenagel condensations. EWG = electron withdrawing group.

(S6.1g). The number of different catalysts that have been used is relatively large, but some of the more common are piperidine, diethyl amine, triethylamine, ethanolic ammonia, DABCO, piperazine, pyridine, imidazole, DBU, NH_4OH, KO^tBu, NH_4OAc, piperidinium acetate, β-alanine, and others, as well as *mixtures* of these and other catalysts.

In many cases, removal of the water by-product can be important to obtain good yields since the reaction is reversible. The water can be removed by using a Dean–Stark trap, by using agents such as molecular sieves, $MgSO_4$ or Na_2SO_4, and/or alumina to absorb the water. Additionally, reagents like anhydrides that more-or-less irreversibly react with and thereby remove water under the reaction conditions may be used. In particular, acetic anhydride alone can sometimes affect the transformation without addition of other catalysts, which may be especially effective with strong acceptors, carbonyl substrates, or when products that might be sensitive to nucleophilic amines or bases are formed. Other chemical agents such as silanes may be used to irreversibly react with the water by-product.[2]

In addition to the many bases, acids, and dehydrating agents that have been used, there are a number of processing and chemical techniques that have been applied to the Knoevenagel condensation including the use of ionic liquids,[3,4] microwave assistance,[5] and solvent free ball milling.[6] It is often the case that combinations of catalysts, dehydrating reagents, and/or chemical/processing techniques may be necessary to discover optimal reaction conditions in a somewhat Edisonian manner. Often, running the reaction in as concentrated a solution as possible, even avoiding solvent if feasible, in combination with water removal may be key to success, especially with microwave irradiation; however, the sheer number of possibilities of catalysts, dehydrating regents, and specialized techniques may lead to an "optimization swamp" that may prove fruitless, and the researcher is many times well served by stopping "incomplete" reactions short of complete conversion of the starting materials in order to isolate a moderate yield of the product, even if conversion is relatively low since the starting materials may be recovered and reused (this would be properly reported as a % yield based on a % conversion). Forcing the reaction, or rather the equilibrium, to "completion" such that starting materials are entirely consumed may result in decomposed product and/or may shift the equilibrium to starting materials if decomposition releases water. As such, trying to force the reaction to a point of complete starting material consumption may ultimately prove counterproductive.

Some common conditions for Knoevenagel condensations are illustrated by the reactions shown in Scheme 6.2. The first example is a classic condensation of a thiophene aldehyde 1 with malononitrile 2, which is an activated methylene compound that has two electron withdrawing groups, catalyzed by piperidine to give the corresponding acceptor-substituted product 3 in high yields.[7] Piperidine as the base, or even as the base and solvent, can be used in Knoevenagel condensations of many acceptors if the substrate and product are relatively stable to nucleophiles. Generally less nucleophilicity can be achieved by using piperidinium acetate as the catalyst, such as with the condensation of aldehyde 4 with the "TCF" acceptor 5 to give product 6 in high yield in refluxing ethanol.[8] Using less basic media/catalysts may be preferable with stronger acceptors that may give nucleophile-sensitive products or may be nucleophile-sensitive themselves. Another option

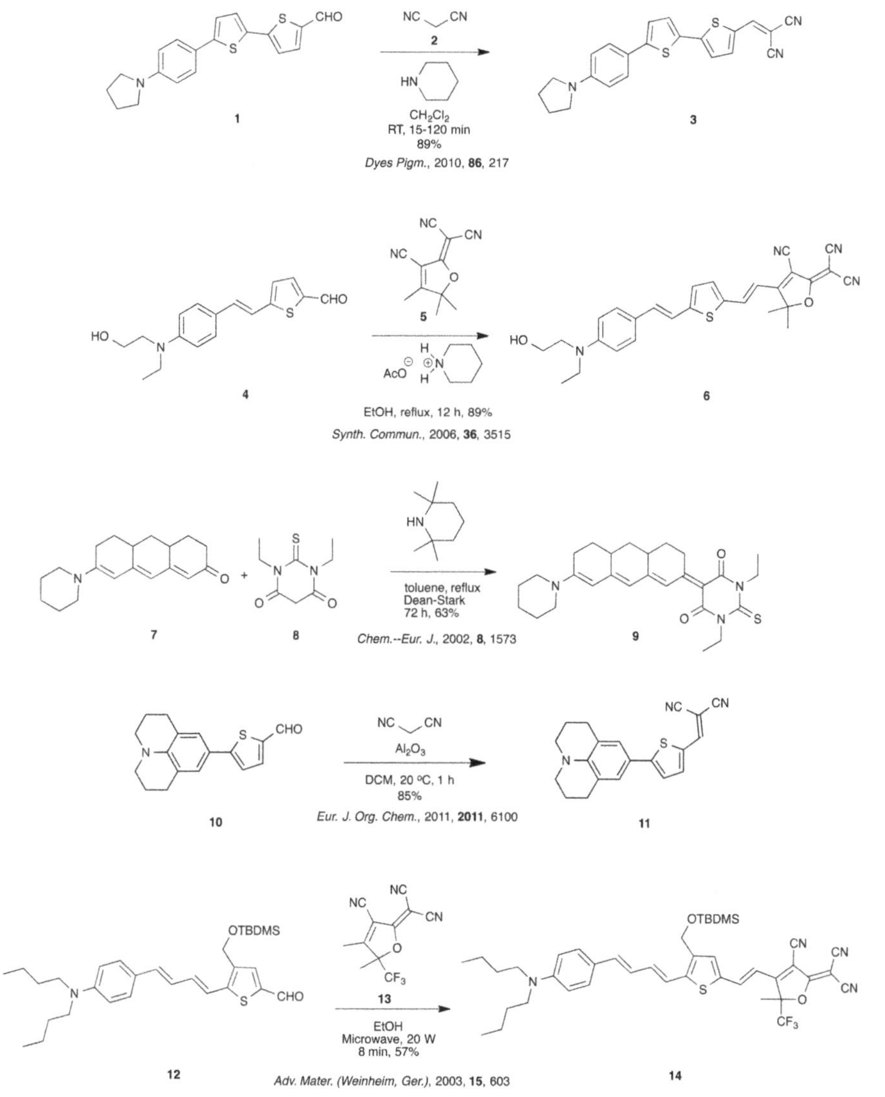

Scheme 6.2 Knoevenagel condensations of aldehyde and ketone substrates.[7–9,119,120]

to reduce nucleophilicity without reducing base strength is to use a sterically hindered base, such as with the condensation of ketone **7** with diethylthiobarbituric acid **8** catalyzed by tetramethylpiperidine (TMP) to give, with azeotropic removal of water, the corresponding product **9** in moderate yield.[9] These conditions are particularly effective since ketones in general are less reactive than aldehydes in Knoevenagel condensations (and addition reactions in general), and ketone

7 is deactivated further toward addition by being in conjugation with a relatively strong aliphatic, cyclic amine donor. A "base free" Knoevenagel condensation is shown for the reaction of aldehyde 10 with 11 in high yield catalyzed by Al_2O_3, which likely also acts as a desiccant, but Al_2O_3 itself may be slightly basic, neutral, or slightly acidic. Finally, very strong acceptors may not require the addition of catalyst, such as with the microwave-assisted reaction of aldehyde 12 with fluorinated TCF acceptor 13 to give the corresponding product 14 in moderate yield. In the case of very strong acceptors, the pK_a of the methylene protons may be sufficiently low that the reaction may be considered an acidic self-catalyzed condensation.

The Knoevenagel condensation offers the possibility of making a given acceptor a *stronger* acceptor that in turn can be reacted with other substrates. Examples of this strategy are shown in Scheme 6.3. Knoevenagel condensation of barbituric acid 15 with pyranone 16 in acetic anhydride gives compound 17 that can be considered a stronger acceptor than pyranone 16 and can be condensed with aldehyde 18 to give product 19 in low yield,[10] perhaps partially due to the possibility of 17 undergoing double condensation. Such tandem Knoevenagel with a double condensation is illustrated by the reaction of pyranone 16 with malononitrile to give acceptor 20 and then reaction of both active methyl groups on 20 with aldehyde 21 to give compound 22 with extended π-conjugation in moderate yield.[11] Tandem Knoevenagel condensations have also been accomplished in one pot, such as with the reaction of ketone 23 with malononitrile to give acceptor 24 and then in one pot condensation of 24 with aldehyde 25 to give product 26 in low yield, although other aromatic amino aldehydes gave greater than 40% yield.[12] A final example is the condensation of diketone 27 with malononitrile catalyzed by β-alanine to give acceptor 28, which in turn was coupled with distannane 29 to give the corresponding product 30 in high yield.[13] Acceptor 28, which is substituted with multiple strong electron withdrawing groups, would generally be a highly activated substrate for Pd coupling chemistry.

Examples of double Knoevenagel condensations on dicarbonyl substrates are illustrated in Scheme 6.4. The first is condensation of donor dialdehyde 31 with malononitrile catalyzed by ammonium acetate to give the corresponding double condensation product 32 in high yield.[14] A double condensation was also possible on the acceptor–donor–acceptor (A–D–A) triad dialdehyde 33 with malononitrile and Al_2O_3 catalysis to give the corresponding product 34 in low-to-moderate yield.[15] The reaction also has been used on the donor diketone 35 to give the corresponding product 36 in moderate

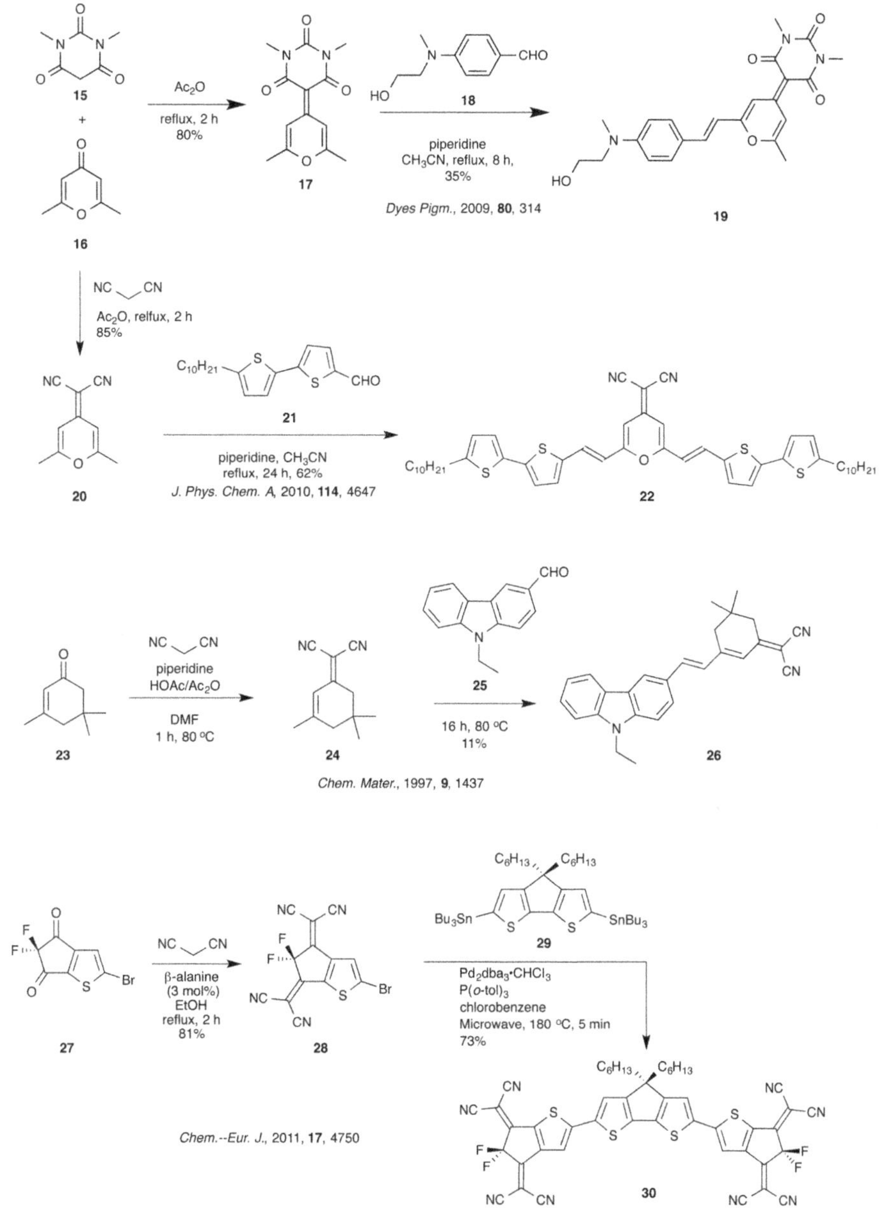

Scheme 6.3 Tandem Knoevenagel condensations in the formation and coupling of acceptors.[10–13]

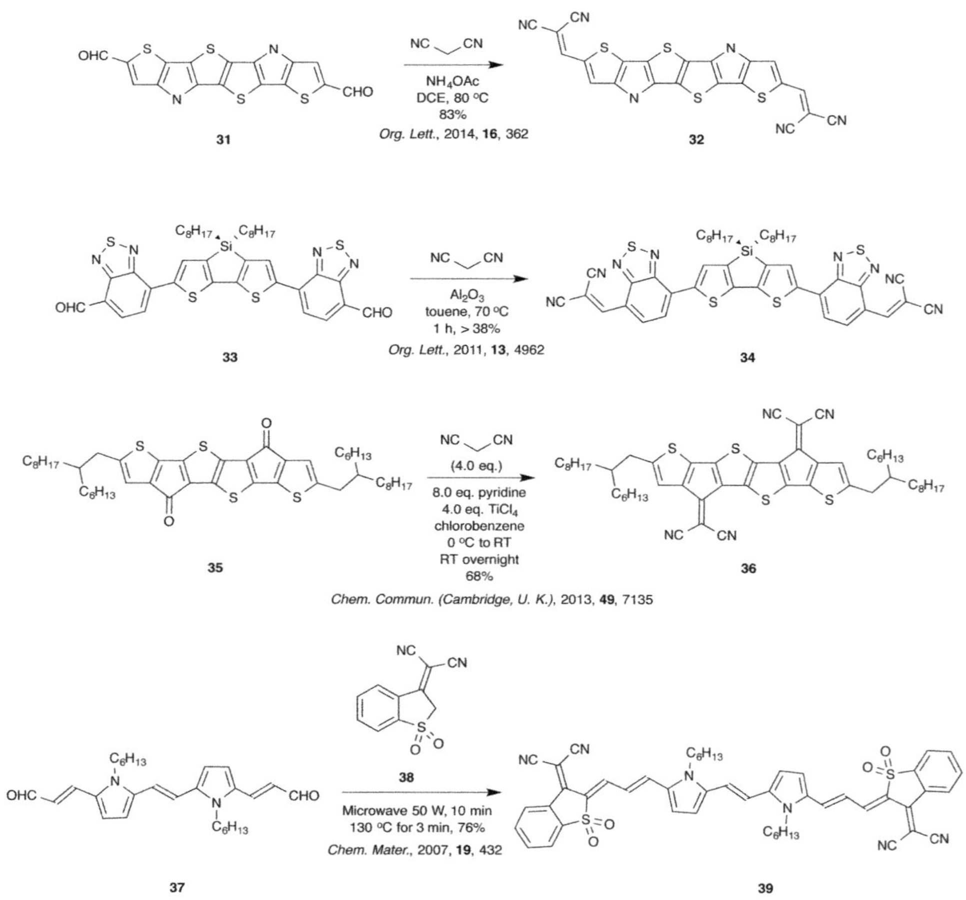

31 → **32**
NC—CN
NH$_4$OAc
DCE, 80 °C
83%
Org. Lett., 2014, **16**, 362

33 → **34**
NC—CN
Al$_2$O$_3$
toluene, 70 °C
1 h, >38%
Org. Lett., 2011, **13**, 4962

35 → **36**
NC—CN
(4.0 eq.)
8.0 eq. pyridine
4.0 eq. TiCl$_4$
chlorobenzene
0 °C to RT
RT overnight
68%
Chem. Commun. (Cambridge, U. K.), 2013, **49**, 7135

37 → **38** → **39**
Microwave 50 W, 10 min
130 °C for 3 min, 76%
Chem. Mater., 2007, **19**, 432

Scheme 6.4 Multiple Knoevenagel condensation on aldehyde and ketone substrates.[14–16,18]

yield[16] using $TiCl_4$ as a Lewis acid catalyst that irreversibly absorbs water, which may allow Knoevenagel condensations on substrates that react with difficulty.[17] Finally, the double condensation is also possible with very strong acceptors in base-free conditions with microwave assistance, as with the conversion of dialdehyde **37** with the "Sandoz" acceptor (**38**) to give the corresponding doubly condensed product **39** in high yield.[18]

Most of the Knoevenagel reactions in Schemes 6.2–6.4, and indeed many in the synthesis of π-conjugated materials, employ aryl aldehydes or diaryl ketones as the reactive substrate. One issue that may arise with the use of α,β-unsaturated aldehydes as a substrate is illustrated in Scheme 6.5, which also underscores the importance of different conditions on reaction outcomes. Reaction of **40** with the TCF acceptor using standard basic piperidine catalysis gave product **41** in 47% yield *with one less alkene* than would be normally expected in the Knoevenagel condensation ($n = 0$); however, reaction with acidic ammonium acetate-catalyzed conditions gave the expected Knoevenagel product **42** ($n = 1$) in 67% yield.[19] Compound **41** with one less alkene can be explained by the mechanism proposed in the

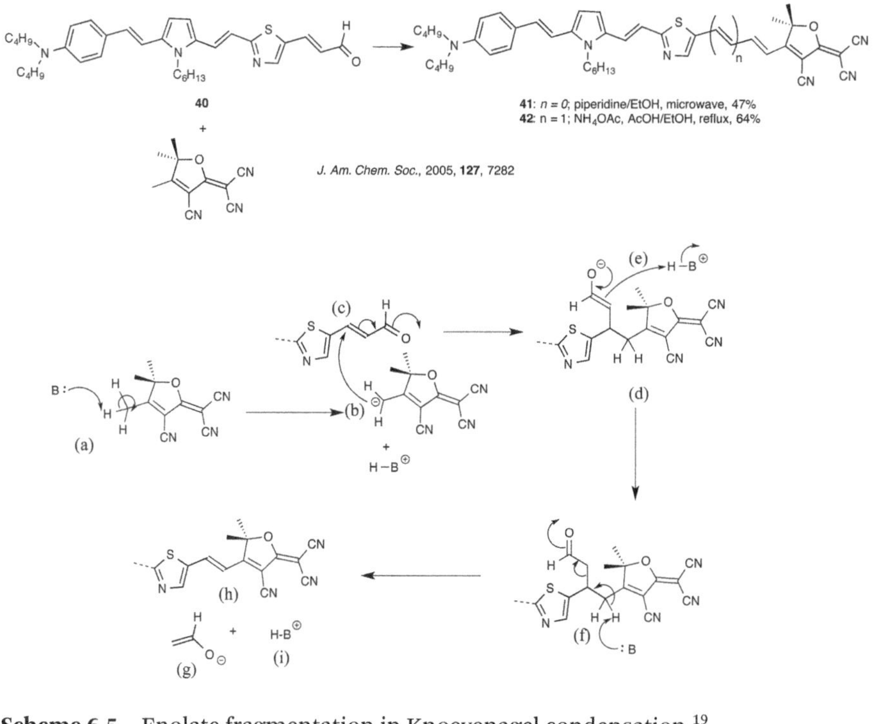

41: $n = 0$; piperidine/EtOH, microwave, 47%
42: $n = 1$; NH$_4$OAc, AcOH/EtOH, reflux, 64%

J. Am. Chem. Soc., 2005, **127**, 7282

Scheme 6.5 Enolate fragmentation in Knoevenagel condensation.[19]

lower half of Scheme 6.5. Deprotonation (S6.5a) of the TCF acceptor to give the corresponding anion (S6.5b) (note: other resonance structures not shown) followed by nucleophilic Michael addition of the anion (S6.5c) to the α,β-unsaturated aldehyde to give adduct (S6.5d) that protonates (S6.5e), and then deprotonation of the most acidic protons on the active methylene eliminates (S6.5f) an acetaldehyde enolate (S6.5g) and the one-alkene product (S6.5h). The acetaldehyde enolate can then be protonated by the conjugate acid (S6.5i) of the base to give acetaldehyde, which is a gas at 20 °C and may volatilize out of the reaction mixture. When acidic conditions are used, Michael addition can be avoided and the expected product **42** can be obtained in moderate yield.

The examples in Schemes 6.2–6.5 demonstrate the synthetic diversity and utility of the Knoevenagel condensation. In particular, a number of catalysts may be used (piperidine, piperidinium acetate, tetramethylpiperidine, Al_2O_3, ammonium acetate, and pyridine in Schemes 6.2–6.5) and removal of the water by-product, possibly along with high reactant concentrations, is often key to obtaining satisfactory yields; however, the nature of the reaction as an equilibrium is important to remember since trying to "force" the equilibrium to completion may result in decomposition of the starting materials and/or product.

6.3   POLYMETHINE DYES

The active methylene compounds presented in Section 6.2 for Knoevenagel condensation derive their acidity from the anion of the conjugate base being delocalized onto one or more *neutral* electron withdrawing groups. Organic *cationic* salts are another class of active methylene compounds that may have sufficient acidity to undergo Knoevenagel condensation with aldehydes and imines and their synthetic equivalents. Such cationic salts (Figure 6.1) generally comprise either a 5-membered heteroaromatic ring or a 6-membered heteroaromatic ring with at least one heteroatom bearing a positive charge, which are represented by the general formulas F6.1a and F6.1b, respectively. These structures can be annulated with aliphatic or aromatic rings, but benzannulation is most common. Cationic salts derive their acidity largely from the anion of the conjugate base delocalizing onto the positively charge heteroatom that results in a "methylidene" form (F6.1c, F6.1d) with an exocyclic double bond in conjugation with a lone pair of electrons on the heteroatom. The methylidene form can be thought of mechanistically as attacking the electrophilic aldehyde or imine, or any equivalent, that initiates the Knoevenagel condensation (F6.1e).

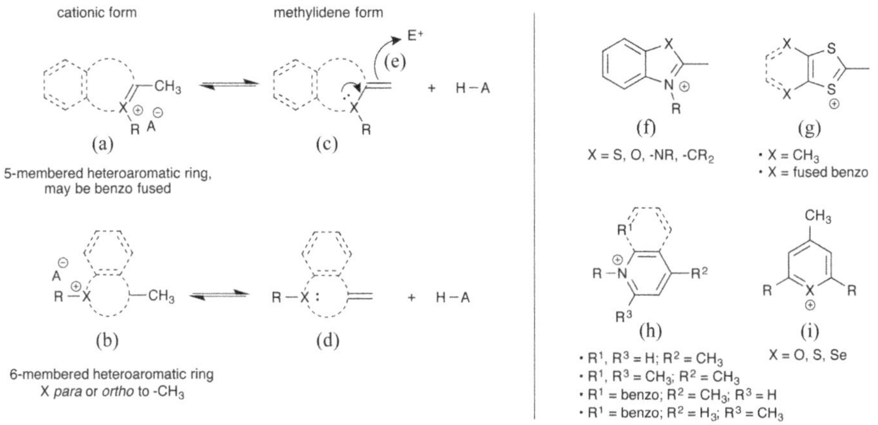

Figure 6.1 General illustration and examples of cationic active methylene organic salts for Knoevenagel condensation.

In some cases, the methylidene form is sufficiently stable for isolation and purification, particularly for cationic salts that retain aromaticity when deprotonated, such as with indoles (*e.g.*, F6.1f, X = CR$_2$). Isolating the methylidene form can be advantageous when Knoevenagel coupling partners or final products are not stable to the basic conditions required to deprotonate the cationic salt. Examples of cationic salts are presented in Figure 6.1 and include: (1) 5-membered benzannulated rings with positively charged nitrogen (F6.1f) such benzothiazoles (X = S) and indoles (X = CR$_2$); (2) 5-membered benzannulated rings with positively charged sulfur (F6.1g); (3) 6-membered heterocycles with a positively charged nitrogen (F6.1h) such as *N*-methyl picoliniums (R^1, R^3 = H; R^2 = CH$_3$ and R^1, R^3 = CH$_3$; R^2 = CH$_3$) and *N*-methyl quinoliums (R^1 = benzo); and (4) 6-membered heterocycles (F6.1i) such as pyriliums with positively charged oxygen (X = O) and thiopyriliums with positively charged sulfur (X = S). In most cases, any methyl group that is *ortho* or *para* to the positively charged heteroatom will have sufficient acidity to react in a Knoevenagel condensation under typical conditions. Additionally, it should be noted that structures F6.1f to F6.1i represent classes of cationic salts that have been commonly utilized, but many other cationic salts are known, including variants of F6.1f to F6.1i with additional annulated rings and variants such as benzo[*c,d*]indolium,[20,21] quinolizinium,[22] and azaazulenium.[23] Further examples of cationic salts are included in recent reviews.[24–27]

Organic cationic salts are key intermediates in the synthesis of polymethine chromophores, which have been examined for second order and third order nonlinear optical applications and as strongly absorbing chromophores in dye sensitized solar cells and organic photovoltaics.

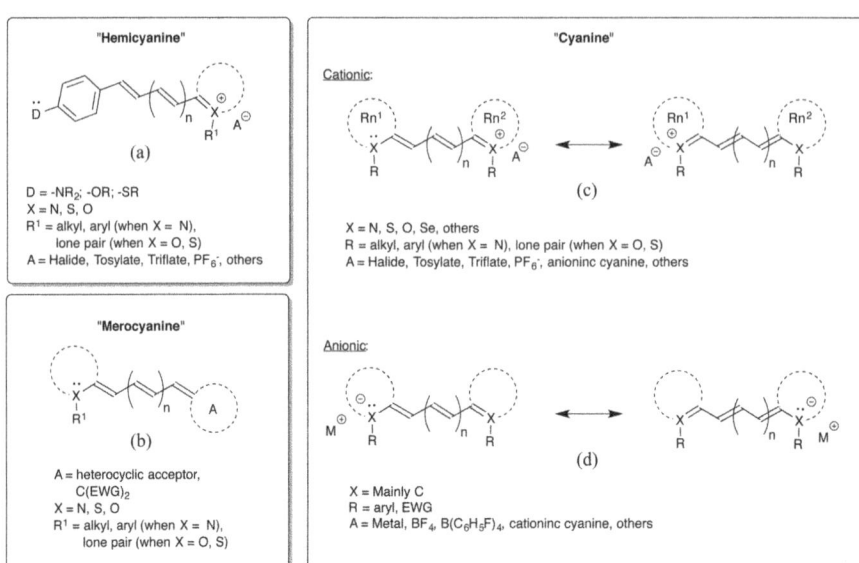

Figure 6.2   General illustration of polymethine dye classes: (a) hemicyanines; (b) merocy anines; (c) cationic cyanines and (d) anionic cyanines.

Polymethine dyes are illustrated in Figure 6.2 with general formulae and include hemicyanines (F6.2a), merocyanines (F6.2b), and cyanines (F6.2c, d).[28,29] Of these, hemicyanines are positively charged, cyanines can be either positively charged or negatively charged, and merocyanines are neutral. The main distinguishing characteristic of hemicyanines is the presence of an aromatic ring between the π-bridge and the donor heteroatom, which is typically a dialkyl or diaryl amine. The main distinguishing characteristics of merocyanines are that the donor is in direct conjugation with the π-bridge and that the acceptor is neutral. The main distinguishing characteristic of cyanines is that two resonance forms exist such that each ring (Rn^1 and Rn^2) bears 1/2 the positive charge and the dye is "electronically" symmetrical even though in some cases the actual chemical structure of the rings may be slightly different to accommodate functionalization at only one end. In some cases, the charge can be localized on one end group and the cyanine is said to be "symmetry broken," and certain properties may be more characteristic of "push–pull" hemicyanines and merocyanines.[30,31]

6.3.1   Hemicyanines

Hemicyanines are typically synthesized by a Knoevenagel condensation between an aldehyde and a cationic salt (Scheme 6.6). The cationic salt is often synthesized by alkylation of an sp^2 nitrogen with an

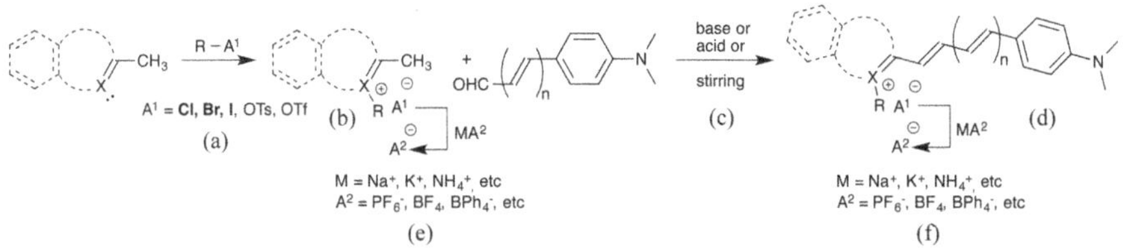

Scheme 6.6 General illustration of the synthesis of hemicyanines from cationic active methylene organic salts.

alkylating agent (S6.6a), which gives the salt with the leaving group as a counter anion (S6.6b). In many cases, conditions similar to those illustrated in Section 6.2 for neutral active methylene compounds may also be utilized in hemicyanine synthesis (acid, base, just stirring if the cationic salt is sufficiently acidic, S6.6c); however, solubility of either the cationic salt or the cationic final product (S6.6d) may require special attention to reaction solvent. In many cases, the organic cation is soluble only in polar solvents and the product, which is a cation with a higher degree of organic content, is substantially less soluble in the polar reaction solvent. Such solubility differences may allow the final product to crystallize from solution, which aids the isolation and purification, or may hinder the progress of the reaction if precipitation occurs to a sufficient extent that starting materials are entrapped. In many cases, the counter anion of the organic salt is a halide since synthesis of the salt intermediate usually involves alkylation of an sp^2 nitrogen in the last step and thus the counter anion of the product is also a halide. Typically, these halide salts are amongst the least soluble of organic salts in common organic solvents; however, solubility of the organic salt intermediate and the final product can be modified by anion metathesis to larger, more polarizable, and/or less coordinating anions, which can be done on either the salt intermediate (S6.6e) or the final product (S6.6f), and depends on matching the relative softness of the organic salt and relative hardness of the halide with the softness of the new counter anion and the hardness of the counter cation (from the new counter anion), respectively. This typically involves stirring the organic halide salt with a sodium (or similar) salt of the softer anion in an organic solvent (such as halogenated solvents) so that the sodium halide salt precipitates from solution and the organic cation with the softer anion remains soluble. Changing the counter anion allows finding a suitable recrystallization solvent for many hemicyanines; however, since the product is cationic, purification by column chromatography may be more difficult and require addition of much more polar solvents (methanol, ethanol, isopropanol) than are traditionally utilized. In some cases, use of reverse-phase silica gel or polystyrene size exclusion beads with a low molecular weight cut off may also prove effective.

Examples of hemicyanine synthesis are shown in Scheme 6.7. The first example is a Knoevenagel condensation of extended aldehyde **43** with *N*-methyl-γ-picolinium PF₆ **44** with catalytic piperidine in methanol to give the corresponding hemicyanine **45** in moderate yield.[32] The second example illustrates a double condensation of a dimethyl pyridinium iodide **46** with 4-(*N,N*-dimethylamino)benzal-dehyde with piperidine and a methanol/chloroform mixture to give

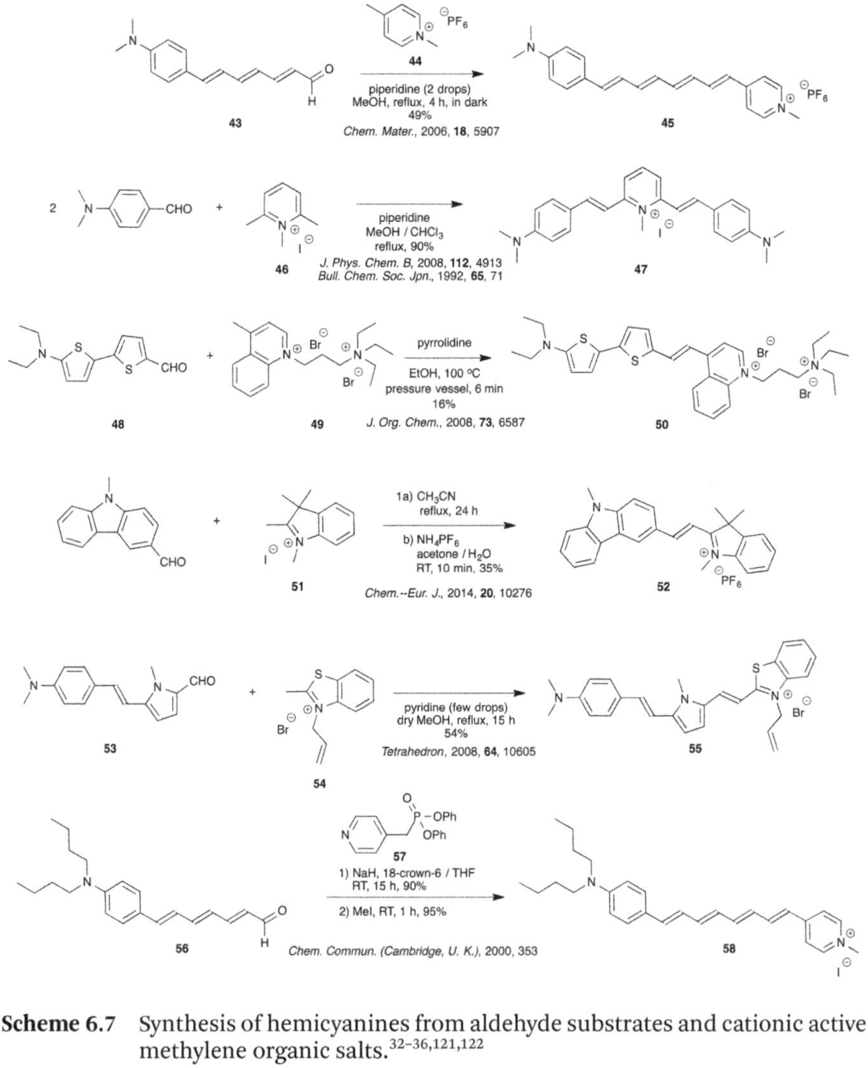

Scheme 6.7   Synthesis of hemicyanines from aldehyde substrates and cationic active methylene organic salts.[32–36,121,122]

the corresponding double condensation product **47** in high yield.[33] In the case of the reaction of salt **46**, the solvent mixture was critical since precipitation during the reaction in methanol resulted in low yields in the previously reported synthesis of **47**.[34] Aldehyde **48**, which is deactivated by the strongly electron-donating aminobithiophene, reacts with a quinolium bromide **49** in rather forcing conditions of pyrrolidine in ethanol at 100 °C to give the corresponding hemicyanine **50**, which was water soluble due to being a dication.[35] An example with an indolium salt is presented for the reaction of indole **51** with 9-methylcarbazole-3-carboxaldehyde without a catalyst in refluxing acetonitrile followed by removal of solvents and anion metathesis with NH_4PF_6 in acetone/H_2O to give the hemicyanine PF_6 salt **52** in 35% yield. Benzothiazoliums are another important class of hemicyanine acceptors that are illustrated by the condensation of the deactivated pyrrole aldehyde **53** with benzothiazolium salt **54** with catalytic pyridine in methanol to give hemicyanine **55** in moderate yield. It should be noted that in some cases synthesis of hemicyanines from active methylene salts might prove impractical due to insolubility and/or side reactivity. In these cases, and alternative route may be possible that relies on forming the cationic acceptor in the last step. Such a strategy is shown in the last example in Scheme 6.6 with the reaction of aldehyde **56** with phosphonate **57** to give a high yield of an intermediate dibutylaminophenyl-pyridinyl polyene followed by *N*-alkylation of the pyridine to give pyridinium hemicyanine **58** in good yield overall.[36] In this case, reaction with picolinium iodide salts similar to **44** resulted in the formation of shorter analogues than expected, likely due to Michael addition mechanism similar to that in Scheme 6.5.

6.3.2 π-Extension of Acceptors

As presented in Section 4.2.4 and summarized in Scheme 6.8a, aryllithium reagents can be treated with vinylogous esters and amides to give the aryl moiety with an extended π-conjugated aldehyde. Similar reactions exist for the π-extension of active methylene compounds through Knoevenagel condensations as presented in Scheme 6.8. Reaction of a variety (S6.8b) of vinylogous amino imines, amino aldehydes, and esters with either neutral active methylene (S6.8c) or cationic active methylene (S6.8d) compounds effectively results in π-extended neutral (S6.8e) and cationic (S6.8f) acceptors. It should be noted that when acetic anhydride is used for the Knoevenagel condensation with extension reagents having a secondary amino group

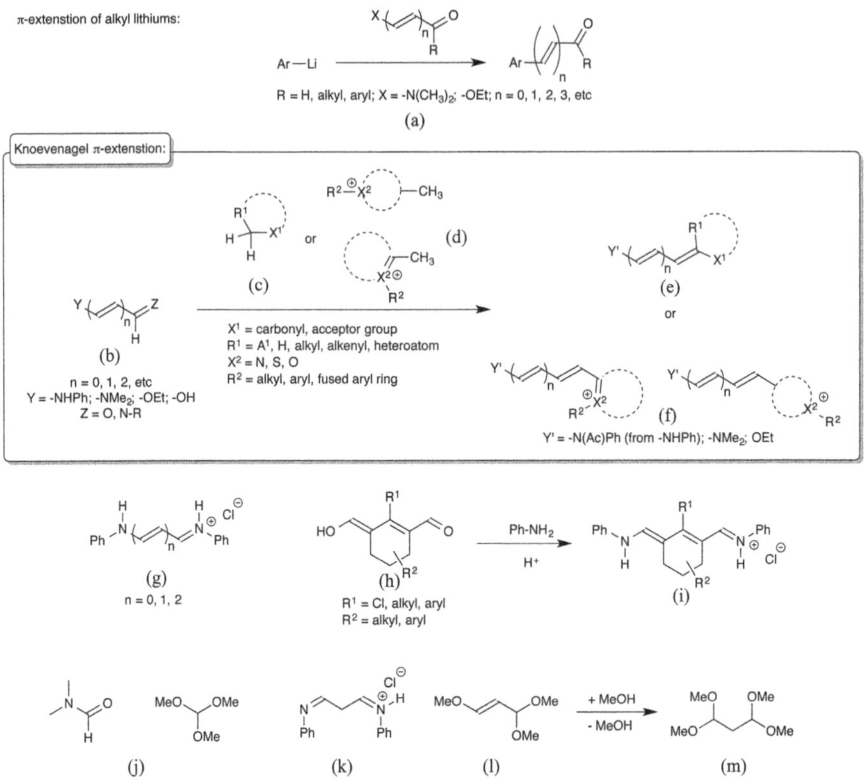

Scheme 6.8 General illustration of π-extension of active methylene compounds *via* Knoevenagel condensation.

(*e.g.*, –NHPh), the secondary amino group is typically acylated to an amide during the reaction. The amide, amines, or ethers present at the end of the π-extended chain, represented by Y′ in S6.8e and S6.8f, may then be replaced by nucleophilic reagents in a manner analogous to a nucleophilic addition–elimination to an acid chloride. In fact, these extended acceptors shown by the general formulas S6.8e and S6.8f are often key synthetic intermediates in the synthesis of merocyanines (Section 6.3.3) and cyanines (Section 6.3.4). Some reagents that have been utilized for π-extension of acceptors are shown in Scheme 6.8. Anilino imines (S6.8g) have been used frequently with 1-carbon formaldehyde (*n* = 0), 3-carbon malonaldehyde (*n* = 1), and 5-carbon glutaconaldehyde (*n* = 2) equivalents known. A cyclic 5-carbon extension reagent has been used in both a hydroxy-aldehyde (S6.8h) and, more widely, in an anilino imine form (S6.8i). Other 1-carbon (S6.8j) and 3-carbon (S6.8k, S6.8l, S6.8m) extension reagents are also known. In some cases, the extension reagent may be particularly reactive

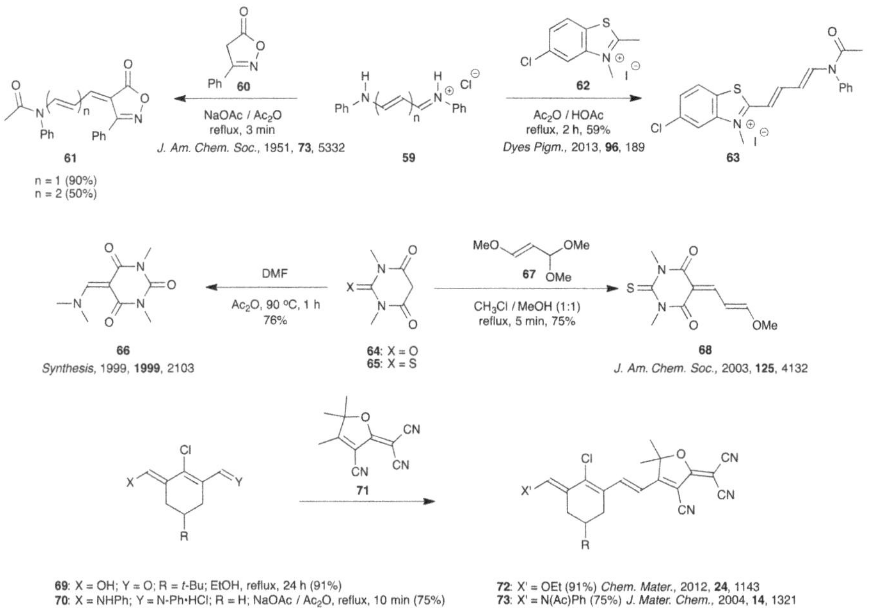

Scheme 6.9 π-Extensions of neutral and cationic active methylene compounds *via* Knoevenagel condensation.[37–42]

towards hydrolysis or other nucleophilic decomposition and should be prepared *in situ*.

Examples of π-extension of acceptors are presented in Scheme 6.9 for compounds that are sufficiently stable to be isolated. Anilino imines **59** can be condensed with neutral acceptors such as isoxazoline **60** with NaOAc in acetic anhydride to give the corresponding 3- and 5-carbon π-extended acceptor amides **61** in yields of 90% ($n = 1$) and 50% ($n = 2$).[37] Very similarly, benzothiaolium acceptor **62** may be extended by 3 carbons to the amide **63** in moderate yield.[38] A relatively efficient reagent from DMF and acetic anhydride can extended acceptors by one carbon such as by reaction of barbituric acid **64** at 90 °C for one hour to give enamine-acceptor **66** in 76% yield.[39] The extension reagent 1,3,3-trimethoxypropene **67** has been used frequently to extend by 3-carbons, such as with reaction of thiobarbituric acid **65** in refluxing $CHCl_3$/MeOH to give ether **68** in high yield.[40] Final examples in Scheme 6.9 illustrate the use of the 5-carbon/cyclohexene extender in the hydroxyl-aldehyde form **69** and the anilino-imine form **70** with acceptor **71** to give ether **72**[41] and amide **73**,[42] respectively, in high yields. As mentioned earlier, although the final products presented in Scheme 6.9 are all push–pull dyes in their own right, they are most often used as intermediates in the synthesis of merocyanines and cyanines. Note that in some cases the relatively short reaction times for Knoevenagel condensations in Scheme 6.9 may be consistent with the discussion in Section 6.2 of the importance of equilibrium in Knoevenagel condensations and stopping the reaction before full consumption of the starting material.

6.3.3 Merocyanines

Merocyanines are typically synthesized by reaction of cationic active methylene compound in the cationic or methylidene form (Figure 6.1) with a π-extended acceptor as presented in Section 6.3.2. Mechanistically, this can be viewed as analogous to an addition–elimination reaction of a nucleophile and an acid chloride, as illustrated in Scheme 6.10. The first step in Scheme 6.10 from the cationic salt (S6.10a) is deprotonation (S6.10b) to the methylidene form (S6.10c). The methylidene form can be generated *in situ* with a base, typically a tertiary amine or pyridine, or in some cases can be separately isolated and purified. The methylidene form can be viewed as nucleophilically attacking the π-extended acceptor compound (S6.10d) in a step that delocalizes electrons onto the acceptor (S6.10e) that then eliminates the leaving group (Y). Deprotonation of the acidic cation (S6.10f) results in the merocyanine. Although in principle any cation salt can be isolated in the methylidene form, the most common compounds that are isolated in the methylidene form are various isomers

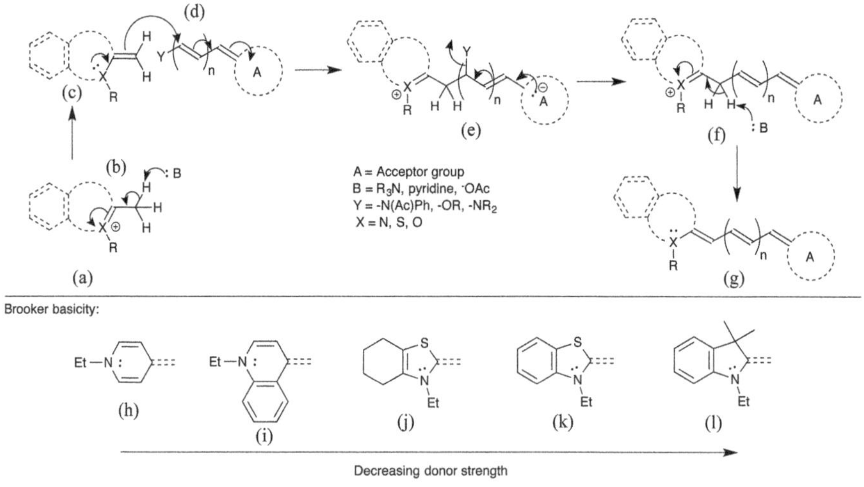

Scheme 6.10 General illustration of merocyanines synthesis of p-extended acceptors and Brooker basicity trend of selected merocyanine donors.

of indoles. Note that whereas in hemicyanines (Scheme 6.7) the moiety from the organic salt retains the positive charge and typically functions as an acceptor, in merocyanines the moiety from the organic salt typically functions as a donor. The donor strength of some typical merocyanine donors, which is a small subset of those known, is given in Scheme 6.10 as the "Brooker basicity". The Brooker basicity is measure of the donor's ability to release electrons into the polymethine chain, and was determined mainly by a donor's ability to red shift the absorption of comparable merocyanines as introduced in Section 2.3.4.[29,37] The basic principles of rationalizing donor strength introduced in Section 2.3 may be applied for the donors in Scheme 6.10: (1) the strongest electron releasing ability of donors S6.10h (from an *N*-ethyl picolinium salt) and S6.10i (from an *N*-ethyl quinolinium salt) are largely influenced by aromaticity gained from donation of the electrons; (2) delocalization of the lone pair of S6.10k onto the benzene ring decreases its donor strength compare to S6.10j; and (3) indole S6.10l is the least Brooker basic due to only one donor atom being in conjugation with the polymethine chain compared to S6.10k.

Examples of merocyanine synthesis are shown in Scheme 6.10. Reaction of extended acceptor **74** with the 1,3-benzodiazolium salt **75** in refluxing pyridine with triethylamine gives merocyanine **76** in 36% yield after five minutes.[43] Another example includes the isolation of a methylidene form by deprotonation of indole **77** with KOH to give methylidene **78** in 65% yield followed by a three-component condensation including acceptor **79** that forms the extended acceptor **80** *in situ* from ethyl ortho formate and then reaction with **78** occurs over one hour to give merocyanine **81** in 60% yield.[44] Vinylogous ester **82** can be reacted with the internal organic salt **83** (*i.e.*, the R—SO$_3^-$ counter anion is covalently attached to the cation) with the weak NaOAc base in a mixture of CHCl$_3$/MeOH at reflux to give the corresponding merocyanine **84** in 67% yield after 30 minutes (Scheme 6.11).[40,45] Two final examples illustrate the reaction of acceptors with longer π extension are shown in Table 6.1 and include the conversion of **85** to the corresponding product **86** with quinolium salt **87** and pyrilium salt **88** under similar conditions in 44% and 76% yields, respectively.[42,46]

6.3.4   Cyanines

The synthesis of cyanines is effectively a combination of π-extension (Section 6.3.2) of a cationic acceptor to give a π-extended intermediate and then merocyanine-like reaction (Section 6.3.3) of the

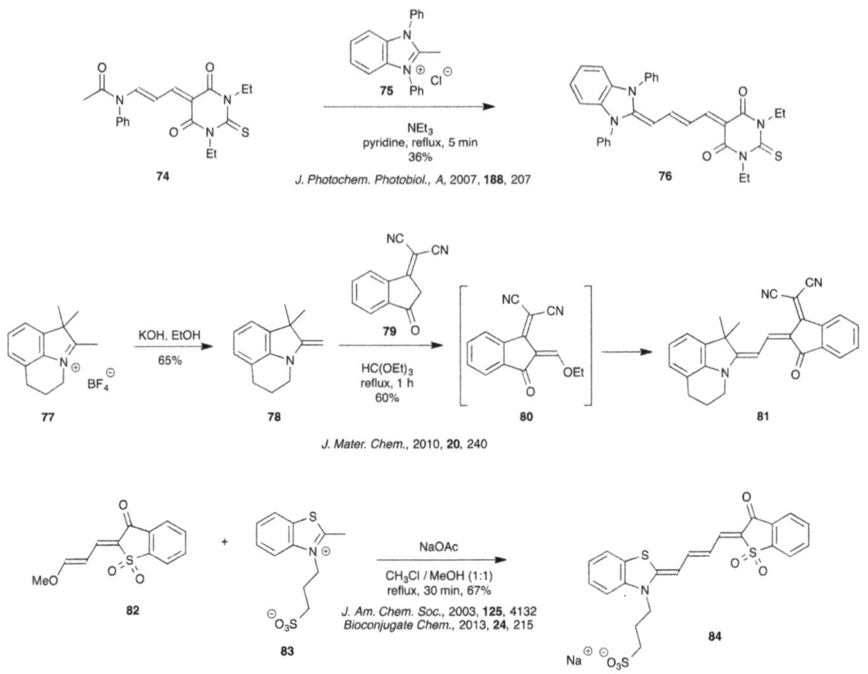

Scheme 6.11 Synthesis of merocyanines from cationic and methylidene forms of cationic active methylene organic salts.[40,43–45]

Table 6.1 Synthesis of merocyanines with 6-carbon bridge.

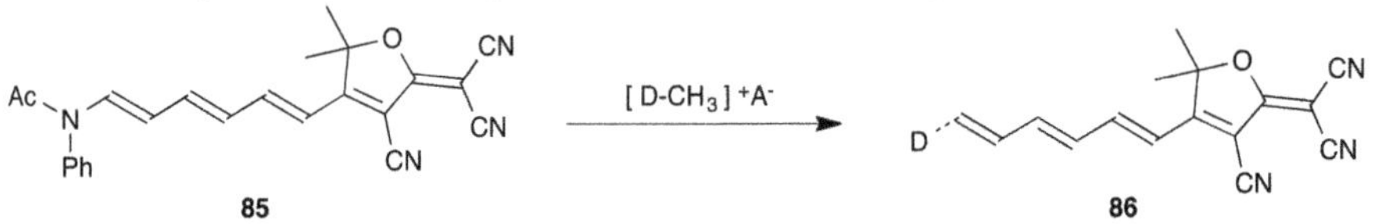

[D-CH₃]⁺A⁻	D	Conditions	Reference
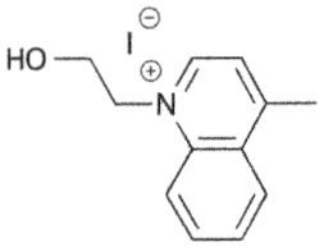 **87**	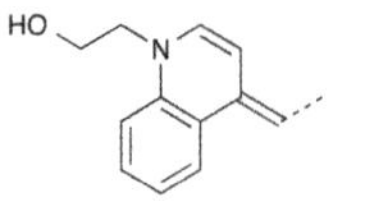	NEt₃/MeOH reflux, 10 min, 44%	42
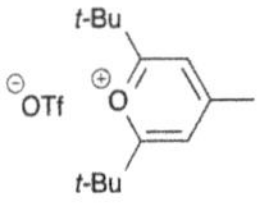 **88**	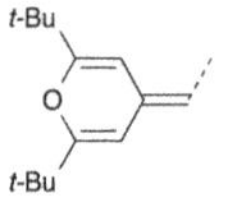	Pyridine (cat.) NEt₃/EtOH reflux, 30 min, 76%	46

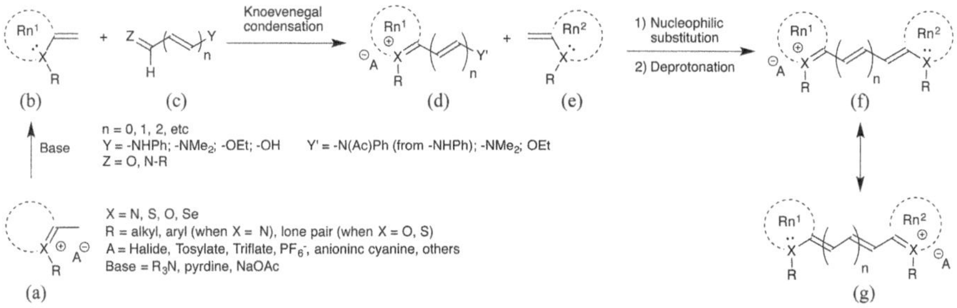

Scheme 6.12   General illustration of the synthesis of cyanines from active methylene cationic salts. Rn^x = cyclic ring end group.

π-extended intermediate to give the cyanine. The sequence is outlined in Scheme 6.12, and includes deprotonation of a cationic active methylene salt (S6.12a) to give the methylidene form (S6.12b) that reacts with the π-extension reagent (S6.12c) *via* a Knoevenagel condensation to give the extended intermediate (S6.12d), which can react with another equivalent of a methylidene compound (S6.12e) in a manner analogous to merocyanine nucleophilic addition–elimination (Scheme 6.10) to give the cyanine (S6.12f), which has end groups (Rn^1, Rn^2) that are equivalent by resonance (S6.12g). The steps of the synthesis may be accomplished in one pot from a single active methylidene compound such that Rn^1 = Rn^2 or in a stepwise sequence from two different active methylidene compounds such that Rn^1 and Rn^2 may be different. Often, differentiating Rn^1 and Rn^2 is done to provide some nonsymmetrical functionalization to the molecule. Differences in Brooker basicity between Rn^1 and Rn^2 will determine if the compound is more "cyanine-like" or more "merocyanine-like". Dyes with end groups having similar electron-releasing ability can be expected to be more cyanine-like while larger differences in electron-releasing ability will tend to promote more merocyanine-like structures. Spectral differences can often be predicted based on Brooker basicity,[29] and the end group that is more electron releasing can be expected to exist in the cation form (Rn^1 in S6.12f).

Examples of symmetrical cyanine synthesis are presented in Scheme 6.13. Reaction of the thiopyrilium PF$_6$ salt **89** with amino-imine **90** in acetic anhydride with the weak base NaOAc provides cyanine **91** in high yield, although in this case reaction temperature is critical as temperatures above 100 °C cause decomposition.[47] Note the PF$_6$ counter anion of **90**, which is less common than a chloride counter anion for this amino-imine, and is most likely used to prevent a mixture of counter anions in product **91**. Another example utilizes

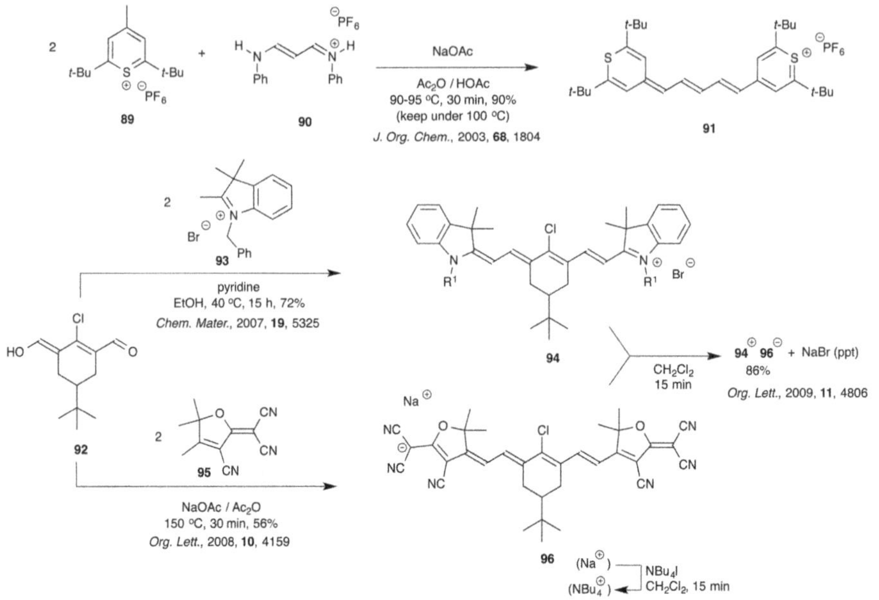

Scheme 6.13 Synthesis of symmetrical cyanines from active methylene cationic salts and selected cation methatheses.[47–50]

hydroxy-aldehyde **92** (a glutaconaldehyde equivalent) and indolium bromide **93** with pyridine in ethanol to give the corresponding indole cyanine **94** in high yield.[48] Alternatively, **92** can be reacted with acceptor **95** to give the anionic cyanine **96** in 56% yield, and note that the Na$^+$ counter cation can be exchanged with NBu$_4^+$ by a metathesis step.[49] One interesting aspect of cyanines with different charges is that they can be used as each other's counterions, which is demonstrated by the metathesis of the Br$^-$ salt of **94** and the Na$^+$ salt of **96** to give the mixed cyanine salt **94$^+$96$^-$**.[50]

The synthesis of unsymmetrical cyanines is illustrated in Scheme 6.14. Condensation of **97** with **98** utilizing NEt$_3$/EtOH is followed by metathesis of the iodide anion to B(C$_6$H$_5$F)$_4^-$ to give unsymmetrical cyanine **99** in 22% yield.[38] Another example in Scheme 6.14 illustrates the use of solid phase synthesis and is a π-extension of indolium salt **100** with malonaldehyde diimine **101** to give the π-extended imine intermediate (isolated in 24% yield) that can be reacted with sulfonylchloride functionalized beads (**102**) to give the solid-supported intermediate **103** as a sulfonamide with a loading of 96%. Subsequent reaction of beads with benzothiazolium **104** gives final product **105** in 67% yield.[51] Isolating **103** is accomplished in this case by washing

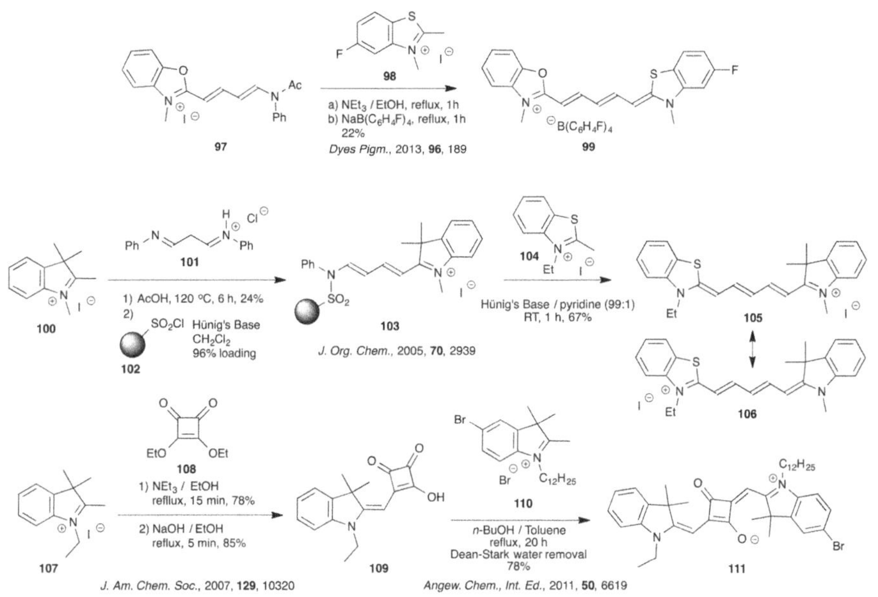

Scheme 6.14 Synthesis of unsymmetrical cyanines.[38,51–53]

away the by- and side-products of the reaction since the beads are insoluble in most solvents, and in general solid phase synthesis may reduce the number of steps and complexity of isolating compounds. Since benzothiazole has significantly greater electron releasing ability than indole, resonance form **106** could be reasonably predicted to predominate in solution and the chromophore may be described as more merocyanine-like that cyanine-like. A final example in Scheme 6.14 presents the synthesis of a squaraine dye, which is a class of dyes closely related to cyanines and distinguished by being zwitterionic and having a 4-membered ring in the π-bridge. Reaction of indolium salt **107** with the squaric acid diethyl ester (**108**) followed by hydrolysis gives half-squaraine **109** in good yield overall.[52] This addition to the squaric acid diethyl ester is analogous to the π-extension presented in Scheme 6.9. Reaction of **109** with indolium salt **110** gives the unsymmetrical squaraine **111** in 78% yield.[53] Note that in the second step, Dean–Stark water removal is used to aid the condensation reaction. Although the squaraine synthesis in Scheme 6.14 is unsymmetrical, one-pot syntheses of symmetrical squaraines are well known, and a large number of end groups have been utilized.[54] One interesting feature of squaraines is that the 4-membered ring can be modified in a number of ways that both effectively retain and substantially modify the cyanine character.

6.4 TCNE-BASED ACCEPTORS

Tetracyanoethylene (TCNE) is an alkene with a high electron affinity that can be used to synthesize relatively strong acceptors. Example of acceptors from TCNE are illustrated in Scheme 6.15. As a high EA material, TCNE can undergo nucleophilic attack from aryllithiums or electron rich compounds *via* an addition–elimination mechanism to give the relatively strong tricyanovinyl (TCV) acceptor. One of the cyano groups of TCNE acts as a leaving group in the reaction, and the reaction is generally faster with aryllithiums, but may not result in a higher yield. One example of the reaction of TCNE is the formation of an aryllithium from **112** by deprotonation followed by addition of TCNE to give the corresponding product **113** in high yield.[55] The nucleophilic reaction with a neutral compound is shown in the reaction of TCNE with the thiophene of **114** in DMF for 24 h to give

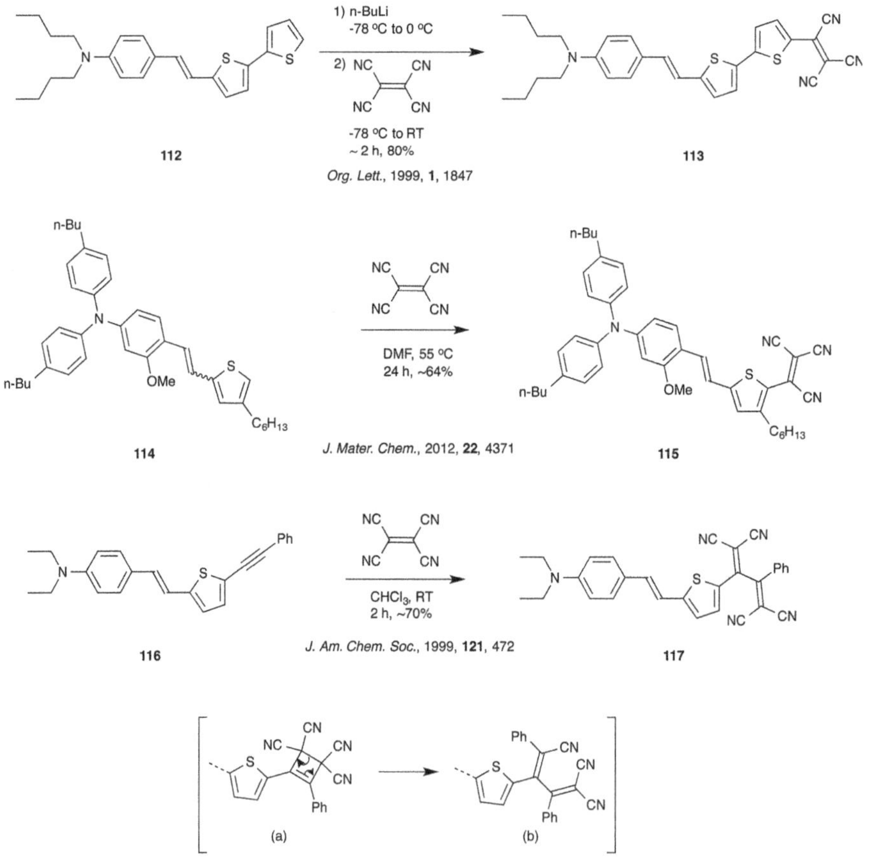

Scheme 6.15 Acceptors synthesized from tetracyanoethylene (TCNE).[55–57]

Table 6.2 Property comparison of TCV and TCBD acceptors synthesized from TCNE.

Entry	A	R	T_d (°C)	λ_{max} (nm)
1	TCV	$-C_4H_9$	249	640
2	TCBD	$-C_4H_9$	250	623
3	TCV	$-Ph$	308	611
4	TCBD	$-Ph$	343	575

the corresponding product **115** with the TCV acceptor in moderate yield.[56] Another acceptor from TCNE results from reaction with alkynes to give a tetracyanobutadiene (TCBD) acceptor. The reaction occurs relatively quickly under mild conditions, usually in high yield. An example of this is the reaction of alkyne **116** with TCNE to give compound **117** with the TCBD acceptor.[57] The reaction may proceed by stepwise addition and then cyclization to give a cyclobutene ring (S6.15a) that opens to give the TCBD acceptor (S6.15b). Compared to the TCV acceptor, one cyano is effectively replaced with a dicyanovinyl-R group in TCBD, where the R is usually phenyl. In addition to the milder reaction conditions for TCBD synthesis compared to TCV synthesis, TCBD chromophores tend to be more chemically stable with respect to reaction with nucleophiles. Other properties are compared briefly in Table 6.2. In addition to higher chemical stability, TCBD tends to be more thermally stable than TCV when the acceptor determines thermal stability (*i.e.*, when the acceptor is the *weakest link*) in comparable chromophores (Table 6.2, Entry 3 *vs.* Entry 4); however TCBD tends to be a less strong acceptor than TCV for intramolecular charge transfer (compare the blue shifted TCBD dyes to the analogous TCV dyes in Table 6.2, Entry 1 to Entry 2 and Entry 3 to Entry 4).

6.5 HETEROCYCLIC ACCEPTORS

There have been numerous acceptors that have been incorporated into π-conjugated materials that cannot be installed *via* Knoevenagel condensation. Most of these acceptors are heterocycles that can be incorporated into π-conjugated materials by reaction of functionalized parts of the heterocyclic or heteroaromatic core or peripheral aromatic moieties; however, the synthesis of many of these acceptors as well as their chemistry is often too specific to be efficiently

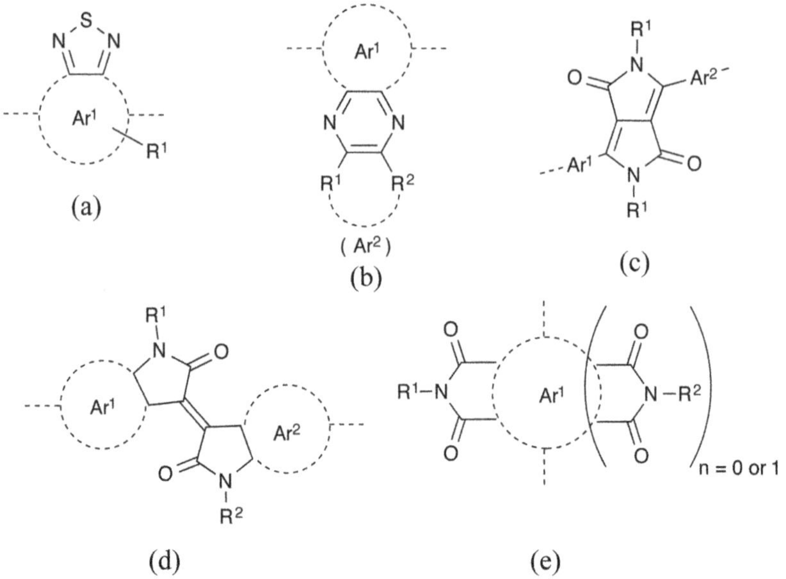

Figure 6.3 General illustration of acceptor classes: (a) thiadiazole; (b) pyrazine; (c) diketopyrrolopyrrole; (d) isoindigo; and (e) imide(s).

covered as broader concepts. Fortunately, there are classes of hetero-cyclic acceptors that have proven to be important for many classes of π-conjugated materials and have been functionalized and structurally diversified by synthetic means that may be generalized. Five of these acceptor classes are illustrated in Figure 6.3 and include: 2,1,3-thiadi-azole-based acceptors (F6.3a); pyrazine-based acceptors (F6.3b); dike-topyrrolopyrrole acceptors (DPP) (F6.3c); isoindigo acceptors (F6.3d); and imide-based acceptors (F6.3e). Each of these acceptor classes has at least one site to attach alkyl groups (R^1) for solubility/processing and all acceptors have aryl group variability (Ar^1, Ar^2) to change features such as through-conjugation and properties such as electron affinity (EA) and ionization energy (IE). Various general reaction conditions for the synthesis of thiadiazole, pyrazine, DPP, isoindigo, and imide acceptors are described below.

6.5.1 2,1,3-Thiadiazole-Based Acceptors

As illustrated in Scheme 6.16, an aryl or heteroaryl core having two amines (S6.16a) on adjacent carbons may undergo 2,1,3-thiadiazole heteroannulation to 2,1,3-thiadizole acceptors (S6.16b) with the use of a sulfur transfer agent and an amine base. Typically, the aromatic core is benzene and the sulfur transfer agent is thionyl chloride or, much

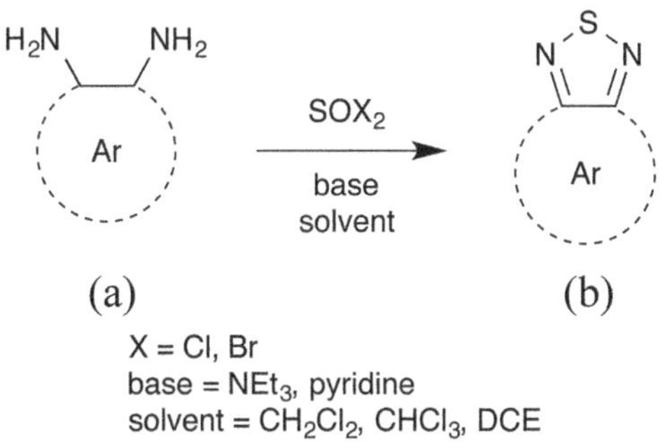

Scheme 6.16 General illustration of synthesis of 2,1,3-thiadiazole-based acceptors from: (a) a diamine to (b) the thiadiazole acceptor.

less often and in special cases, thionyl bromide. Other sulfur transfer agents have been used for diamino benzene and other diamino aromatic and heteroaromatic cores include S_2Cl_2 [58] and N-sulfinylaniline (Ph−N=S=O).[59] Two amine bases that have been used almost exclusively are triethylamine and pyridine. The method is reasonably broad in that many aromatic and heteroaromatic cores can be utilized if they can be functionalized with 1,2-amines.

Scheme 6.17 provides a number of conditions for the synthesis of 2,1,3-benzothiadiazole acceptors and subsequent functionalization chemistry. Diamines of benzene (**118**),[60] dialkoxybenzene (**119**),[61] mono-(**120**)[62] and difluoro benzenes (**121**),[63] and dibromobenzene (**122**)[64] all give their respective 2,1,3-benzothiadiazole (BT) products (**123–127**) mostly in high yields, with the exception of the relatively low IE dialkoxydiamine **119**, which may suffer from oxidation side-reactions under the electrophilic conditions. All these preparations of 2,1,3-benzothiadiazole derivatives more-or-less use the same conditions with changes only in halogenated solvent used, which is either CH_2Cl_2 or $CHCl_3$. The substrates range from those with relatively low IE (**119**) to medium IE (**118**) to relatively higher IE (**122**, **120**, and **121**), which demonstrates a reasonable scope to the reaction. Parent 2,1,3-benzothiadiazole (**123**) can be brominated with activated conditions to give 4,7-dibromo-2,1,3-benzothiadiazole (**128**) in high yield[60] and the monobromo product **129** in low yield[65] with Br_2/HBr. The low yield of **129** is partially due to over-bromination to **128** since the bromine in the 4 position may activate the ring slightly. However, the electron-withdrawing fluoro substituents in the 5- (**126**) and 5,6-positions (**127**) result in low bromination yields of 33% for monofluoro **130**[66] and 40% for dibromo **132**[67] even with highly forcing conditions of Br_2/acid/heat with longer reaction times. Both diiodo versions have been synthesized as well for the monofluoro **131**[62] and the difluoro **133**,[63] each

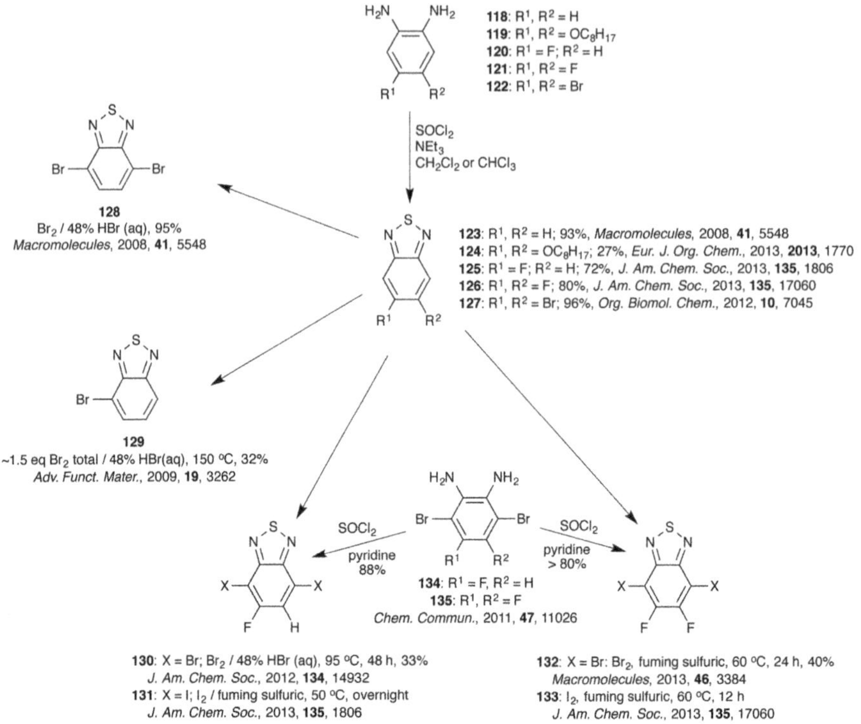

Scheme 6.17   Synthesis   of   2,1,3-benzothiadiazole   acceptors   from diaminobenzenes.[60–68]

with forcing conditions, and each were used directly in other reactions without purification due to instability, so reaction yields were not reported. Due to the difficulty of electrophilic bromination of the high IE molecules **125** and **126**, both dibromo **130** and **132** were made in high yields by treating the dibromodiamines **134** and **135**, respectively, with thionyl chloride and pyridine.[68] In this route, electrophilic bromination on the relatively high IE fluorinated BT compounds **125** and **126** can be avoided. It should be noted that the harsh conditions required to brominate the high IE 2,1,3-benzothiadiazole derivatives **125** and **126** illustrated in Scheme 6.17 encourages the direct arylation of BT compounds discussed in Section 5.5, given that the bromines are added almost exclusively for subsequent coupling reactions.

Examples of 2,1,3-thiadiazole heteroannulation on other heterocycles are illustrated in Scheme 6.18. These include the heteroannulation of diamino quinone **136** to **137** in high yield from thionyl chloride without the use of an amine base.[69] The thienobenzothiophene **138** can also be heteroannulated under base free conditions using S_2Cl_2 in DMF to give thiadiazole derivative **139** in high yield.[58] One

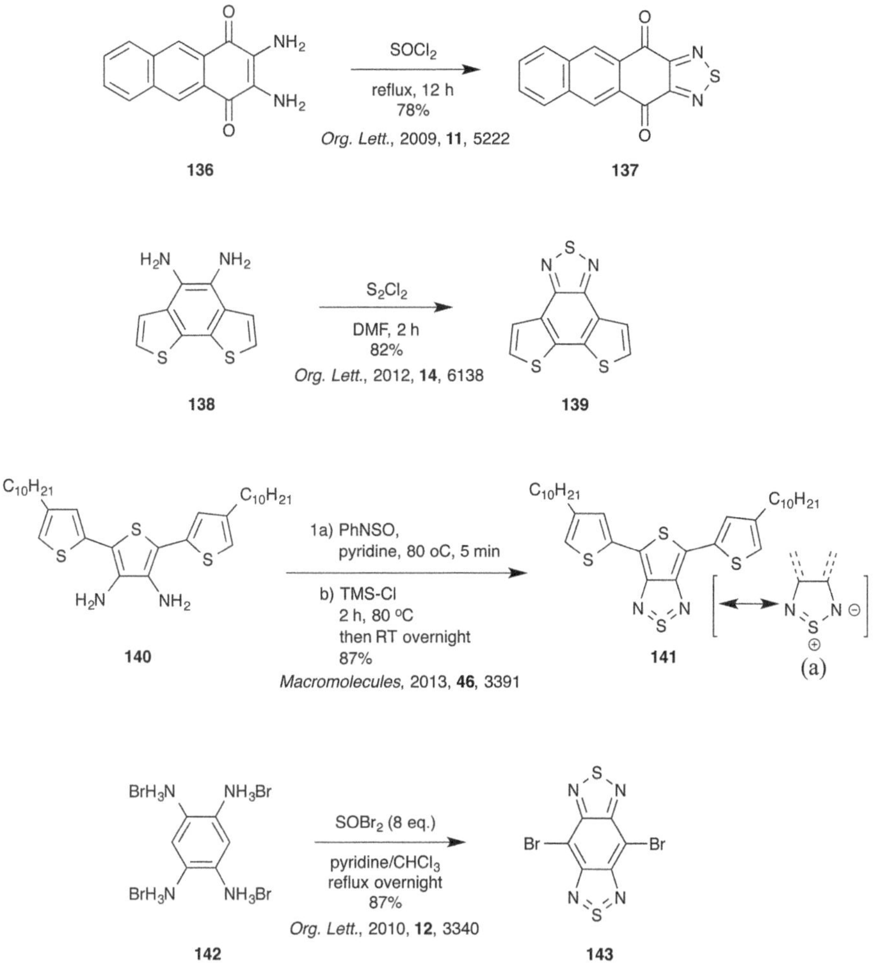

Scheme 6.18 Synthesis of aryl 2,1,3-thiadiazole acceptors from aryl-*o*-diamines.[58,59,69,72]

example using the *N*-sulfinylaniline reagent is the heteroannulation of 3,4-diaminothiophene **140** to give the biheterocyclic **141** in high yield.[59] Biheterocycle **141** is a special case of a 2,1,3-thiadiazole ring that contains a hypervalent sulfur atom, although such compounds more likely correspond to the ylidic resonance form S6.18a,[70] which can result in relatively strong acceptors due mainly to a high partial positive charge on sulfur.[71] Another example of such bonding and the use of thionyl bromide is illustrated by the reaction of the tetraaminobezene hydrogen bromide salt **142** with excess thionyl bromide to give dibromo compound **143** in high yield.[72] Acceptor **143** has been included in π-conjugated materials that have high electron affinity.[73]

6.5.2 Pyrazine-Based Acceptors

The synthesis of pyrazine-based acceptors is illustrated in Scheme 6.19 and generally involves the double condensation of a 1,2-aromatic amine (S6.19a) with a 1,2-diketone (S6.19b) in a polar protic solvent, which is typically ethanol, to give the pyrazine ring (S6.19c). In some cases, when the diketone is less active to addition because of steric hindrance or because of conjugation, an acid catalyst may be used to decrease reaction time and improve yields. Many times, this has been acetic acid used as a co-solvent with ethanol. The 1,2-diketone can be dialkyl or diaryl, or mixed (*i.e.*, R^1 = alkyl and R^2 = aryl), in addition to being either acyclic or cyclic. Many aromatic and heteroaromatic 1,2-diamines have been used; however, sometimes the 1,2-diamine is unstable due to low IE from the amine and/or other donors, in which case a protecting group may be used before the pyrazine heteroannulation step. One of the most effective and widely used strategies is reaction of the 1,2-diamine (S6.19d) with a sulfur transfer agent (S6.19e) to form a 2,1,3-thiadiazole (S6.19f, and discussed in Section 6.5.1). Deprotection is accomplished using a reducing agent (S6.19g) to give the 1,2-diamine, which can then be condensed (S6.19h) with a diketone to give the pyrazine. In some cases, the deprotection and condensation may occur in one pot. Another potential advantage of the 2,1,3-thiadiazole protection strategy is that reactivity towards electrophilic substitution is reduced compared to a 1,2-diamine, which may allow more controlled derivatization. Indeed, the ability to control electrophilic substitution regiochemistry on the aromatic ring is often the main reason for forming a 2,1,3-thiadiazole in the synthetic sequence of a pyrazine acceptor.

Some examples of pyrazine ring synthesis are illustrated in Schemes 6.20 and 6.21. In these Schemes, the structures of the diketones are

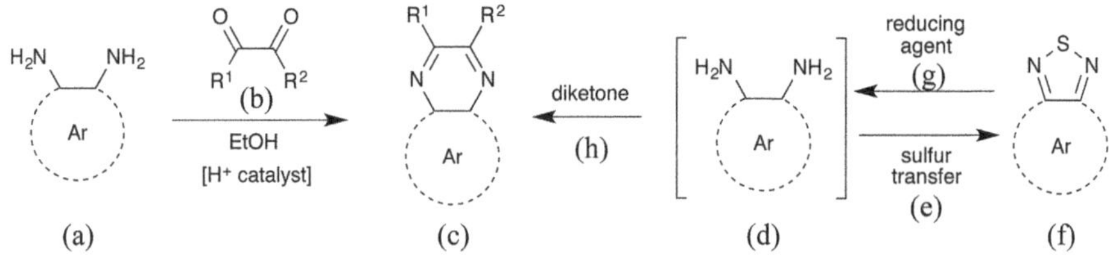

Scheme 6.19 General illustration of the synthesis of pyrazine-based acceptors: (a) a diamine; (b) a 1,2-diketone; (c) a pyrazine acceptor; (d) a diamine that has been (e) protected to a (f) thiadiazole and (g) deprotection of a thiadiazole; (h) synthesis of pyrazine.

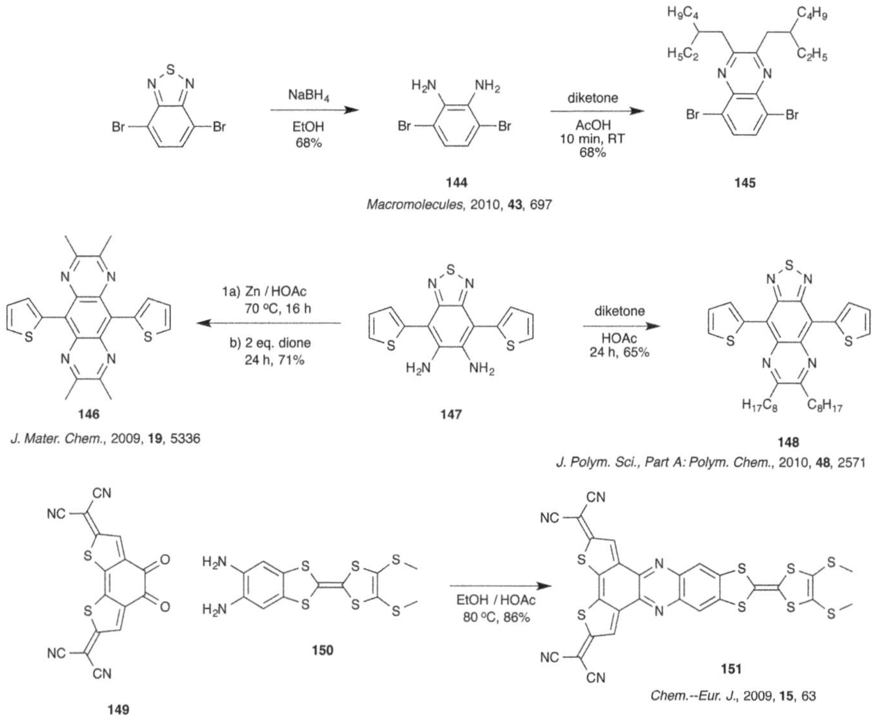

Scheme 6.20 Synthesis of several pyrazine-based acceptors from aryl-*o*-diamines.[74,76,77,123]

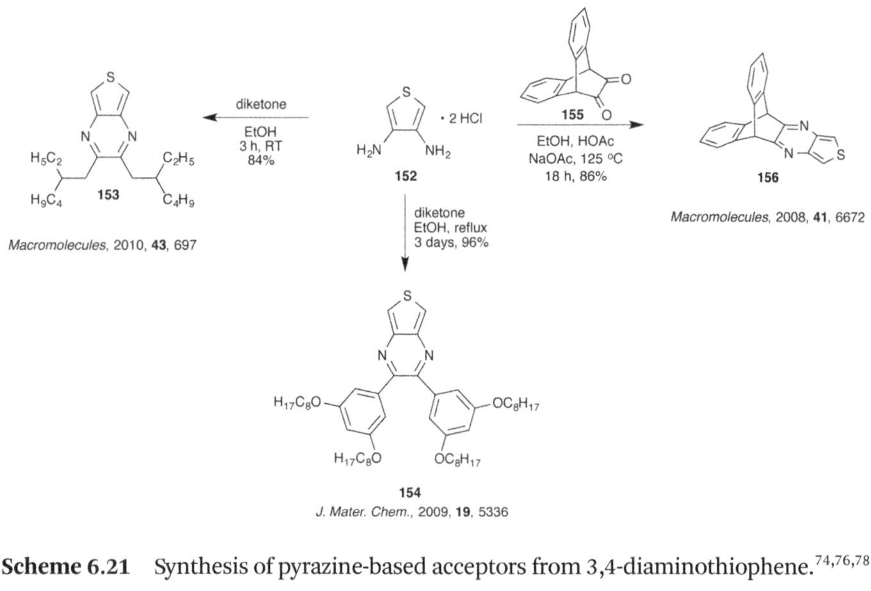

Scheme 6.21 Synthesis of pyrazine-based acceptors from 3,4-diaminothiophene.[74,76,78]

generally not given due to space constraints and since the diketone structure can be easily inferred from the structure of the final product. The first example shows the deprotection of 4,7-dibromobenzo[2,1,3]-thiadiazole with the mild reducing agent sodium borohydride (NaBH$_4$) to give 1,2-diamine **144** in 68% yield.[74] In this case, synthesis of **144** by brominating 1,2-diaminobenzene would be challenging due to over bromination and 4,5-regiochemistry instead of 3,6.[75] Diamine **144** can then be condensed with a diketone to give the pyrazine based acceptor **145** in moderate yield.[74] Pyrazine based acceptors such as **144** with one benzannulated ring are often referred to as "quinoxalines", which is the common name of the heterocyclic parent. An example with a one-pot deprotection–condensation is illustrated by the reaction of **147** with zinc reducing agent to give the tetra-amine followed by condensation with two equivalents of 3,4-butanedione to give dipyrazine acceptor **146** in 71% yield.[76] Thiadiazole diamine **147** may also react directly with a diketone to give the thiadiazole pyrazine acceptor **148** in 65% yield with acetic acid as the solvent.[77] Another example showing the relative mildness of the reaction is condensation of the relatively high EA diketone **149** with the relatively low IE diamine **150** to give donor–acceptor pyrazine **151** in 86% yield.[77] The moderate-to-high yields of the condensations given in Scheme 6.20 to form the pyrazine ring are typical.

The synthesis of thiophene-based pyrazine acceptors is illustrated in Scheme 6.21. 3,4-Diaminothiophene (**152**) is a low IE compound that is susceptible to oxidation, and is more stable and commercially supplied as the di(hydrochloride salt); however, this high reactivity allows for the condensation of diamine **152** with a variety of diketones including dialkyl diketones to give pyrazines such as **153**,[74] diaryl diketones to give pyrazines such as **154**,[76] and even bicyclic diketones such as **155** to give pyrazines such as **156**,[78] all in high yield. The increased nucleophilicity of the amines of **152** due to the electron donating thiophene ring may often give high yields of the corresponding pyrazines. Overall, the examples in Schemes 6.20 and 6.21 demonstrate the relatively broad reactivity and substrate variability possible in the synthesis of pyrazine-based acceptors.

6.5.3 Diketopyrrolopyrrole Acceptors (DPP)

Two routes for the synthesis of diketopyrrolopyrrole (DPP) acceptors are outlined in Scheme 6.22. The first route is reaction of two equivalents of an aryl nitrile (S6.22a) with 1,4-alkyl diester (S6.22b) and an alkoxide (MOR) to give the DPP-diNH acceptor (S6.22c). Alternatively,

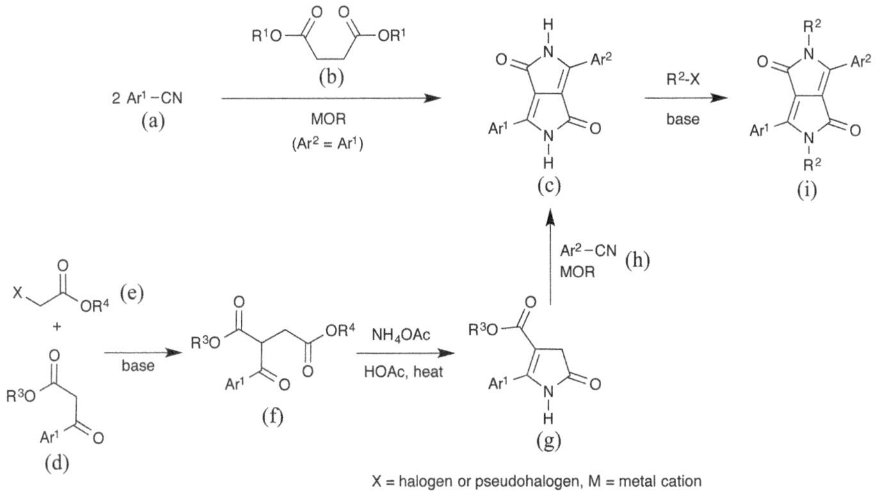

Scheme 6.22 General illustration of the synthesis of diketopyrrolopyrrole (DPP) acceptors.

a stepwise synthesis may be accomplished by first alkylating a 3-aryloxo-ester (S6.22d) with a haloester (S6.22e) to give a 2-acyl-1,4-alkyl diester (S6.22f) that can be cyclized to a 2*H*-pyrrol-2-one (S6.22g), which is essentially a half-DPP, that can then be reacted with an equivalent of an aryl nitrile (S6.22h) to give DPP-diNH (S6.22c). The stepwise cyclization through a half-DPP (S6.22g) allows synthesis of DPPs with different aryl substituents (Ar^1, Ar^2). The DPP-diNH can then be alkylated with a variety of alkyl groups to give the final DPP acceptor (S6.22i). In the synthesis of π-conjugated materials, the aryl groups (Ar^1 and Ar^2) have been used to incorporate the acceptors into the π-conjugated backbone of the polymer or small molecule and the *N*-alkyl groups (R^2) have been use to aid solubility and/or processability or to provide functional groups. The various esters that are used in the synthetic strategies are typically methyl, ethyl, and isopropyl esters. The metal oxides used are typically bulky bases such as *t*-butoxides or *t*-amyloxides. In general, *t*-amylalcohol (2-methyl-2-butanol) is preferred as the solvent for the nitrile cyclization reactions.[79] When *t*-amylalcohol (*t*-AmylOH) is used, generation of *t*-AmylONa in the reaction as the hindered base by addition of sodium metal to *t*-AmylOH may be preferred.

Table 6.3 summarizes some of the variations in the synthesis of DPP acceptors. The aryl groups range from bromophenyl (T6.3, Entry 1)[80] and cyanophenyl (T6.3, Entry 3)[81] to a phenyl acetal (T6.3, Entry 4)[82] to electron donor thiophenes (T6.3, Entry 5 and 6)[83,84] to an

Table 6.3 Synthesis and *N*-alkylation of symmetrical DPP acceptors. X = halogen or tosylate; MOR = metal alkoxide.

Entry	Ar	R^2	Conditions	Reference
1	Br—⟨⟩---	---C_8H_{17}	(1) KO*t*-Bu/HO*t*-Amyl and dimethylsuccinate, 80% (2) KO*t*-Bu/NMP, then 1-bromooctane	80
2	MeO—⟨⟩--- (MeO)	OEt / OEt	(1) NaO*t*-Amyl (FeCl₃(cat.))/HO*t*-Amyl and di(isopropyl)succinate, 18% (2) K₂CO₃/DMF, then bromoacetaldehyde diethyl acetal, 48%	86
3	NC—⟨⟩---	None	(1) KO*t*-Amyl/HO*t*-Amyl and diethylsuccinate, 77%	81
4	(dioxolane)—⟨⟩---	---(⟨O⟩)₃	(1) NaO*t*-Amyl (FeCl₃(cat.))/HO*t*-Amyl and di(isopropyl)succinate, 30% (2) K₂CO₃/DMF, Me(OCH₂CH₂)₃-OTs, 65%	82
5	thiophene---	---$C_{16}H_{33}$	(1) KO*t*-Bu/HO*t*-Amyl and dimethylsuccinate, 82% (2) K₂CO₃/DMF then 1-bromohexadecane, 88%	83
6	thienothiophene---	$C_{10}H_{21}$ / C_8H_{17}	(1) NaO*t*-Amyl (FeCl₃(-cat.))/HO*t*-Amyl and di(isopropyl)succinate, 73% (2) K₂CO₃/DMF/18-crown-6 then 9-(bromomethyl)nonadecane, 34%	84
7	thiazole---	C_4H_9 / Et	(1) NaO*t*-Amyl and diethylsuccinate, 73% (2) NaH/DMF, then 2-ethylhexylbromide, 22%	85

electron acceptor thiazole (T6.3, Entry 7).[85] The *N*-alkyl groups vary from straight and branched chain alkyls to an acetal (T6.3, Entry 2)[86] to PEG (T6.3, Entry 4). Other *N*-alkyl groups not presented in Table 6.3 include groups having alkenes and siloxanes,[87] azides for "Click" reactions,[88] alkyl and phenyl cyano groups,[89,90] and alkyl halides that may then be converted to trimethylammonium salts for water solubility.[91] Bases in the cyclization step include NaO*t*-Bu, KO*t*-Bu and NaO*t*-Amyl, and bases for the alkylation step are typically K_2CO_3 but in some cases are NaH (T6.3, Entry 7) and others. Yields are variable for each of the two steps. When amyl alkoxide is generated and used in the cyclization (*e.g.*, T6.3, Entry 2–4, 6, and 7), $FeCl_3$ is often used as a catalyst for formation of the alkoxide.

Examples of the synthesis of differently aryl-substituted DPP acceptors are summarized in Table 6.4. Many of the cyclization reaction conditions are similar to those in Table 6.3 for the symmetrically substituted DPPs, and the references presented in Table 6.4 also have preparation of the half-DPP compound. Note that the differential substitution includes diaryl (T6.4, Entry 1)[92] and bromo-functionalized diaryl (T6.4, Entry 2),[93] bromoaryl-thiophene (T6.4, Entry 3),[94] and thiophene-bromopyridine (T6.4, Entry 4).[95] Again, most reported

Table 6.4 Stepwise synthesis of unsymmetrical DPP acceptors.

Entry	Ar¹	Ar²	Yield (reference)
1			54% (92)
2			70% (93)
3			63% (94)
4			% Not specified (95)

yields are variable; however, the ability to synthesize acceptors with different aryl substituents and the range of aryl substituents presented in Tables 6.3 and 6.4 along with the ability to install functional groups at both the *N*-alkyl and aryl substituents demonstrates the relatively broad scope of DPP acceptor synthesis. It should also be noted that DPP acceptors with an aryl substituent and an alkyl substituent are also synthetically available by the two-step synthetic method.[79]

6.5.4 Isoindigo Acceptors

The synthesis of isoindigo acceptors is summarized in Scheme 6.23 and is essentially an acid catalyzed Knoevenagel condensation between a 1,2-ketoamide arene (S6.23a) and a cyclic arene amide (S6.23b) to give the isoindigo core (S6.23c). When there are alkyl groups on both amides, acid catalysis (S6.23d) gives the isoindigo core. When at least one amide does not have an alkyl group (*i.e.*, is N–H), then a two-step process is required that involves acid catalyzed condensation followed by alkylation or, less commonly, acylation in a second step (S6.23e). Since both aryl groups and alkyl groups on the amides can be varied, there is a reasonable amount of structural diversity that can be explored with isoindigo acceptors. To date, there has been relatively less variation of the synthetic methodology for isoindigo synthesis.

Some examples of isoindigo synthesis are illustrated in Scheme 6.24. In the first example, 1,2-ketoamide **157** is reacted with amide **158** with hydrochloric acid catalysis to give the isoindigo di-NH **159** in 90% yield followed by deprotonation with NaH and then alkylation with 6-bromohexane to give isoindigo **160** in 70% yield.[96] Another example from **159** illustrates the acylation of the isoindigo di-NH with

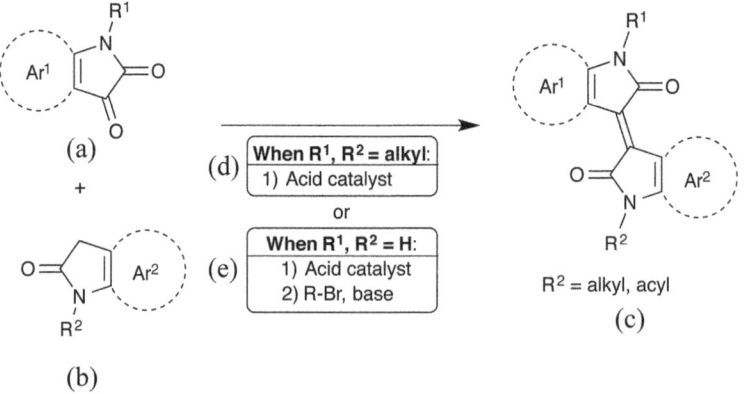

Scheme 6.23   General illustration of the synthesis of isoindigo-based acceptors.

an acyl chloride in pyridine to give **161** in high yield.[97] Formation of an isoindigo with thieno groups is shown in the synthesis of di(thieno) isoindigo **164** from the thieno ketoamide **162** and thieno amide **163** in acetic acid in 66% yield. In this case, no stronger acid catalyst was used other than the acetic acid solvent. Mixed (benzothieno)isoindigo **165** has been prepared in a similar manner from thieno ketoamide **166** and amide **167** to give the (benzothieno)isoindigo mono-NH compound in 51% yield followed by alkylation of the benzo-NH with an alkyliodide and K_2CO_3 to give the final (benzothieno)isoindigo **165** in 61% yield.[98] It should be noted that, for symmetrical isoindigos, homocoupling of a ketoamide using Lawesson's reagent is possible. For example, Lawesson's reagent affected the formation of **164** from **162** with a reported yield of 36%.[99]

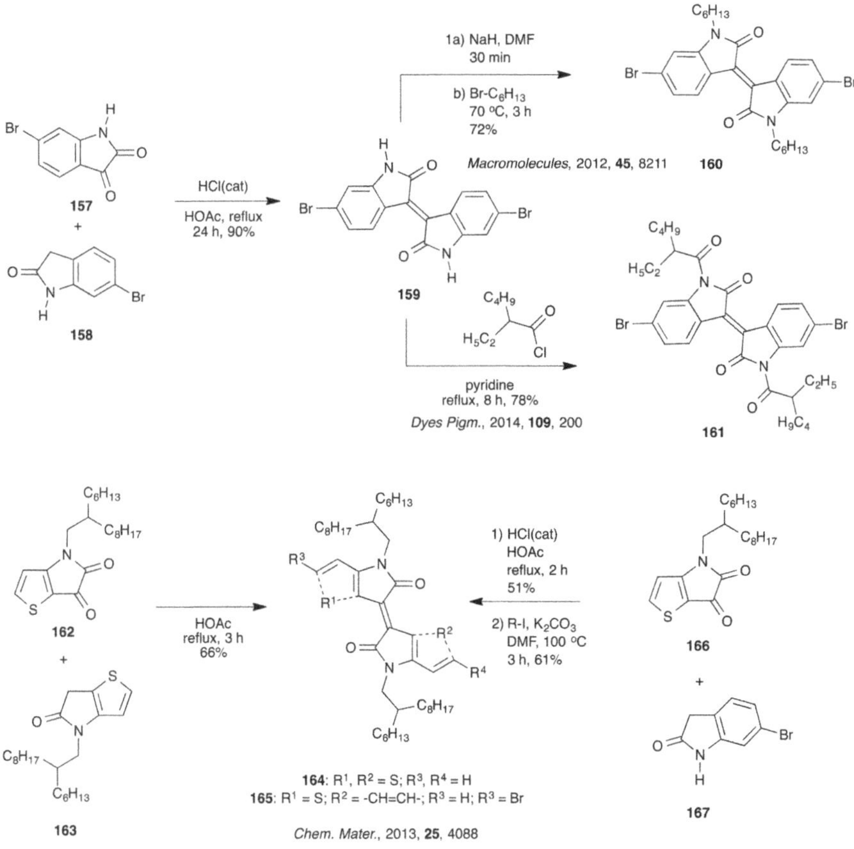

Scheme 6.24 Synthesis of symmetrical and unsymmetrical isoindigo acceptors from arenes and thiophenes.[96,98]

6.5.5 Imide-based Acceptors

A number of imide-based acceptors have been used in π-conjugated materials. Three of the most widely used are shown in Scheme 6.25 and include the monoimide thieno[3,4-*c*]pyrrole-4,6-dione (TPD, S6.25a) and the diimides naphthalene diimide (NDI, S6.25b) and perylene diimide (PDI, S6.25c). The general synthetic route to generate derivatives of the imides is illustrated in Scheme 6.25. Typically, reaction of an anhydride or dianhydride (S6.25e) with an amine (S6.25f) can give the amic acid intermediate (S6.25g) that can then cyclize to the imide (S6.25h) in the presence of heat, acid or Lewis acid catalysis, base, $SOCl_2$, or any combination. Often, trial and error must be used to determine the optimal conditions. The amic acid can be isolated as a stable intermediate in some cases, particularly during TPD synthesis. In other cases, the amic acid cannot be isolated, especially when forcing conditions must be used to react the anhydride with the amine to form the amic acid, which then may quickly undergo condensation to form the imide. This often happens with less nucleophilic phenyl amines, which typically have to be reacted at elevated temperatures to form the amic acid; however, in most cases, isolation of the amic acid is not desirable if condensation to form the imide can occur in the same reaction mixture as addition to the anhydride. With diimides, a second equivalent of a different amine (S6.25i) can be used to provide a differential substitution ($R^1 \neq R^2$) either in one pot or in a stepwise sequence where the monoimide of the first amine addition is isolated.

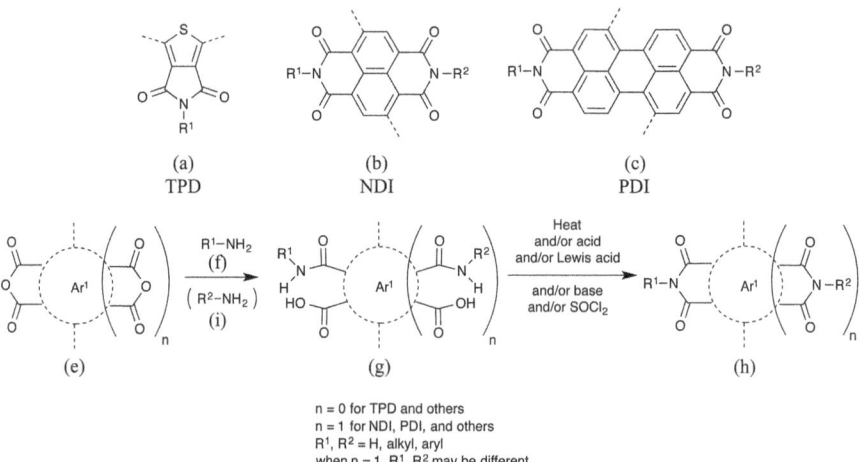

(a) TPD   (b) NDI   (c) PDI

(e)   (g)   (h)

n = 0 for TPD and others
n = 1 for NDI, PDI, and others
R^1, R^2 = H, alkyl, aryl
when n = 1, R^1, R^2 may be different

Scheme 6.25   General illustration of the synthesis of imide-based acceptors.

Examples of formation of TPD imides are shown in Scheme 6.26. Typically, these reactions occur from the commercially available thiophene-3,4-dicarboxylic acid, which can also be synthesized by a more economical route (Scheme 6.26 inset).[100] In one example, formation of TPD imide **168** occurs first with formation of the TPD anhydride with acetic anhydride (Ac$_2$O) followed by formation of the amic acid in toluene and finally condensation to the imide with SOCl$_2$ to give imide **168** in 57% yield over three steps.[101] In this case, the TPD anhydride and TPD amic acid were isolated but not purified. Similar chemistry was employed for the synthesis of dibromo TPD **169** except in this case bromination was carried out first on the thiophene diacid in 51% yield followed by formation of the anhydride in 81% yield and formation of the amic acid and condensation of the with SOCl$_2$ to give dibromo TPD **169** in 84% yield over two steps;[102] however, TPD itself can be directly brominated with relatively high yields such as with conversion of **169** to **170** with NBS/H$_2$SO$_4$/CF$_3$CO$_2$H in 81% yield.[101] TPD imide **170** was synthesized by an alternative route from di(acid chloride) **171** using the corresponding 2-hexyl-1-decylamine as the solvent in 72% yield.[103]

The synthesis of NDI and PDI compounds from the corresponding dianhydrides is similar, although PDI compounds may require

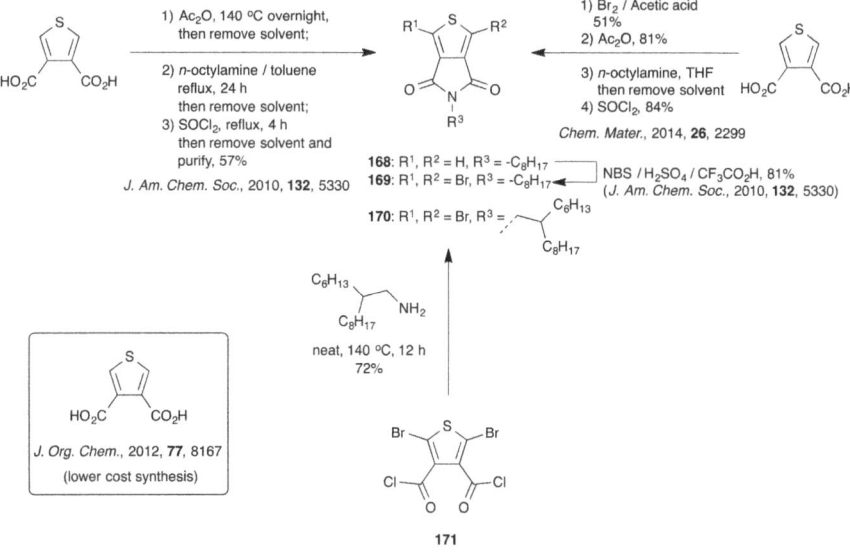

Scheme 6.26   Synthesis of thienodiketopyrrole (TPD) imide acceptors.[101–103]

higher temperatures and different solvents since perylene dianhydride is less soluble than naphthalene dianhydride and PDI
compounds are typically less soluble than their NDI analogues.
A number of conditions are summarized in Table 6.5 for both the
synthesis of NDI ($n = 0$) and PDI ($n = 1$) from the naphthalene or
perylene diimides, respectively. Note that in both cases, the anhydrides can be substituted with bromines. The conditions include
imidazole as a melt at high temperature[104,105] (T6.5, Entry 1 and
9), Zn(OAc) in a basic solvent[106] (T6.5, Entry 2), acidic solvent conditions[107–109] (T6.5, Entry 3, 4, and 8), and approximately neutral
conditions in polar solvents[110–112] (T6.5, Entries 5–7) for both aliphatic and aromatic amines. The yields range from 58% to 90%,
and in general the yields are moderate to high, unless there is
exceptional steric hindrance in the amine, low reactivity, or high
reactivity of the amine that causes decomposition at the high temperatures of the reaction. In some cases it is better to dissolve the
naphthalene or perylene dianhydride at high temperatures in the
solvent (often over 120 °C) and then add the amine so that the mixture is heterogeneous and the amine reacts more rapidly with both
anhydrides. This may be particularly true when the naphthalene or
perylene dianhydride contains aryl bromines that can react with
excess amine, which is effectively the situation when an insoluble
naphthalene or perylene dianhydride reacts with an amine and
is "pulled" into solution where there is an excess of amine if the
remaining solid naphthalene or perylene dianhydride does not
react sufficiently quickly.

Differentially substituted NDIs can be synthesized by: (1) reacting
the naphthalene dianhydride with one equivalent of amine, isolating
the intermediate monoimide, and then reacting with a second equivalent of amine; or (2) reacting the dianhydride with a mixture of the
two amines and then isolating the desired differentially substituted
amine. However, both these routes proceed by a more-or-less statistical reaction that results in a mixture of products. Reaction with one
amine typically provides an ~1:1:1 mixture of starting dianhydride,
monoimide, and diimide while reaction with two amines provides an
~1:1:1 mixture of one product with two R^1, one product with R^1 and R^2,
and one product with two R^2. These mixtures can be difficult to purify
but differences in solubility and/or polarity may allow purification by
crystallization and/or chromatography. Additionally, the ~1:1:1 product ratio means yields are often limited to around 20–40%. Scheme
6.27 illustrates some examples of differential NDI substitution. The
first example is the reaction of dianhydride **172** in a stepwise route to

Table 6.5 Synthesis of symmetrical naphthalene- and perylenediimides (NDI and PDI).

Entry	N	X^1	X^2	Amine	Conditions	Reference
1	1	H	H	C_8H_{17} / C_6H_{13} —NH$_2$	Imidazole, 180 °C (melt), 4 h, 86%	104
2	1	H	H	$C_{12}H_{25}$ / $C_{10}H_{21}$ —NH$_2$	Zn(OAc)$_2$, quinoline, 180 °C overnight, 87%	106
3	1	H	Br	C_4H_9 / C_2H_5 —NH$_2$	CH$_3$CH$_2$CO$_2$H, reflux, 12 h, 90%	107
4	0	Br	Br	C_4H_9 / C_2H_5 —NH$_2$	AcOH at 120 °C until solids dissolved, 58%	108
5	1	H	Br	C_4H_9 / C_2H_5 —NH$_2$	n-BuOH/H$_2$O (1:1), 80 °C, 4 h, 86%	110
6	1	H	Br	C_4H_9 / C_2H_5 —NH$_2$	DMF, 70 °C, 12 h, 72%	111
7	0	H	H	C_4H_9 / C_2H_5 —NH$_2$	DMF, reflux overnight, 56%	112
8	0	H	H	—NH$_2$ (2,3-dimethylphenyl)	AcOH, reflux, 2.5 h, 61%	109
9	1	H	H	—NH$_2$ (2,6-diisopropylphenyl)	Imidazole, 190 °C (melt), 24 h, 66%	105

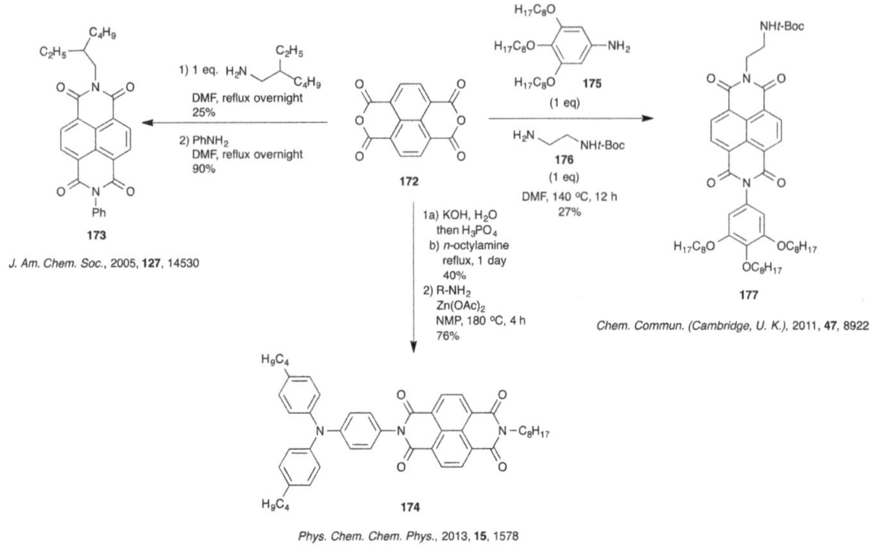

Scheme 6.27 Synthesis of NDIs with differential imide *N*-substitution.[113–115]

give NDI **173** by first reaction with one equivalent of amine and isolation of the monoimide intermediate in 25% yield followed by another imidization with aniline to give the NDI **173** in 90% yield (23% overall).[113] The second example synthesizes **174** using KOH to hydrolyze the anhydride to a dicarboxylic acid followed by neutralization with H_3PO_4 and reaction with the amine in one pot to give the intermediate anhydride-imide in 40% yield followed by imidization with $Zn(OAc)_2$/imidazole to give **174** in 76% yield (30% overall).[114] Another example uses a mixture of aniline **175** and amine **176** to give the corresponding differentially substituted NDI **177** in 27% yield accompanied (presumably) by side-product dimides of **175** and **176**.[115] These examples illustrate that even the statistical reactivity of naphthalene dianhydride can be used to obtain research quantities of differentially substituted NDIs in yields of 20–40%.

Although differentially substituted PDIs have been synthesized in a similar manner to differentially substituted NDIs,[116] the generally lower solubility of the perylene core relative to the naphthalene core can be used to precipitate intermediate perylene imide-anhydrides from the hydrolysis of symmetrical PDIs in relatively high yields. The intermediate perylene imide-anhydride then can be imidized with an equivalent of a different amine to give the differentially substituted PDI. An example of this is illustrated in Scheme 6.28 with the hydrolysis of symmetrical PDI **178** to the monoanhydride **179** in 65% yield

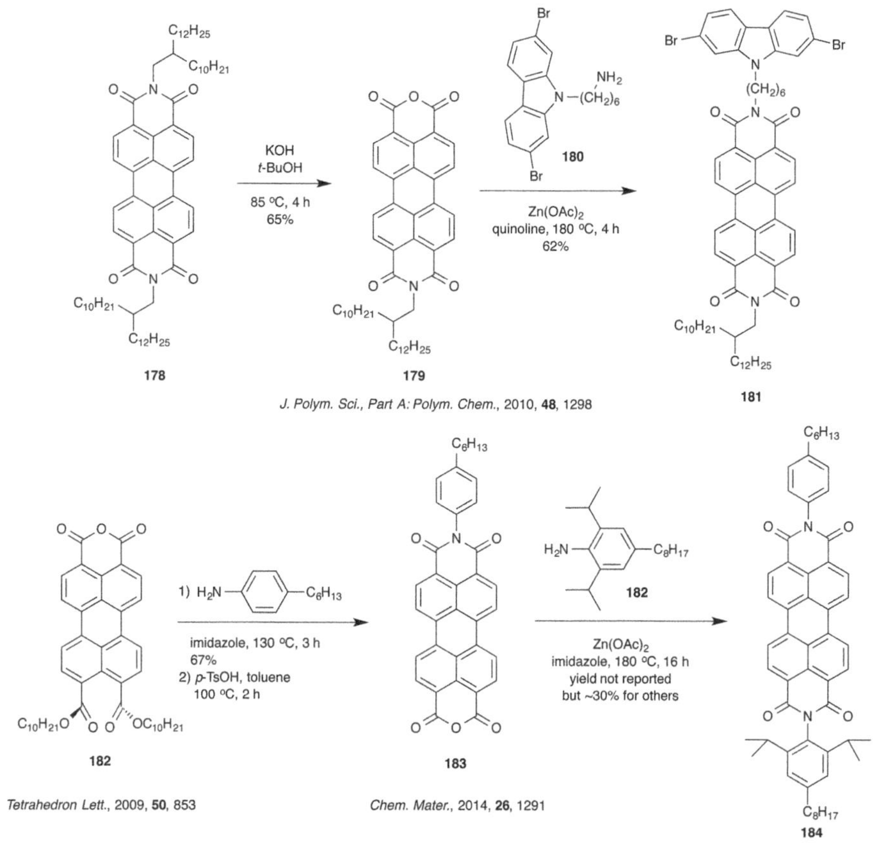

J. Polym. Sci., Part A: Polym. Chem., 2010, **48**, 1298

Tetrahedron Lett., 2009, **50**, 853

Chem. Mater., 2014, **26**, 1291

Scheme 6.28 Synthesis of PDIs with differential imide *N*-substitution.[106,117,118]

followed by imidization of **179** with **180** with Zn(OAc)$_2$ in quinolone to give differentially substituted **181** in 62% yield.[106] Another strategy uses the perylene anhydride-di(decyl)ester **182**, which precipitates from solution when the perylene tetra(didecyl)ester is hydrolyzed.[117] Anhydride-diester **182** can be reacted with an aniline to provide the intermediate imide-diester in 67% yield that is then converted to monoanhydride **183** with *p*-TsOH and reacted without purification with aniline **182** to give differentially substituted **184**.[118]

6.6 CONCLUSION

The synthesis and installation of acceptors in π-conjugated materials is often a critical step in achieving material properties. However, the nature of acceptors as generally having high electron affinity (EA) makes them resistant to electrophilic aromatic substitution and reactive toward nucleophiles, especially lithiated arenes that are intermediates in the synthesis of π-conjugated materials. Because of these reactivity challenges, acceptors are often installed near the end of π-conjugated material synthesis. This is particularly the case with the Knoevenagel condensation in the synthesis of D–π–A and A–D–A or D–A–D chromophores for nonlinear optical applications, and also for small molecule electron acceptor materials. For polymers, halogenated acceptors are often used in the organometallic polymerization *via* Stille and Suzuki coupling to donors or π-bridges in the final step, which is the topic of Chapter 7.

REFERENCES

1. G. Jones, in *Organic Reactions*, ed. R. Adams, A. Blatt, V. Boekelheide, T. Cairns, A. Cope and H. House, Wiley, New York, 1967, vol. 15, p. 204.
2. D. Volochnyuk, S. Ryabukhin, A. Plaskon and O. Grygorenko, *Synthesis*, 2009, **2009**, 3719.
3. T. L. Greaves and C. J. Drummond, *Chem. Rev.*, 2008, **108**, 206.
4. D.-Z. Xu, Y. Liu, S. Shi and Y. Wang, *Green Chem.*, 2010, **12**, 514.
5. P. Lidstrom, J. Tierney, B. Wathey and J. Westman, *Tetrahedron*, 2001, **52**, 9225.
6. B. Rodríguez, A. Bruckmann, T. Rantanen and C. Bolm, *Adv. Synth. Catal.*, 2007, **349**, 2213.
7. C. Herbivo, A. Comel, G. Kirsch, A. M. C. Fonseca, M. Belsley and M. M. M. Raposo, *Dyes Pigm.*, 2010, **86**, 217.
8. M. C. Davis, R. A. Hollins and B. Douglas, *Synth. Commun.*, 2006, **36**, 3515.
9. U. Lawrentz, W. Grahn, K. Lukaszuk, C. Klein, R. Wortmann, A. Feldner and D. Scherer, *Chem.–Eur. J.*, 2002, **8**, 1573.
10. S. Wang and S. Kim, *Dyes Pigm.*, 2009, **80**, 314.
11. S. Mukhopadhyay, R. B. Kanth, S. Ramasesha and S. Patil, *J. Phys. Chem. A*, 2010, **114**, 4647.
12. S. Ermer, S. M. Lovejoy, D. S. Leung, H. Warren, C. R. Moylan and R. J. Twieg, *Chem. Mater.*, 1997, **9**, 1437.
13. Y. Ie, K. Nishida, M. Karakawa, H. Tada, A. Asano, A. Saeki, S. Seki and Y. Aso, *Chem.–Eur. J.*, 2011, **17**, 4750.
14. C. Wetzel, A. Mishra, E. Mena-Osteritz, A. Liess, M. Stolte, F. Wurthner and P. Bauerle, *Org. Lett.*, 2014, **16**, 362.
15. L. Y. Lin, C. W. Lu, W. C. Huang, Y. H. Chen, H. W. Lin and K. T. Wong, *Org. Lett.*, 2011, **13**, 4962.

16. X. Shi, J. Chang and C. Chi, *Chem. Commun.*, 2013, **49**, 7135.
17. B. S. Ong and B. Keoshkerian, *J. Org. Chem.*, 1984, **49**, 5002.
18. S. Zheng, A. Leclercq, J. Fu, L. Beverina, L. A. Padilha, E. Zojer, K. Schmidt, S. Barlow, J. Luo, S.-H. Jiang, A. K. Y. Jen, Y. Yi, Z. Shuai, E. W. Van Stryland, D. J. Hagan, J.-L. Brédas and S. R. Marder, *Chem. Mater.*, 2007, **19**, 432.
19. L. Beverina, J. Fu, A. Leclercq, E. Zojer, P. Pacher, S. Barlow, E. W. Van Stryland, D. J. Hagan, J. L. Bredas and S. R. Marder, *J. Am. Chem. Soc.*, 2005, **127**, 7282.
20. S. H. Sinha, E. A. Owens, Y. Feng, Y. Yang, Y. Xie, Y. Tu, M. Henary and Y. G. Zheng, *Eur. J. Med. Chem.*, 2012, **54**, 647.
21. I. G. Davydenko, A. D. Kachkovsky, M. L. Dekhtyar, Y. L. Slominskii and A. I. Tolmachev, *J. Phys. Org. Chem.*, 2009, **23**, 96.
22. E. Macoas, G. Marcelo, S. Pinto, T. Caneque, A. M. Cuadro, J. J. Vaquero and J. M. Martinho, *Chem. Commun.*, 2011, **47**, 7374.
23. O. V. Przhonska, H. Hu, S. Webster, J. L. Bricks, A. A. Viniychuk, A. D. Kachkovski and Y. L. Slominsky, *Chem. Phys.*, 2013, **411**, 17.
24. V. Shirinian and A. Shimkin, *Top. Heterocycl. Chem.*, 2008, **14**, 75.
25. M. Mojzych and M. Henary, *Top. Heterocycl. Chem.*, 2008, **14**, 1.
26. A. V. Kulinich and A. A. Ishchenko, *Russ. Chem. Rev.*, 2009, **78**, 141.
27. M. Panigrahi, S. Dash, S. Patel and B. K. Mishra, *Tetrahedron*, 2012, **68**, 781.
28. *Heterocyclic Polymethine Dyes*, ed. L. Strekowski, Springer GmbH, Berlin, 2008.
29. J. Fabian and H. Hartmann, *Light Absorption of Organic Chromophores*, Springer-Verlag, Berlin, 1980.
30. R. S. Lepkowicz, O. V. Przhonska, J. M. Hales, J. Fu, D. J. Hagan, E. W. Van Stryland, M. V. Bondar, Y. L. Slominsky and A. D. Kachkovski, *Chem. Phys.*, 2004, **305**, 259.
31. F. Terenziani, O. V. Przhonska, S. Webster, L. A. Padilha, Y. L. Slominsky, I. G. Davydenko, A. O. Gerasov, Y. P. Kovtun, M. P. Shandura, A. D. Kachkovski, D. J. Hagan, E. W. Van Stryland and A. Painelli, *J. Phys. Chem. Lett.*, 2010, **1**, 1800.
32. B. J. Coe, J. A. Harris, J. J. Hall, B. S. Brunschwig, S.-T. Hung, W. Libaers, K. Clays, S. J. Coles, P. N. Horton, M. E. Light, M. B. Hursthouse, J. Garín and J. Orduna, *Chem. Mater.*, 2006, **18**, 5907.
33. X. Xu, W. Qiu, Q. Zhou, J. Tang, F. Yang, Z. Sun and P. Audebert, *J. Phys. Chem. B*, 2008, **112**, 4913.
34. M. Matsui, S. Kawamura, K. Shibata and H. Muramatsu, *Bull. Chem. Soc. Jpn.*, 1992, **65**, 71.
35. P. Yan, A. Xie, M. Wei and L. M. Loew, *J. Org. Chem.*, 2008, **73**, 6587.

36. V. Alain, M. Blanchard-Desce, I. Ledoux-Rak and J. Zyss, *Chem. Commun.*, 2000, 353.

37. L. G. S. Brooker, G. H. Keyes, R. H. Sprague, R. H. VanDyke, E. VanLare, G. VanZandt, F. L. White, H. W. J. Cressman and S. G. Dent, *J. Am. Chem. Soc.*, 1951, **73**, 5332.

38. R. Sun, B.-L. Yan, J.-F. Ge, Q.-f. Xu, N.-J. Li, X.-Z. Wu, Y.-L. Song and J.-M. Lu, *Dyes Pigm.*, 2013, **96**, 189.

39. F. Würthner, *Synthesis*, 1999, **1999**, 2103.

40. A. Toutchkine, V. Kraynov and K. Hahn, *J. Am. Chem. Soc.*, 2003, **125**, 4132.

41. C. Cabanetos, W. Bentoumi, V. Silvestre, E. Blart, Y. Pellegrin, V. Montembault, A. Barsella, K. Dorkenoo, Y. Bretonnière, C. Andraud, L. Mager, L. Fontaine and F. Odobel, *Chem. Mater.*, 2012, **24**, 1143.

42. A. J. Kay, A. D. Woolhouse, Y. Zhao and K. Clays, *J. Mater. Chem.*, 2004, **14**, 1321.

43. A. V. Kulinich, N. A. Derevyanko and A. A. Ishchenko, *J. Photochem. Photobiol., A*, 2007, **188**, 207.

44. H. Bürckstümmer, N. M. Kronenberg, M. Gsänger, M. Stolte, K. Meerholz and F. Würthner, *J. Mater. Chem.*, 2010, **20**, 240.

45. C. J. MacNevin, D. Gremyachinskiy, C. W. Hsu, L. Li, M. Rougie, T. T. Davis and K. M. Hahn, *Bioconjugate Chem.*, 2013, **24**, 215.

46. R. Andreu, L. Carrasquer, S. Franco, J. Garin, J. Orduna, N. Martinez de Baroja, R. Alicante, B. Villacampa and M. Allain, *J. Org. Chem.*, 2009, **74**, 6647.

47. J. Panda, P. R. Virkler and M. R. Detty, *J. Org. Chem.*, 2003, **68**, 1804.

48. P.-A. Bouit, G. Wetzel, G. Berginc, B. Loiseaux, L. Toupet, P. Feneyrou, Y. Bretonnière, K. Kamada, O. Maury and C. Andraud, *Chem. Mater.*, 2007, **19**, 5325.

49. P. A. Bouit, E. Di Piazza, S. Rigaut, B. Le Guennic, C. Aronica, L. Toupet, C. Andraud and O. Maury, *Org. Lett.*, 2008, **10**, 4159.

50. P. A. Bouit, D. Rauh, S. Neugebauer, J. L. Delgado, E. Di Piazza, S. Rigaut, O. Maury, C. Andraud, V. Dyakonov and N. Martin, *Org. Lett.*, 2009, **11**, 4806.

51. S. J. Mason, J. L. Hake, J. Nairne, W. J. Cummins and S. Balasubramanian, *J. Org. Chem.*, 2005, **70**, 2939.

52. J. H. Yum, P. Walter, S. Huber, D. Rentsch, T. Geiger, F. Nuesch, F. De Angelis, M. Gratzel and M. K. Nazeeruddin, *J. Am. Chem. Soc.*, 2007, **129**, 10320.

53. Y. Shi, R. B. Hill, J. H. Yum, A. Dualeh, S. Barlow, M. Gratzel, S. R. Marder and M. K. Nazeeruddin, *Angew. Chem., Int. Ed.*, 2011, **50**, 6619.

54. S. Yagi and H. Nakazumi, *Top. Heterocycl. Chem.*, 2008, **2008**, 133.
55. C. Cai, I. Liakatas, M.-S. Wong, M. Bösch, C. Bosshard, P. Günter, S. Concilio, N. Tirelli and U. W. Suter, *Org. Lett.*, 1999, **1**, 1847.
56. Y. A. Getmanenko, J. M. Hales, M. Balu, J. Fu, E. Zojer, O. Kwon, J. Mendez, S. Thayumanavan, G. Walker, Q. Zhang, S. D. Bunge, J.-L. Brédas, D. J. Hagan, E. W. Van Stryland, S. Barlow and S. R. Marder, *J. Mater. Chem.*, 2012, **22**, 4371.
57. X. Wu, J. Wu, Y. Liu and A. K. Y. Jen, *J. Am. Chem. Soc.*, 1999, **121**, 472.
58. F. A. Arroyave, C. A. Richard and J. R. Reynolds, *Org. Lett.*, 2012, **14**, 6138.
59. J. Zhou, S. Xie, E. F. Amond and M. L. Becker, *Macromolecules*, 2013, **46**, 3391.
60. Y. Jin, Y. Kim, S. H. Kim, S. Song, H. Y. Woo, K. Lee and H. Suh, *Macromolecules*, 2008, **41**, 5548.
61. L. Chen, X. Li, W. Ying, X. Zhang, F. Guo, J. Li and J. Hua, *Eur. J. Org. Chem.*, 2013, **2013**, 1770.
62. A. C. Stuart, J. R. Tumbleston, H. Zhou, W. Li, S. Liu, H. Ade and W. You, *J. Am. Chem. Soc.*, 2013, **135**, 1806.
63. N. Wang, Z. Chen, W. Wei and Z. Jiang, *J. Am. Chem. Soc.*, 2013, **135**, 17060.
64. J. Shao, J. Chang and C. Chi, *Org. Biomol. Chem.*, 2012, **10**, 7045.
65. J. C. Bijleveld, M. Shahid, J. Gilot, M. M. Wienk and R. A. J. Janssen, *Adv. Funct. Mater.*, 2009, **19**, 3262.
66. S. Albrecht, S. Janietz, W. Schindler, J. Frisch, J. Kurpiers, J. Kniepert, S. Inal, P. Pingel, K. Fostiropoulos, N. Koch and D. Neher, *J. Am. Chem. Soc.*, 2012, **134**, 14932.
67. L. Dou, C.-C. Chen, K. Yoshimura, K. Ohya, W.-H. Chang, J. Gao, Y. Liu, E. Richard and Y. Yang, *Macromolecules*, 2013, **46**, 3384.
68. Y. Zhang, S. C. Chien, K. S. Chen, H. L. Yip, Y. Sun, J. A. Davies, F. C. Chen and A. K. Jen, *Chem. Commun.*, 2011, **47**, 11026.
69. A. L. Appleton, S. Miao, S. M. Brombosz, N. J. Berger, S. Barlow, S. R. Marder, B. M. Lawrence, K. I. Hardcastle and U. H. Bunz, *Org. Lett.*, 2009, **11**, 5222.
70. T. Strassner and J. Fabian, *J. Phys. Org. Chem.*, 1997, **10**, 33.
71. D. G. Patel, F. Feng, Y.-y. Ohnishi, K. A. Abboud, S. Hirata, K. S. Schanze and J. R. Reynolds, *J. Am. Chem. Soc.*, 2012, **134**, 2599.
72. T. L. Tam, H. Li, F. Wei, K. J. Tan, C. Kloc, Y. M. Lam, S. G. Mhaisalkar and A. C. Grimsdale, *Org. Lett.*, 2010, **12**, 3340.
73. T. C. Parker, D. G. Patel, K. Moudgil, S. Barlow, C. Risko, J.-L. Brédas, J. R. Reynolds and S. R. Marder, *Mater. Horiz.*, 2014, **2**, 22.

74. C. H. Chen, C. H. Hsieh, M. Dubosc, Y. J. Cheng and C. S. HSu, *Macromolecules*, 2010, **43**, 697.

75. M. Milek, A. Witt, C. Streb, F. W. Heinemann and M. M. Khusni-yarov, *Dalton Trans.*, 2013, **42**, 5237.

76. A. P. Zoombelt, M. Fonrodona, M. G. R. Turbiez, M. M. Wienk and R. A. J. Janssen, *J. Mater. Chem.*, 2009, **19**, 5336.

77. J. S. Song, C. Zhang, C. H. Li, W. W. Li, R. P. Qin, B. S. Li, Z. P. Liu and Z. S. Bo, *J. Polym. Sci., Part A: Polym. Chem.*, 2010, **48**, 2571.

78. Z. H. Chen, J. Bouffard, S. E. Kooi and T. M. Swager, *Macromolecules*, 2008, **41**, 6672.

79. C. J. H. Morton, R. Gilmour, D. M. Smith, P. Lightfoot, A. M. Z. Slawin and E. J. MacLean, *Tetrahedron*, 2002, **58**, 5547.

80. S.-Y. Liu, H.-Y. Li, M.-M. Shi, H. Jiang, X.-L. Hu, W.-Q. Li, L. Fu and H.-Z. Chen, *Macromolecules*, 2012, **45**, 9004.

81. S. Luňák, J. Vyňuchal, M. Vala, L. Havel and R. Hrdina, *Dyes Pigm.*, 2009, **82**, 102.

82. G. Zhang, L. Song, S. Bi, Y. Wu, J. Yu and L. Wang, *Dyes Pigm.*, 2014, **102**, 100.

83. J. R. Matthews, W. Niu, A. Tandia, A. L. Wallace, J. Hu, W.-Y. Lee, G. Giri, S. C. B. Mannsfeld, Y. Xie, S. Cai, H. H. Fong, Z. Bao and M. He, *Chem. Mater.*, 2013, **25**, 782.

84. H. Bronstein, Z. Chen, R. S. Ashraf, W. Zhang, J. Du, J. R. Durrant, P. S. Tuladhar, K. Song, S. E. Watkins, Y. Geerts, M. M. Wienk, R. A. Janssen, T. Anthopoulos, H. Sirringhaus, M. Heeney and I. McCulloch, *J. Am. Chem. Soc.*, 2011, **133**, 3272.

85. B. Carsten, J. M. Szarko, L. Lu, H. J. Son, F. He, Y. Y. Botros, L. X. Chen and L. Yu, *Macromolecules*, 2012, **45**, 6390.

86. M. Grzybowski, E. Glodkowska-Mrowka, T. Stoklosa and D. T. Gryko, *Org. Lett.*, 2012, **14**, 2670.

87. J. Lee, A. R. Han, J. Kim, Y. Kim, J. H. Oh and C. Yang, *J. Am. Chem. Soc.*, 2012, **134**, 20713.

88. M. Castelain, H. Salavagione and J. L. Segura, *Org. Lett.*, 2012, **14**, 2798.

89. Y. Sun, S. C. Chien, H. L. Yip, K. S. Chen, Y. Zhang, J. A. Davies, F. C. Chen, B. P. Lin and A. K. Y. Jen, *J. Mater. Chem.*, 2012, **22**, 5587.

90. Y. L. Han, L. Chen and Y. W. Chen, *J. Polym. Sci., Part A: Polym. Chem.*, 2013, **51**, 258.

91. F. He, L. B. Liu and L. D. Li, *Adv. Funct. Mater.*, 2011, **21**, 3143.

92. S. Luňák Jr, L. Havel, J. Vyňuchal, P. Horáková, J. Kučerík, M. Weiter and R. Hrdina, *Dyes Pigm.*, 2010, **85**, 27.

93. H. Ftouni, F. Bolze, H. de Rocquigny and J. F. Nicoud, *Bioconjugate Chem.*, 2013, **24**, 942.

94. Y. Xu, Y. Jin, W. Lin, J. Peng, H. Jiang and D. Cao, *Synth. Met.*, 2010, **160**, 2135.

95. T. W. Holcombe, J.-H. Yum, Y. Kim, K. Rakstys and M. Grätzel, *J. Mater. Chem. A*, 2013, **1**, 13978.

96. L. A. Estrada, D. Y. Liu, D. H. Salazar, A. L. Dyer and J. R. Reynolds, *Macromolecules*, 2012, **45**, 8211.

97. S. Li, L. Ma, C. Hu, P. Deng, Y. Wu, X. Zhan, Y. Liu and Q. Zhang, *Dyes Pigm.*, 2014, **109**, 200.

98. M. S. Chen, J. R. Niskala, D. A. Unruh, C. K. Chu, O. P. Lee and J. M. J. Fréchet, *Chem. Mater.*, 2013, **25**, 4088.

99. G. W. P. Van Pruissen, F. Gholamrezaie, M. M. Wienk and R. A. J. Janssen, *J. Mater. Chem.*, 2012, **22**, 20387.

100. P. Berrouard, S. Dufresne, A. Pron, J. Veilleux and M. Leclerc, *J. Org. Chem.*, 2012, **77**, 8167.

101. Y. Zou, A. Najari, P. Berrouard, S. Beaupre, B. R. Aich, Y. Tao and M. Leclerc, *J. Am. Chem. Soc.*, 2010, **132**, 5330.

102. J. Warnan, A. El Labban, C. Cabanetos, E. T. Hoke, P. K. Shukla, C. Risko, J.-L. Brédas, M. D. McGehee and P. M. Beaujuge, *Chem. Mater.*, 2014, **26**, 2299.

103. G. Y. Chen, Y. H. Cheng, Y. J. Chou, M. S. Su, C. M. Chen and K. H. Wei, *Chem. Commun.*, 2011, **47**, 5064.

104. Z. Chen, B. Fimmel and F. Wurthner, *Org. Biomol. Chem.*, 2012, **10**, 5845.

105. C. Ramanan, A. L. Smeigh, J. E. Anthony, T. J. Marks and M. R. Wasielewski, *J. Am. Chem. Soc.*, 2012, **134**, 386.

106. M. Yuan, M. Su, M. Chiu and K. Wei, *J. Polym. Sci., Part A: Polym. Chem.*, 2010, **48**, 1298.

107. J. Chen, M. Shi, X. Hu, M. Wang and H. Chen, *Polymer*, 2010, **51**, 2897.

108. X. Guo and M. D. Watson, *Org. Lett.*, 2008, **10**, 5333.

109. H. Langhals and H. Jaschke, *Chem.–Eur. J.*, 2006, **12**, 2815.

110. E. Zhou, K. Tajima, C. Yang and K. Hashimoto, *J. Mater. Chem.*, 2010, **20**, 2362.

111. J. Hou, S. Zhang, T. L. Chen and Y. Yang, *Chem. Commun.*, 2008, 6034.

112. S. Guo, W. Wu, H. Guo and J. Zhao, *J. Org. Chem.*, 2012, **77**, 3933.

113. P. Ganesan, X. Yang, J. Loos, T. J. Savenije, R. D. Abellon, H. Zuilhof and E. J. Sudholter, *J. Am. Chem. Soc.*, 2005, **127**, 14530.

114. R. Rybakiewicz, J. Zapala, D. Djurado, R. Nowakowski, P. Toman, J. Pfleger, J. M. Verilhac, M. Zagorska and A. Pron, *Phys. Chem. Chem. Phys.*, 2013, **15**, 1578.

115. A. Das and S. Ghosh, *Chem. Commun.*, 2011, **47**, 8922.

116. Y. Che, X. Yang, G. Liu, C. Yu, H. Ji, J. Zuo, J. Zhao and L. Zang, *J. Am. Chem. Soc.*, 2010, **132**, 5743.
117. C. M. Xue, R. K. Sun, R. Annab, D. Abadi and S. Jin, *Tetrahedron Lett.*, 2009, **50**, 853.
118. W. Benjamin, D. Veit, M. Perkins, E. Bain, K. Scharnhorst, S. McDowall, D. Patrick and J. Gilbertson, *Chem. Mater.*, 2014, **26**, 1291.
119. J. Shao, S. Ji, X. Li, J. Zhao, F. Zhou and H. Guo, *Eur. J. Org. Chem.*, 2011, **2011**, 6100.
120. S. Liu, M. A. Haller, H. Ma, L. R. Dalton, S. H. Jang and A. K. Y. Jen, *Adv. Mater. (Weinheim, Ger.)*, 2003, **15**, 603.
121. J. Garcia-Amoros, S. Swaminathan, S. Sortino and F. M. Raymo, *Chem.–Eur. J.*, 2014, **20**, 10276.
122. M. Zajac, P. Hrobárik, P. Magdolen, P. Foltínová and P. Zahradník, *Tetrahedron*, 2008, **64**, 10605.
123. X. Guegano, A. L. Kanibolotsky, C. Blum, S. F. Mertens, S. X. Liu, A. Neels, H. Hagemann, P. J. Skabara, S. Leutwyler, T. Wandlowski, A. Hauser and S. Decurtins, *Chem.–Eur. J.*, 2009, **15**, 63.

Polymerization

7.1 INTRODUCTION

Many π-conjugated materials for organic electronic and photonic applications are polymers, which due to their mechanical strength, elasticity, and solution processability may be used for printed and flexible devices and modules.[1] Most of these polymers are π-conjugated through the backbone by linking aryl moieties through carbon–carbon single bonds, alkenes, or alkynes. Some polymers with non-conjugated backbones with π-conjugated moieties as pendant side-chains have been synthesized for hole- and electron-transporting host polymers for OLEDs,[2,3] hole-transporting materials,[4] electron-transporting polymers,[5] and nonlinear optical polymers.[6,7] However, this chapter presents the synthesis of main-chain π-conjugated polymers given their prevalence in several research and application areas.[8,9] Additionally, many of the reactions that have been used to form main-chain π-conjugated polymers have been discussed in previous chapters and include the Wittig and Horner–Wadsworth–Emmons (HWE) reactions, Knoevenagel condensations, and, in particular, the organometallic coupling reactions presented in Chapter 5. Since much of the terminology, mechanism, and general synthetic issues for these reactions have already been presented, this chapter focuses on polymerization conditions that have been optimized for certain substrates and reaction types. *Emphasis on optimized conditions for polymerization reactions are critical to the synthesis of π-conjugated polymers*, as will be discussed in Section 7.2 below.

Synthetic Methods in Organic Electronic and Photonic Materials: A Practical Guide
By Timothy C. Parker and Seth R. Marder
© Timothy C. Parker and Seth R. Marder, 2015
Published by the Royal Society of Chemistry, www.rsc.org

This chapter focuses *concisely* on the modern synthetic aspects of the main classes of π-conjugated polymers, which is supported by references having conditions and substrate variations. More in-depth treatments of π-conjugated polymer design and synthesis for nearly all the current classes of conducting and semiconducting polymers have been published and are highly recommended.[8,9] Additionally, a number of books are available on the general synthesis and characterization of polymer molecules, including Boyd and Phillips[10] and Odian.[11] In this chapter, a working knowledge of much of the terminology and the measures that define a polymer molecule and a polymer sample are assumed. Throughout this chapter, the *number average molecular weight* (M_n) of a polymer sample is used as a measure of a "successful" polymerization (*i.e.*, the degree of polymerization) since the more monomers that are incorporated into longer polymer chains results in higher M_n. Moreover, physical quantities/qualities of polymers that are important to the general processability, flexibility, and stability of organic electronic and photonic devices—such as the polymer's melting point, stiffness, and fracture stress—typically increase with increasing M_n.[12] Additionally, in organic electronics, charge carrier mobilities can increase[13,14] or decrease[15] by up to 10^4 as M_n increases depending on the device and polymer studied. It should be noted that other measures of a polymer sample including the *weight average molecular weight* (M_W) and the *polydispersity index* (PDI = M_W/M_n) may be more important in certain situations and the reader should be familiar with them as well.[12]

7.2 CRITICAL ISSUES IN POLYMERIZATION

Most π-conjugated polymers are synthesized *via* step polymerization. Scheme 7.1 illustrates a prototypical step polymerization between two monomers (S7.1a, S7.1b) that have complementary reactive functional groups, such as aryl ditins and aryl dihalides, aryl diboronates and aryl dihalides, or aryl dialdehydes and aryl diphosphonates. The first step is reaction of the monomers to give a dimer (S7.1c) followed by a "second step" of reaction of the dimer with both another dimer to give a tetramer (S7.1d) and with either of the monomers to give trimers (S7.1e, S7.1f). A "third step" reacts the dimer (S7.1g) and trimers (S7.1h, S7.1i) with other monomers, dimers, or trimers to provide a mixture of growing polymer chains up to an octamer. Further steps results in a distribution of polymer chains with different molecular weights having an even (S7.1j) and odd number of monomers (S7.1k, S7.1l). If the degree of polymerization is sufficiently high, the polymer chains with even and odds numbers likely have substantially the same

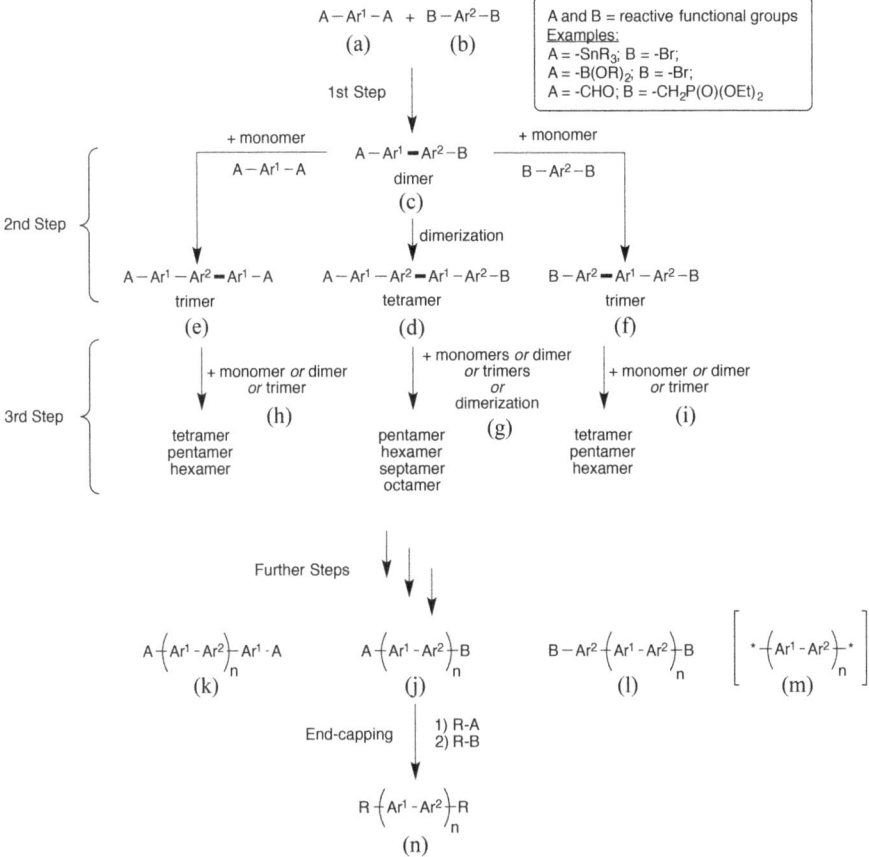

Scheme 7.1 General illustration of a step-growth polymerization process including optional end-capping.

properties so the mixture may be abbreviated as a polymer with exact monomer balance with asterisks denoting a mixture of end groups (S7.1m). End-capping may be accomplished by quenching the polymerization in a two-step sequential process with reactants that each have only one functionality to give an end-capped polymer (S7.1n). Such end-capping may be important since end groups may affect polymer electronic or photonic properties.[16–23] Note that as the condensation reaction begins and proceeds through intermediate completion, there is generally a much higher concentration of monomers and shorter oligomers.[12] Because of this, and since longer polymer chains are formed from reaction of longer oligomers with each other, achieving a high degree of polymerization and hence high M_n generally occurs only toward the end of the polymerization.

In 1936, Carothers formulated that the dependence of polymer chain length in a step polymerization depends on the extent of the reaction as defined by the overall consumption of reactive groups.[24] The Carothers equation (Figure 7.1a) relates the fraction of functional groups remaining in the reaction (*i.e.*, the extent of the reaction) to

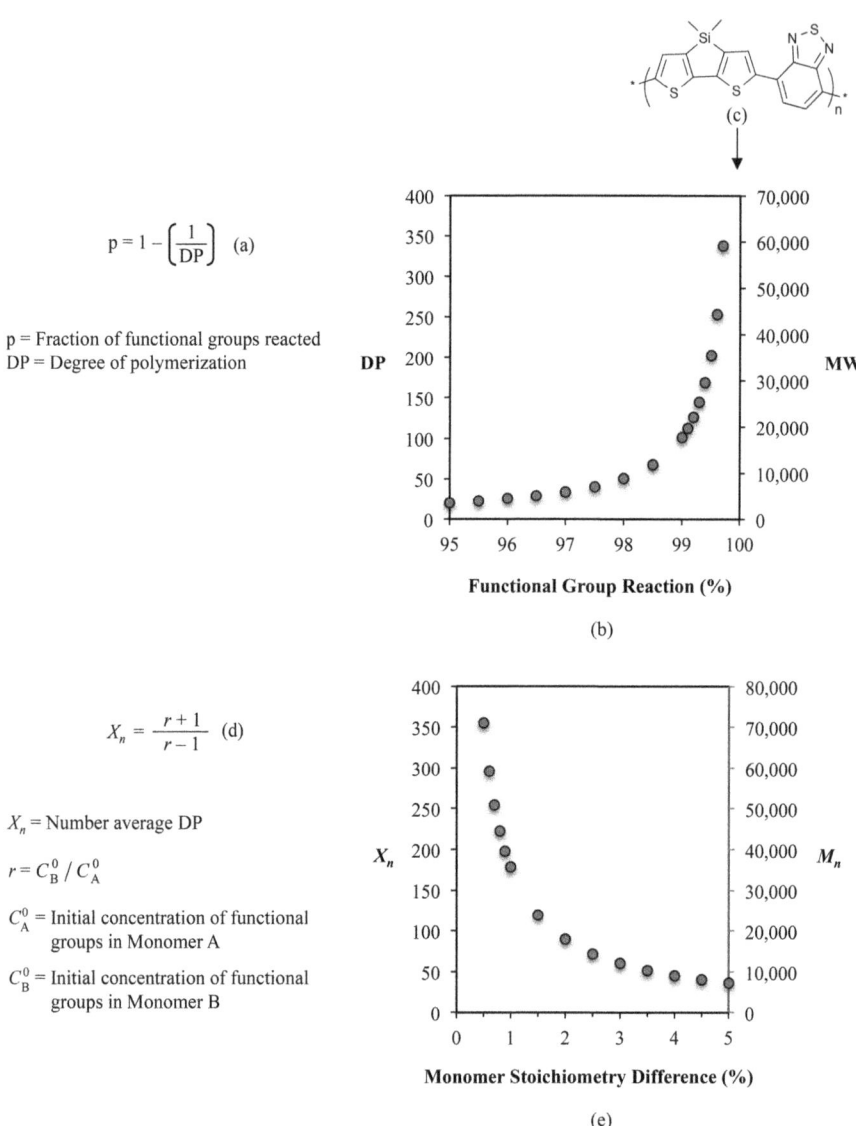

$$p = 1 - \left(\frac{1}{DP}\right) \quad \text{(a)}$$

p = Fraction of functional groups reacted
DP = Degree of polymerization

$$X_n = \frac{r+1}{r-1} \quad \text{(d)}$$

X_n = Number average DP

$r = C_B^0 / C_A^0$

C_A^0 = Initial concentration of functional groups in Monomer A

C_B^0 = Initial concentration of functional groups in Monomer B

Figure 7.1 Graphical representation of the Carothers equation and dependence of step polymerization on stoichiometry balance: (a) the Carothers equation; (b) graph of Carothers equation; (c) DTS-BT polymer structure; (d) A-B polymerization stoichiometry equation; and (e) graph of stoichiometry equation.

the degree of polymerization (DP). The DP obtained from the Carothers equation as a function of the percentage of functional groups reacted is presented graphically in Figure 7.1b. Note that to achieve a DP of 100 monomers in a polymer chain, at least 99% of the functional groups need to react (*i.e.*, the *conversion* of the reaction is 99%). The MW of the polymer chain for a given DP is shown on the right axis of Figure 7.1b for a prototypical donor–acceptor polymer (F7.1c). Another important aspect of step polymerizations is the *balance* of monomers and hence the balance of reactive functional groups. For a step polymerization of two monomers, Monomer A and Monomer B, each with two complementary reactive functional groups, eqn (F7.1d) relates the *number average degree of polymerization* (X_n) to the initial concentration of functional groups when all the functional groups of Monomer A are consumed, which is when the polymerization stops.[12] Eqn (F7.1d) is represented graphically in Figure 7.1e with the prototypical M_n for polymer F7.1c on the right axis.

Graphs F7.1b and F7.1e help illustrate two *critical* issues in any step polymerization:

(1) The reaction chosen to polymerize the monomers should be highly efficient with respect to conversion of the starting material/s into products. A reaction that would proceed with 99+% conversion to the specific product in a simple dimerization of two starting materials is often a reasonable *starting point* for developing a polymerization reaction. Additionally, reactions with by-products or side-products that might retard high conversion should be avoided. Reagents and solvents typically should be of the highest practical purity and extra care should be taken in steps such as degassing. Note that the conversion efficiency of a reaction is different from the isolated yield of the reaction, since product in the reaction can be lost during workup and purification. Thus, a reaction with an isolated yield of 70% may still be adequate for polymerization if an analytical technique such as gas chromatography shows a 99+% conversion to the product;

(2) The monomers ratio must be closely balanced (1.00 equivalent : 1.00 equivalent) to achieve a high M_n, which *requires careful and accurate weighing and dispensing of monomer materials*. Practically, having monomers that are highly pure is a necessary starting point, which may require multiple monomer purifications with crystallization and/or high performance liquid chromatography to purity *at least* within the American Chemical Society guidelines for combustion analysis (C, H, N within 0.4%). In some cases, a monomer may be difficult

to purify to the ideal extent, particularly if the monomer is a viscous oil that retains solvent or is unstable to purification conditions; in these cases, establishing the actual monomer content of the mixture may be possible or varying the *presumed* equivalents of a certain monomer over several reactions may be necessary to find a correct amount for an optimized polymerization, which in effect is a "stoichiometric correction"[25] (*e.g.*, 1.03 equivalents "excess" of a monomer of 97% purity may actually be 1.00 equivalents in the reaction). Such a stoichiometric correction typically requires that the impurity does not degrade the conversion efficiency of the reaction and thereby limit achievable M_n even though the corrected monomer ratio may be ideal.

It should be noted that the functional group conversion efficiency and monomer ratio balance are merely two of many important issues that affect the outcome of a step polymerization. Other issues include differential reactivity of the functional groups on one or both monomers compared to the reactivity of the functional groups on the growing polymer chain, which can affect reaction times, degree of polymerization, and PDI of the polymers.[26] Differences in reactivities between functional groups on the polymer chain and the monomers may result from differences in steric availability and/or differences in electronic effects.

Taken as a whole, all these factors often necessitate optimization of polymerization conditions including monomer purity and stoichiometric accuracy, catalyst efficiency and loading, solubility of all reagents and growing polymer chains, additives including phase transfer catalysts, temperatures and temperature gradients, concentrations, *etc.* The optimization processes may require many reaction variations to increase M_n by an order-of-magnitude or more. Many of the polymerizations reported in π-conjugated polymer literature may not have been optimized to an exhaustive extent; thus, caution should be exercised when comparing polymerization conditions since a report of Conditions A for a reaction and an M_n of ~1000 may not have been optimized whereas another report of Conditions B for a similar reaction and an M_n of ~10 000 may have been serendipitously optimized, and thoroughly optimizing Conditions A may ultimately result in greater M_n than Conditions B. Fundamentally, step polymerizations often require much more meticulous technique, attention to detail, and thorough reaction exploration than the synthesis of small molecules or monomers.

Some π-conjugated polymers are synthesized *via* chain growth polymerization. One advantage to chain growth polymerization is that the M_n is built up more rapidly since monomers react with the end of the growing polymer chain instead of the (relatively) slow process of gradual M_n buildup through lower oligomers in step polymerization. One disadvantage is that most often the monomers are identical or nearly so such that donor–acceptor type polymers are difficult to synthesize. Polythiophenes have been synthesized *via* chain growth polymerization and one in some cases the polymerization is of a living nature, *i.e.*, there are very few chain-transfer and interchain terminations during the polymerization. Living polymerization often results in a relatively narrow distribution of polymer chain lengths in the sample (*i.e.*, a narrow PDI).[27–29] Indeed, controlled chain-growth polymerization of π-conjugated polymers is an active and important field of research.[30]

Throughout this chapter, yields are reported for a sample of the polymer with a specific M_n. Most often, these values are in the range of 30–60% and should not be confused with the overall functional group reaction conversion as referenced in Figure 7.1b. It is often the case that both the highest molecular weight polymers and lower molecular weight polymers and oligomers are lost or excluded during purification by precipitation/filtration, Soxhlet extraction, slurry filtration, or size exclusion chromatography. In some examples in this chapter, more than one fraction of a particular polymer is referenced along with the percentage yield of each fraction.

7.3   ALKENE- AND ALKYNE-LINKED POLYMERS

Alkene- and alkyne-linked polymers have been synthesized by a variety of the methods that have been presented in this book including Wittig and Horner–Wadsworth–Emmons (HWE) coupling (Section 4.3); Knoevenagel condensation (Section 6.2); and the Heck (Section 5.4.6), Stille (Section 5.4.2), and Sonogashira (Section 5.4.3) couplings, among others. The effects of the alkene, alkyne, and direct single bond linkages on properties such as ionization energy (IE), electron affinity (EA), absorption wavelength (λ_{max}), *etc.* in polymers are similar to those seen in small molecules (Section 2.3.4); however, twisting of the polymer backbone due to steric torsions between the aryl subunits (Figure 2.15) or aryl subunits and alkenes can result in greater solubility than for analogous polymers with alkyne linkages or polymers that have little steric torsional twisting of the backbone and thus are substantially linear and planar.

7.3.1 Wittig and Horner–Wadsworth–Emmons Polymerization

The Wittig reaction has been used in many examples for the synthesis of aryl vinylene homopolymers and copolymers. Examples are illustrated in Scheme 7.2 from the common Wittig salt **1**. Polymerization of **1** with dialdehyde **2** using common Wittig conditions of an alkoxide base in an alcohol solvent gives homopolymer **3** (poly[2-methoxy-5-(2-ethylhexyloxy)-1,4-phenylenevinylene], or "MEH-PPV") with a M_n = 10 200 in 44% yield.[31] Similar conditions for the polymerization of **1** with fluorene dialdehyde **4** give the corresponding copolymer **5** with a M_n = 6600 in 31% yield.[32] Another example, using THF as the solvent, and biphenyl dialdehyde **6** gives the corresponding copolymer **7** with a M_n ~ 3900 in 52% yield.[32] It should be noted that in the Wittig synthesis of aryl vinylene polymers, *cis* double bonds may occur along the polymer backbone as a "defect" that generally

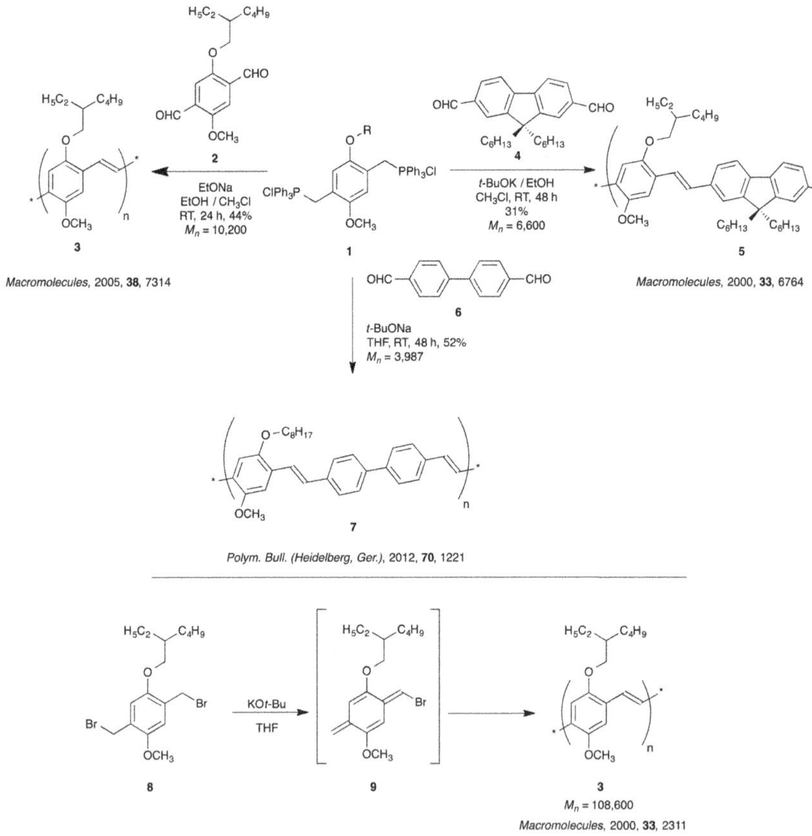

Scheme 7.2 Wittig polymerization and the Gilch route to poly(*p*-phenylenevinylenes).[31,32,36,102]

reduces π-conjugation and may limit device performance and lifetime.[33] Another issue with Wittig polymerization is that the triphenylphosphine oxide by-product may be difficult to remove and require lengthy solvent extraction procedures and/or repeated precipitation. Also, note that the synthesis of MEH-PPV, a polymer studied widely as being a relatively easily processible PPV polymer,[34] may be achieved by a variety of methods[35] including the Gilch route.[33] An example of the Gilch route is shown in the Scheme 7.2 and involves the polymerization of a 1,4-dibromomethyl monomer **8** to give the MEH-PPV polymer **3** with a $M_n \sim 108\,000$,[36] which is approximately an order-of-magnitude higher M_n than the M_n reported in the Wittig synthesis of MEH-PPV from **1** and **2** in Scheme 7.2. The actual monomer of the Gilch polymerization appears to be quinodimethane **9**,[33] generated by 1,6-dehydrohalogenation of **8**. However, one drawback to the Gilch and similar routes is that achieving regular *copolymers* is difficult. Additionally, Gilch-route PPV polymers may have defects, in addition to *cis* defects seen in Wittig reactions, that may reduce device efficiency and lifetime, for example, in the case of OLEDs.[33]

In Scheme 7.2, the Wittig reagent from **1** would be relatively reactive with the presence of the electron donating alkoxy substituents, and the dialdehydes are either relatively less reactive due to the presence of electron donating groups (**2**) or are relatively neither activated nor deactivated (**4** and **6**). Scheme 7.3 illustrates examples of Wittig polymerizations where the reactant pairs have different reactivity

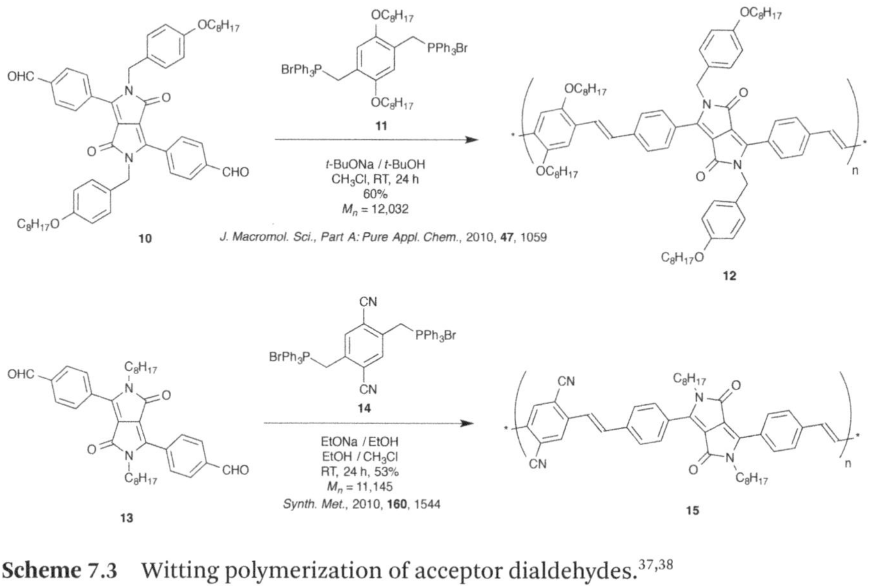

Scheme 7.3 Witting polymerization of acceptor dialdehydes.[37,38]

than the examples in Scheme 7.2. The first example is the reaction of a relatively nucleophilic Wittig reagent generated from Wittig salt **11** with relatively electron accepting—and more reactive—diketopyrrolopyrrole (DPP) dialdehyde **10** to give the corresponding donor–acceptor (D–A) copolymer **12** with $M_n \sim 12\,000$ in 60% yield.[37] Similar conditions with electron accepting DPP monomer **13** and the relatively less reactive Wittig reagent generated from Wittig salt **14**, which has electron withdrawing cyano groups on the phenyl core, give the corresponding copolymer **15** with an $M_n \sim 11\,000$ in 53% yield.[38] Thus, although Wittig polymerizations may give *cis* defects and may be relatively difficult to purify, the method has certain utility due to the wide range of monomers that can be used to synthesize both homopolymers and copolymers.

As mentioned in Section 4.3.2, the Horner–Wadsworth–Emmons (HWE) reaction has certain advantages over the Wittig reaction, including higher nucleophilicity and reactivity of the phosphonate reagent, large reduction or practical elimination of *cis* isomers, and easier purification of the final product from the water-soluble dialkylphosphate by-product. Examples of HWE polymerization are shown in Scheme 7.4. The first example polymerizes diphosphonate **16** with dialdehyde **2** to give MEH-PPV **3** with an $M_n = 9100$ and in a 66% yield.[39] Note that, compared to the comparable Wittig reaction of **1** with **2** (Scheme 7.2), the HWE polymerization is completed in 5 h instead of 24 h, largely due to the generally higher reactivity of HWE reagents. Another example couples diphosphonate **16** to the mixed donor–acceptor–donor (D–A–D) dialdehyde **17** to give the corresponding copolymer **18** with an $M_n = 6400$ and in 72% yield.[40] Another example with a mixed D–A–D dialdehyde is the coupling of **19** with the relatively nucleophilic and reactive thiophene diphosphonate **20** to give the corresponding copolymer **21** with an $M_n = 13\,300$ in a yield of 50% after 3.5 h of reflux.[41] Finally, coupling of the electron accepting, less reactive diphosphonate **22** with the electron donating, less reactive aldehyde **23** gives the corresponding D–A copolymer **24** with an $M_n \sim 3400$ in relatively high yield, albeit with reflux and a relatively long reaction time of 24 h for a HWE coupling.[42]

7.3.2 Knoevenagel Polymerization

Scheme 7.5 illustrates a general Knoevenagel condensation polymerization with an aryl monomer having at least two methylene groups substituted with electron-withdrawing groups (EWG) (S7.5a) and an aryl dialdehyde monomer (S7.5b) that react under basic and/or catalytic conditions to give the corresponding vinylene polymer (S7.5c).

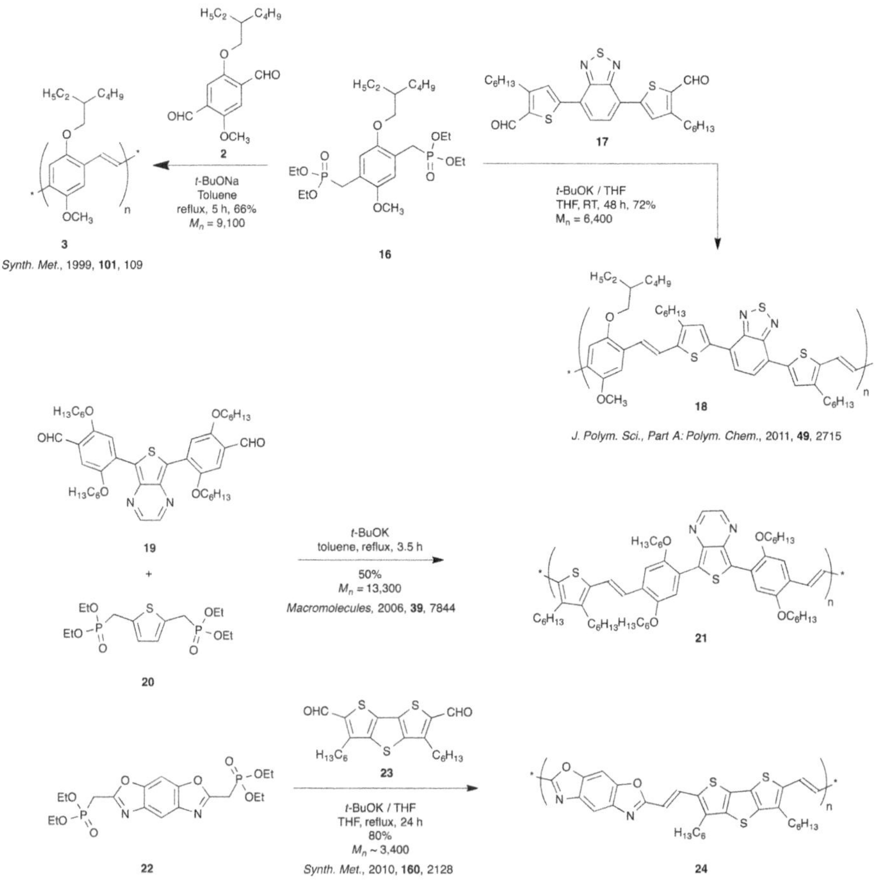

Scheme 7.4 Horner–Wadsworth–Emmons (HWE) synthesis of donor-acceptor polymers.[39–42]

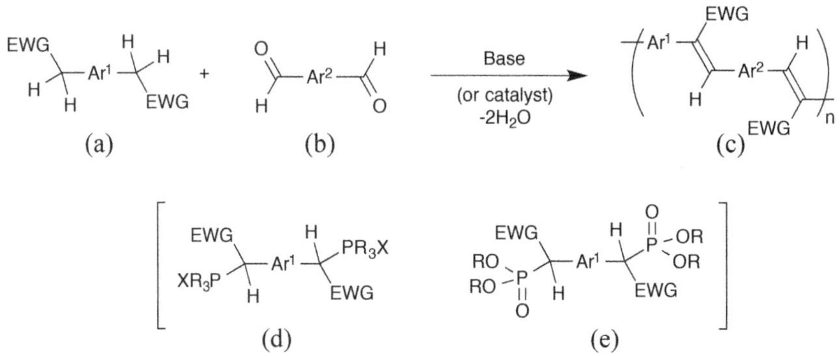

Scheme 7.5 General illustration of Knoevenagel condensation polymerization.

Many synthetic applications of the Knoevenagel condensation were discussed in Section 6.2. One advantage of Knoevenagel polymerization is that an electron-withdrawing group generally is installed on the alkene as a result of the reaction, which is another site for functionalization and property modification in addition to the aryl core of the monomers. Indeed, Knoevenagel polymerization is perhaps the preferred choice of reaction for installation of an electron-withdrawing group on the alkene linkage; especially as compared to the Witting or HWE condensation (Section 7.3.2 above), which would require a Wittig salt (S7.5d) or phosphonate (S7.5e) substituted with an electron-withdrawing group that would generally reduce the reactivity of the reagent significantly. One limitation in a typical Knoevenagel condensation polymerization is that, since the carbon atom bearing the two acidic protons must be substituted with the aryl group, there is only *one* electron-withdrawing group bonded to the active methylene group that affects acidity, which generally results in stronger bases/catalysts being used in a Knoevenagel polymerization than might be used in a Knoevenagel condensation with active methylene compounds having two electron-withdrawing groups. Because of this lower acidity, many known Knoevenagel polymerizations use relatively strong bases, and examples are shown in Table 7.1 for di(cyanomethyl) aryl monomers and aryl dialdehydes. Note that most of the examples use either KOt-Bu (Entry 1[43] and Entry 3[44]) or Bu$_4$NOH (Entry 2[45] and Entry 4[46]) as strong bases, with Entry 5 employing a C−H activation catalysis with a Ru(II) catalyst under relatively mild conditions.[47] Also note that, in the examples in Table 7.1, M_n ranges from ~37 000 to 4000 and yields range from moderate to high (with the exception of Entry 4).

7.3.3 Heck Polymerization

The Heck reaction has been used to synthesize a large variety of PPV polymers carrying functional groups as side-chains or in the conjugated backbone including nonlinear optical chromophores, metalloporphyrins, Ru(bpy)$_3$ and other complexes, and N-hydroxysuccinimide (NHS) esters.[48] Often the optimized conditions include Pd(OAc)$_2$/ P(o-tol)$_3$ and triethylamine and/or DMF as the solvent, which are generally high yielding conditions for the Heck reaction, as discussed in Section 5.4.6. In some cases, the solvent ratio may need to be adjusted to favor solubility of the longer polymer chains. Potential drawbacks with Heck polymerization is that divinyl aromatic monomers may be unstable and that strong electron donor divinyl aromatic monomers may result in cross-conjugated 1,1-coupled defects (Scheme 5.26)

Table 7.1 Synthesis of polymers from various monomers *via* the Knoevenagel condensation.

Entry	Ar^1	Ar^2	Conditions	Reference
1	(OC$_{12}$H$_{25}$, OC$_{12}$H$_{25}$ substituted benzene)	(thiophene/dioxepine, H$_{13}$C$_6$ C$_6$H$_{13}$)	t-BuOK/t-BuOH/ THF, 70 °C, 4 h, 67% $M_n = 17\,400$	43
2	(fluorene, H$_{13}$C$_6$ C$_6$H$_{13}$)	(biscarbazole, (H$_2$C)$_6$ (CH$_2$)$_6$, fluorene)	Bu$_4$N$^+$OH/THF/ MeOH, reflux, 12 h, 81% $M_n = 37\,000$	45
3	(carbazole, C$_2$H$_5$ C$_4$H$_9$)	(OC$_8$H$_{17}$, OC$_8$H$_{17}$ substituted benzene)	t-BuOK/t-BuOH/ THF, RT, 24 h, 68% $M_n = 7900$	44
4	(carbazole, H$_{17}$C$_8$ C$_8$H$_{17}$)	(benzothiadiazole bithiophene)	Bu$_4$N$^+$OH/THF/ MeOH, 70 °C, 2 days, 12% $M_n = 4000$	103
5	(benzene)	(OC$_8$H$_{17}$, OC$_8$H$_{17}$ substituted benzene)	RuH$_2$(PPh$_3$)$_2$ (3 mol%), dppe (6 mol%) THF, 60 °C, 30 min, 90% $M_n = \sim 7000$	47

in the polymer backbone. Examples of the Heck polymerization are shown in Scheme 7.6. Divinyl fluorene monomer **25** may be coupled with the electron accepting, activated dibromo oxadiazole monomer **26** to give the corresponding donor–acceptor (D–A) polymer **27** with an $M_n = 11\,300$ in 57% yield while reaction under nearly identical conditions with electron donor, deactivated dibromo triarylamine **28** gives the corresponding copolymer **29** with an $M_n = 7500$ in 49% yield.[49] A similar divinyl fluorene monomer **30** was polymerized with donor–acceptor–donor (D–A–D) dibromo monomer **31** to give the corresponding D–A polymer **32** with an $M_n = 6600$ in ~64% yield.[50] Finally, the electron donating divinyl monomer **33** was coupled with

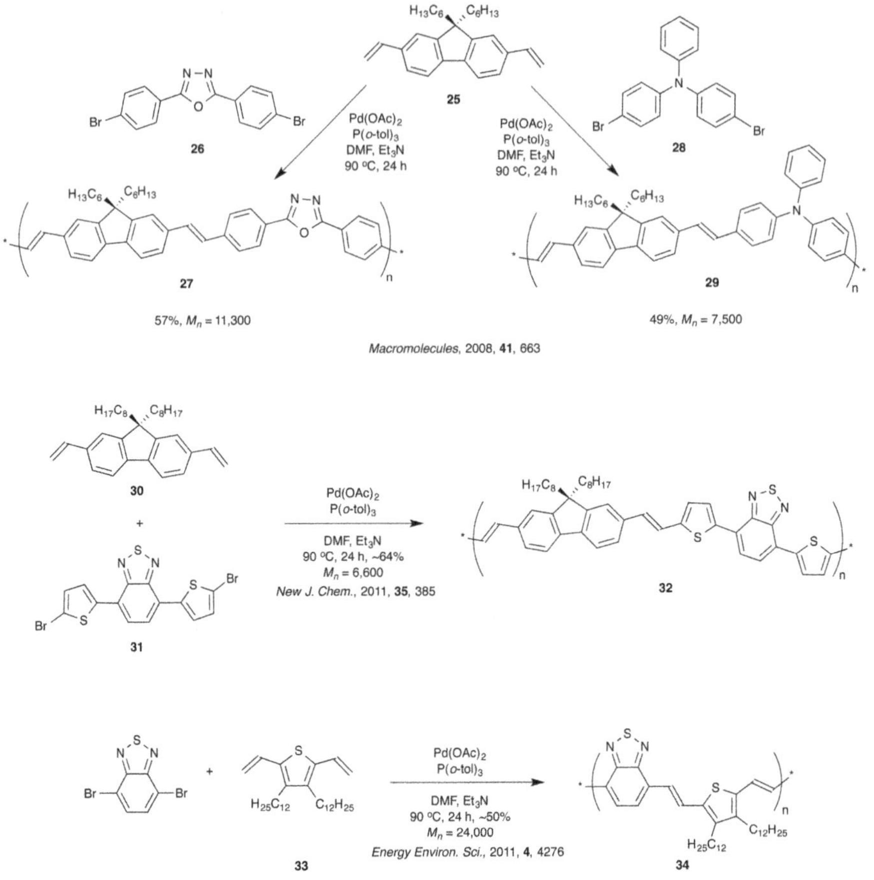

Scheme 7.6 Heck polymerization examples.[49-51]

4,7-benzothiadiazole to give D–A copolymer **34** with an $M_n = 24\,000$ in ~50% yield.[51] In the case of donating monomer **33**, no 1,1-coupled isomeric defects were reported in the backbone, although in many cases, 1,1-coupled defects can be difficult to detect. Note that all the conditions are identical to a first order of approximation, so the Pd(OAc)$_2$/P(o-tol)$_3$/NEt$_3$/DMF conditions may be considered a good starting point in Heck polymerization.

7.3.4 Stille Vinyl and Alkyne Polymerization

The Stille coupling has been used for the synthesis of a large array of π-conjugated polymers, but the vast majority are direct aryl–aryl polymers (Section 7.4.1 below) and not vinyl and alkyne linked polymers.[52] This may be because alkyne linked polymers are more easily prepared by the Sonogashira coupling (Section 7.3.5 below) and vinyl tin reagents may be more challenging to prepare[53] compared to corresponding divinyl aromatic monomers that are used in Heck polymerization. However, a few specialized Stille reagents have been utilized in the synthesis of homopolymers and examples are illustrated in Scheme 7.7. Electron donor dithienothiophene (DTT) **35** may be coupled with vinyl distannane **36** to give homopolymer **37** with an $M_n \sim 21\,000$ in high yield.[54] Note the use of aqueous KF as the last step in the reaction, which may decompose the tributyltin bromide by-product and, if reacted for an extended period of time, the vinyl distannane **36** to water-soluble products. DTT **35** may also be reacted with an analogous alkyne distannane **38** to give the corresponding homopolymer **39** with an $M_n \sim 60\,000$ and in 90% yield.[54] Distannanes **36** and **38** also have been utilized in the synthesis of the electron acceptor containing polymer such as the polymerization **36** with electron accepting perylene diimide (PDI) dibromide **40** to give homopolymer **41**[55] and the polymerization of **38** with **42** to give homopolymer **43**.[56]

7.3.5 Sonogashira Polymerization

The Sonogashira coupling has been used for the synthesis of π-conjugated copolymers having alkyne links between different aryl monomers. Many of the conditions that can be used in the Sonogashira coupling as well as factors that affect the efficiency of the reaction

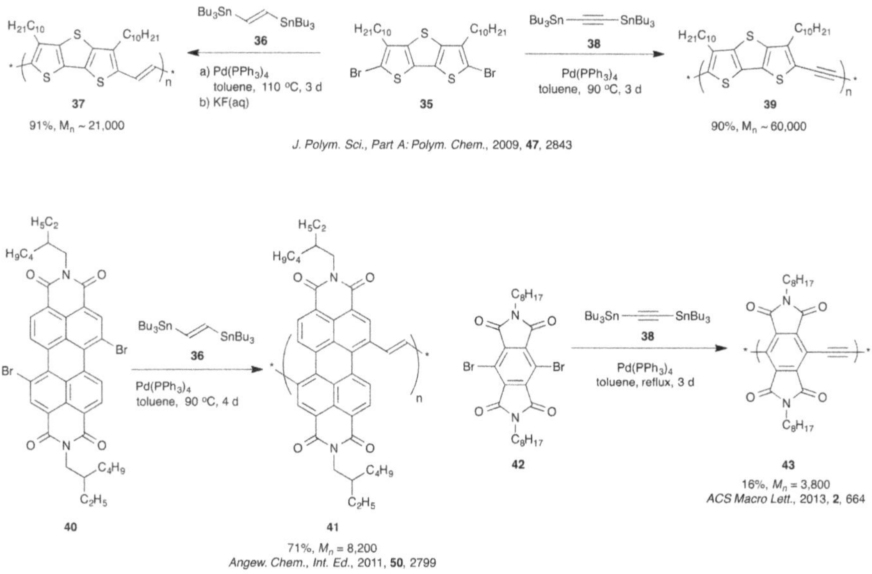

Scheme 7.7 Poly(arylvinylenes) and poly(arylalkynes) from Stille polymerization.[54–56]

were presented in Section 5.4.5. As with the synthesis of small molecule materials, Sonogashira polymerization has been reported using a variety of conditions. Scheme 7.8 provides some examples of Sonogashira polymerization of donor–acceptor (D–A) polymers. The first example includes an electron accepting dialkyne **44** and an electron donor diiodo thiophene **45** that are polymerized under common Sonogashira conditions to provide the corresponding polymer **46** with an $M_n = 52\,000$ in 52% yield.[51] Note that in this case, since the thiophene is deactivated towards Pd insertion, use of a diiodo thiophene is preferred; however, more electron accepting monomers may undergo

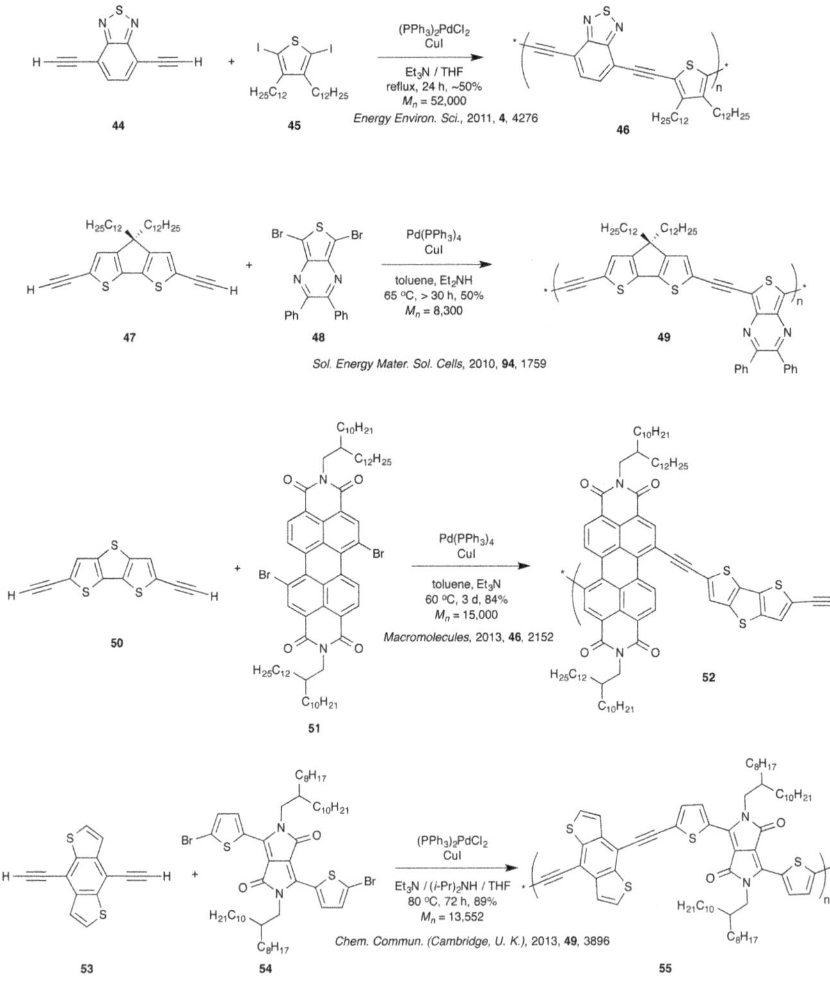

Scheme 7.8 Sonogashira polymerization of various donor and acceptor monomers.[51,57–59]

efficient coupling as dibromides, which is the case for the other examples in Scheme 7.8. These include coupling of donor **47** to acceptor **48** with Pd(0) to give copolymer **49**;[57] coupling of strong dithienothiophene (DTT) donor **50** to strong perylene diimide (PDI) acceptor **51** to give the copolymer **52**;[58] and coupling of donor **53** with acceptor **54** using an amine mixture of NEt_3 and i-Pr_2NH for solubility to give the corresponding copolymer **55**.[59] Note that the yields are moderate to high and the M_n ranges from 8300 to 52 200 for these examples, indicating moderate to moderately high efficiency in the polymerization. Also note that in all the examples either THF or toluene is used as either a co-solvent or the main solvent for the polymerization, which is most likely to maintain solubility of the growing polymer chains.

7.4 ARYL–ARYL POLYMERS

Although many alkene- and alkyne-linked π-conjugated polymers have been synthesized and studied in organic electronics and photonics, in particular poly(*p*-phenylene vinylene) (PPV) and derivatives, aryl–aryl polymers generally show higher thermal and operational stability in a range of electronics and photonics applications.[60] Aryl–aryl polymers largely have been synthesized using Stille (Section 7.4.1), Suzuki (Section 7.4.1), and Kumada (Section 7.4.3) cross-coupling and homo-coupling reactions. More recently, direct arylation (Section 7.4.4) has been used to synthesize both homopolymers and copolymers. Many of the reaction conditions illustrated in this section have been optimized for polymerization, and thus may be somewhat different from the small-molecule organometallic coupling conditions presented in Chapter 5. In particular, many polymerization reactions may employ solvents that solubilize the growing polymer chains and catalysts that work well in those solvents (*e.g.*, toluene, chlorobenzene, dichlorobenzene, *etc.*). As mentioned in Section 7.2, in step-polymerizations, which are operative in nearly all donor–acceptor (D–A) copolymer syntheses, care should be taken to purify monomers and/or properly balance the monomer stoichiometry. Additionally, non-optimized Pd cross-coupling reactions may result in polymers with a varying amount of homocoupled defects that were shown to significantly affect a material's properties.[61]

As mentioned in Section 5.4.1, π-conjugated polymers synthesized with Pd catalysts may be contaminated with Pd nanoparticles, which can affect the electrical or photophysical properties significantly.[62] Moreover, the Pd nanoparticles may be difficult to detect by normal characterization techniques such as ^{1}H NMR, HPLC, and UV/vis spectroscopy.[63]

In these cases, Pd scavengers such as *N,N*-diethylphenylazothiofor-mamide,[64] diethylammonium diethyldithiocarbamate,[65] or silica supported scavengers[63] may be used as part of the reaction workup or as a separate purification step to remove traces of Pd, which has been shown to reduce the concentration of Pd in π-conjugated polymers to less the 0.1 ppm.[63]

7.4.1 Stille Polymerization

The factors that affect Stille couplings are presented in Section 5.4.2 along with many examples of small molecule and monomer synthe-ses. The Stille coupling has been utilized for many polymerizations in the synthesis of π-conjugated polymers; however, the Stille coupling partner is often limited to electron-donating monomers due mostly to the general difficulty in preparing, relative instability of, and slug-gish reactivity of Stille reagents with high electron affinity. Although the low reactivity of acceptor Stille reagents might be tolerated in the synthesis of small molecules, slow monomer conversion rate in step polymerization may result in monomer decomposition or substantial competing side-reactions to an extent sufficient to result in unaccept-ably low M_n in polymers in accordance with the Carothers and stoichi-ometry equations in Figure 7.1. Practically, this means electron donor Stille reagents are used in combination either with electron donor halides or, most often, with electron acceptor halides in the synthesis of π-conjugated polymers.

Scheme 7.9 illustrates several examples of Stille polymerizations from the dithienosilole (DTS) Stille reagent 56. The first example cou-ples DTS distannane 56 to the electron donor DTS dibromide 57 with Pd(II)PPh₃Cl₂ followed by end-capping with tributyltin thiophene and bromothiophene to give the corresponding DTS homopolymer 58 with an $M_n \sim 8200$ in 85% yield.[66] Note that, with the use of a Pd(II) cat-alysts, conversion of the Pd(II) to Pd(0) through homocoupling of the organometallic species (Section 5.4.2, Scheme 5.14) would produce at least a dimer of 56 and hence modify the stoichiometry of monomer 56 and 57 *in situ*, which will potentially reduce the otherwise achiev-able M_n in accordance with Figure 7.2e above. Additionally, in cases of copolymer synthesis with a Pd(II) catalyst source, for example with a donor distannane (R₃Sn–D–SnR₃) and an acceptor dihalide (X–A–X), one could also anticipate at least some dimer defects in the alternating polymer chain (*e.g.*, –D–A–D–A–**D–D**–A–). Homocoupled defects can substantially affect the properties and performance of π-conjugated copolymers.[61] Thus, Pd(II) catalyst sources in Pd(0) polymerizations

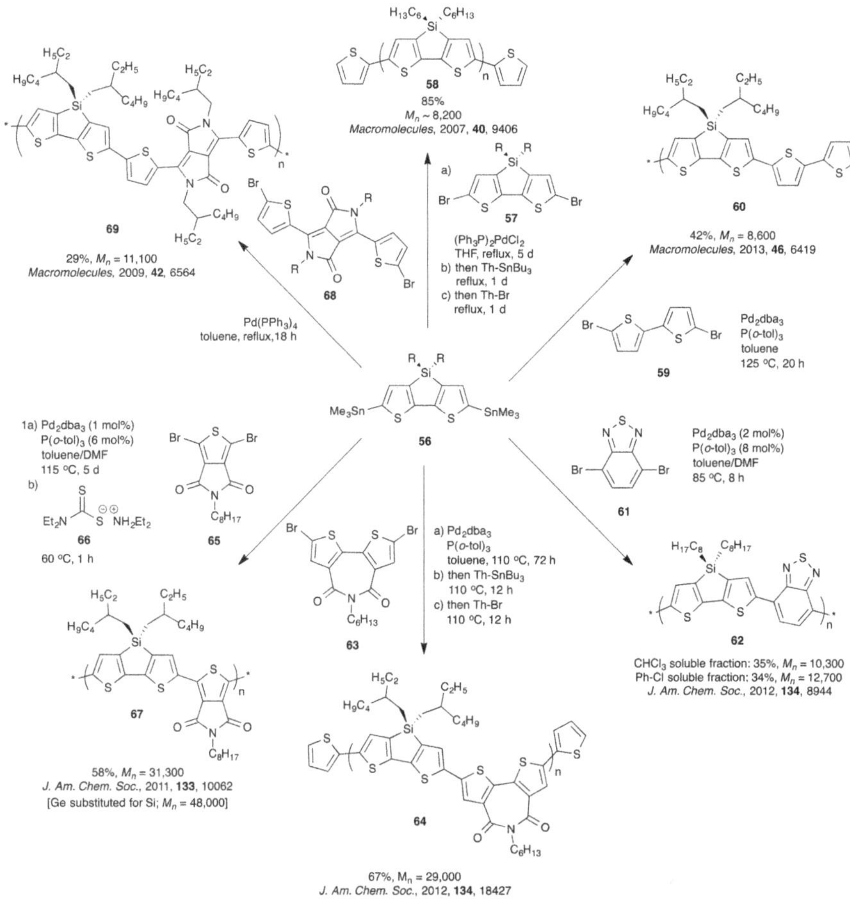

Scheme 7.9 Synthesis of polymers *via* Stille polymerization from an electron donor Stille monomer.[66–71]

should be used sparingly, especially in copolymerizations. Most of the examples in Scheme 7.9 utilize Pd(0) catalyst sources, including Pd_2dba_3, an air-stable source of Pd(0) that is also relatively active since the dba ligands readily dissociate in the reaction. Typical conditions are shown for the polymerization of DTS **56** with donor dibromide **59** using the ligand $P(o\text{-tol})_3$ and toluene as the solvent to give copolymer **60** with M_n = 8600 in 42% yield.[67] Another example of polymerization with **56** illustrates the use of an electron accepting dibromo benzothiadiazole (BT) monomer **61** using DMF as co-solvent to give the corresponding D–A copolymer **62** with two isolated fractions of M_n = 12 700 and M_n = 10 300.[68] End-capping conditions are illustrated with the copolymerization of **56** with dibromo monomer **63** followed

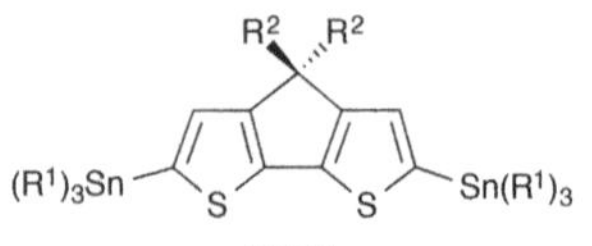

CPDT

R¹ = Me; R² = -C₆H₁₃. *Macromolecules*, 2007, **40**, 1981
R¹ = n-Bu; R² = -C₈H₁₇. *Chem. Mater.*, 2012, **24**, 1434

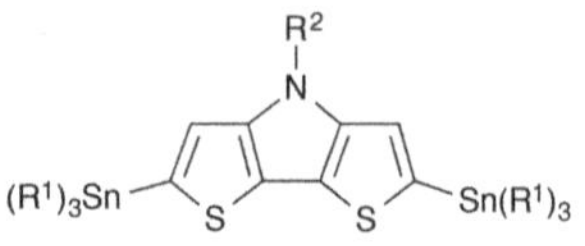

DTP

R¹ = Me; R² = -C₁₂H₂₅. *Macromolecules*, 2008, **41**, 8953
R¹ = n-Bu; R² = -C₁₂H₂₅. *Synth. Met.*, 2010, **160**, 1438

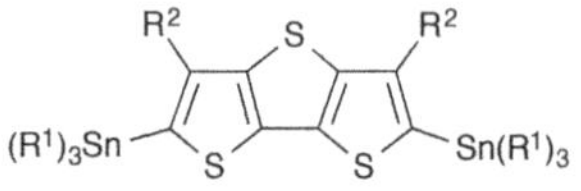

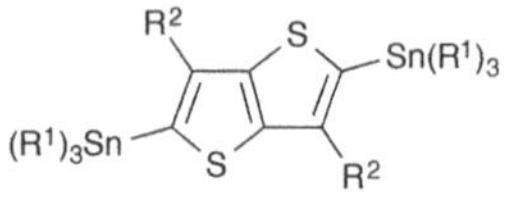

TT

R¹ = Me; R² = -C₁₅H₃₁. *J. Mater. Chem.*, 2009, **19**, 4938
R¹ = n-Bu; R² = -H. *Chem. Mater.*, 2012, **24**, 1434

DTT

R¹ = Me; R² = -H. *Energy Environ. Sci.*, 2012, **5**, 6857 (recrystallized from CH₃CN)
R¹ = n-Bu; R² = -C₁₀H₂₁. *J. Polym. Sci., Part A: Polym. Chem.*, 2009, **47**, 2843

Figure 7.2 Additional electron donor Stille monomers.[54,72–77]

by reaction with tributyltin thiophene and bromothiophene to give copolymer **64** with an M_n = 29 000 in 67% yield.[69] Another example polymerizes **56** with the thienopyrroledione (TPD) acceptor dibromo monomer **65** followed by use of the palladium scavenger **66** to give the corresponding copolymer **67** with M_n ~ 31 000 in 58% yield.[70] A final example in Scheme 7.9 uses Pd(PPh₃)₄ as the Pd(0) source to copolymerize **56** with diketopyrrolopyrrole (DPP) acceptor dibromo monomer **68** in toluene to give copolymer **69** with M_n ~ 11 000 in 29% yield.[71] Although not shown in Scheme 7.9, microwave assistance has been demonstrated to effectively aid the Stille polymerization.[25] Although Scheme 7.9 is limited to examples with the donor monomer DTS, many other Stille donor monomers have been used in the synthesis of π-conjugated polymers, with examples shown in Figure 7.2 for the CPDT,[72,73] DTP,[74,75] DTT,[54,76] and TT[73,77] donors. Note that Stille polymerization is often the preferred method for DTS monomers since the silicon atom may be a source of base and acid sensitivity.

Stille polymerization may also be used to form homopolymers from a single monomer as outlined in Scheme 7.10. Reaction of an aryl dihalide monomer (S7.10a) with one equivalent of (hexaalkyl) ditin and Pd(0) can give a mixture of a haloaryl stannane (S7.10b), an aryl distannane (S7.10c), and unreacted aryl dihalide monomer (S7.10d), which has a 1:1 ratio of aryl halide and aryl stannane functional groups. All of the components of this mixture can enter a step polymerization with the Pd(0) catalyst to give the homopolymer (S7.10e). Examples of this strategy are also illustrated in Scheme 7.10 including

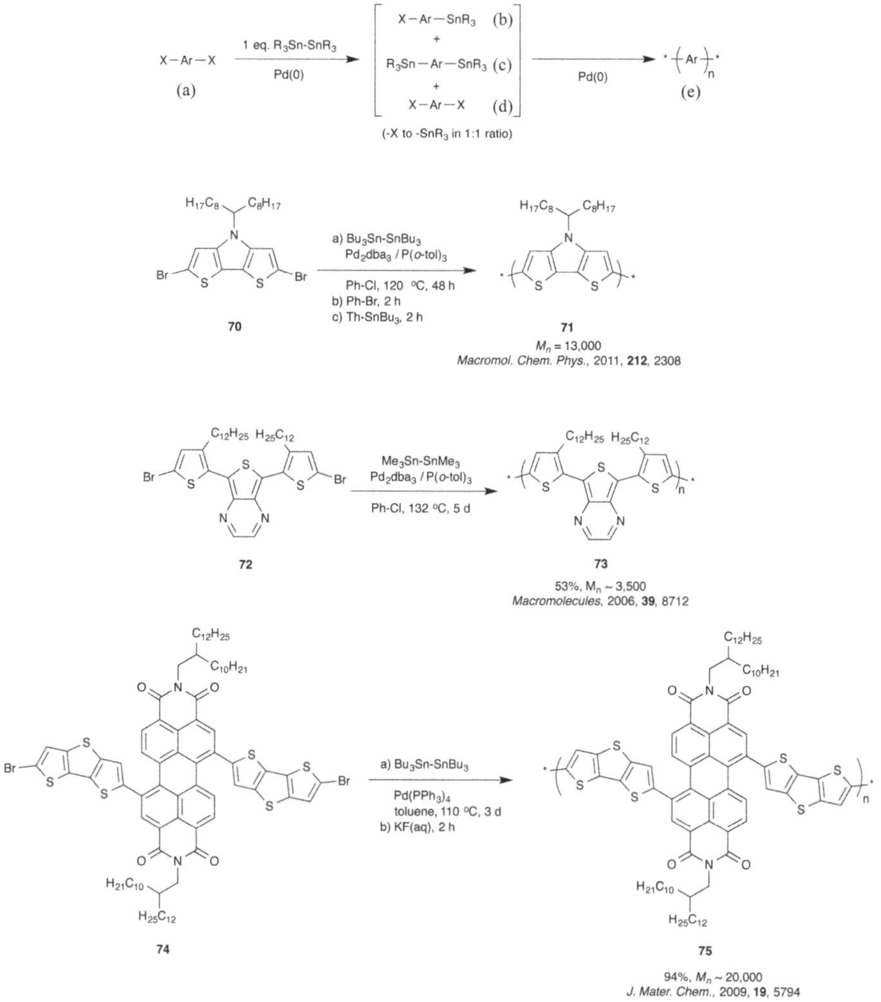

Scheme 7.10 General illustration and examples of Stille homopolymerization from aryl dihalide monomers and hexa(alkyl) ditin reagents.[104–106]

the polymerization of strong electron donating monomer **70** with hexabutyl ditin and Pd$_2$dba$_3$/P(o-tol)$_3$, followed by end-capping *in situ*, to give corresponding homopolymer **71** with an M_n = 13 000. The method has also been applied to the polymerization of donor–acceptor–donor D–A–D monomers such as the polymerization of **72** to **73** with Pd$_2$dba$_3$ ($M_n \sim 3500$ in 53% yield) and **74** to **75** with Pd(PPh$_3$)$_4$ ($M_n \sim 20 000$ in 94% yield). Note that the chlorobenzene in the polymerization of **72** is the solvent and not an end-capping reagent since the reactivity of arylchlorides is generally far lower than arylbromides.

Also note that the aqueous KF added at the end of the reaction in the polymerization of **74** is used to decompose the tributyltin bromide by-product to tributyltin fluoride, although it may also act to remove tributyltin groups from the end of polymer chains in certain cases.

7.4.2 Suzuki Polymerization

The Suzuki coupling has been used for the synthesis of a wide array of π-conjugated polymers. The general conditions and issues that affect Suzuki coupling are presented in Section 5.4.3. Many features of the Stille coupling that are critical for polymerization—in particular the potentially high conversion of starting materials to products, ability to purify the organometal component, and functional group tolerance—also apply to the Suzuki coupling for polymerization. In addition, electron-acceptor Suzuki reagents are often more stable than electron-acceptor Stille reagents and since the boron must be activated by base (Section 5.4.3, Scheme 5.16), electron-acceptor Suzuki reagents often have sufficiently high reactivity for polymerization; because of this, both electron-donor *and* electron-acceptor Suzuki reagents can be utilized in polymerizations, which broadens the scope of the Suzuki polymerization compared to Stille polymerization. In many cases, end capping is employed, most likely from concerns that boronate/boronic acid end groups may deleteriously affect device performance by acting as a charge trap.[16] One potential disadvantage to Suzuki polymerization is that the reaction is often biphasic, and longer polymer chains may precipitate from the organic phase at the organic/aqueous interface. Varying the stirring speed to increase/decrease the organic/aqueous interfacial surface area, decreasing the amount of water, and/or using a phase transfer catalyst may help alleviate early precipitation in some cases. As with most polymerizations, optimization of the reaction conditions is often necessary to increase the monomer conversion to a level that allows a high degree of polymerization.

Scheme 7.11 presents some examples of Suzuki polymerization with a donor diboronate **76** and several different dibromo substrates. In the first example, a carbazole monomer **76** is co-polymerized with electron donor dibromide **77** with Ph(PPh$_3$)$_4$ in toluene/Na$_2$CO$_3$(aq) to give the corresponding homopolymer **78** with M_n = 3400 and M_n = 7800 for toluene-soluble and chloroform-soluble fractions, respectively, in high overall yield.[78] Most of the other examples in Scheme 7.11 use very active and relatively air-stable Pd$_2$dba$_3$ as the source of Pd(0) for the catalyst along with varying base and solvent conditions.

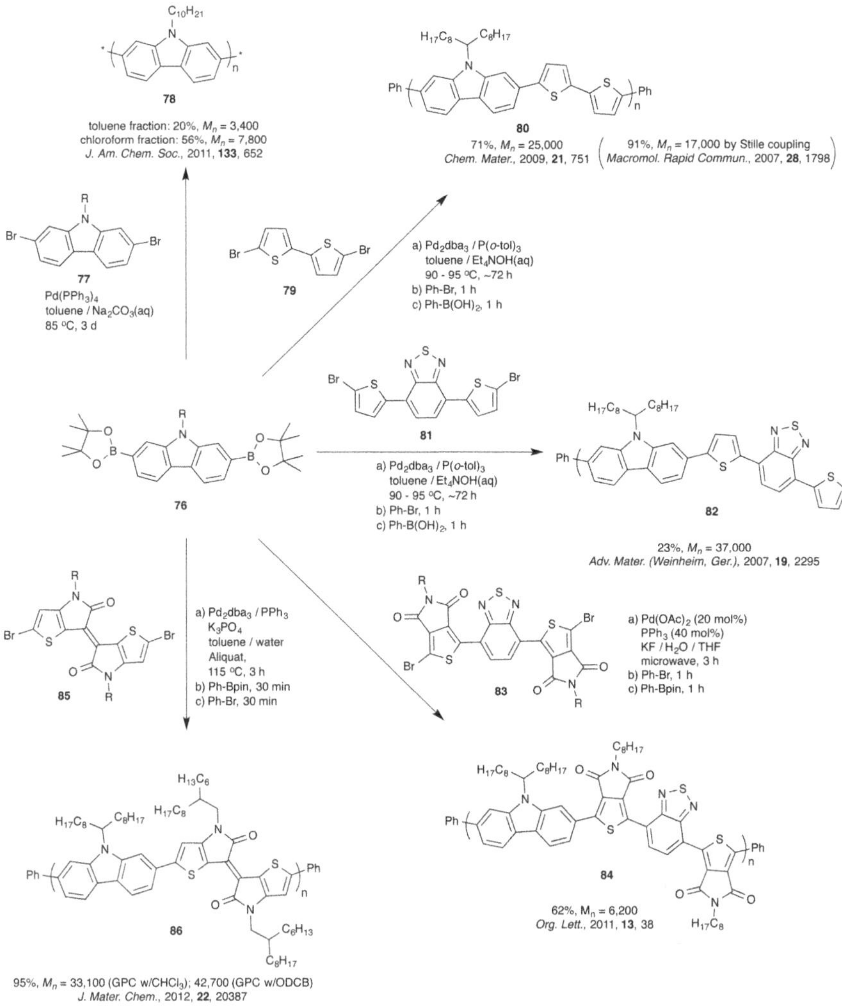

Scheme 7.11 Suzuki polymerization from an electron donor carbazole diboronate.[78-83]

Polymerization of **76** with electron donor bithiophene dibromide **79** using the P(o-tol)$_3$ ligand and the phase transfer base Bu$_4$NOH followed by end-capping with bromobenzene and phenylboronic acid gives corresponding donor copolymer **80** with an M_n = 25 000 in 71% yield,[79] which can be compared to a similar results of M_n = 17 000 and 91% yield for the same polymer *via* a Stille polymerization.[80] Similar to conditions used for the synthesis of **80**, except with the D–A–D monomer **81**, give the corresponding D–A copolymer **82** with an M_n = 37 000 in 23% yield.[81] A more electron accepting monomer **83** may be polymerized with **76** using Pd(OAc)$_2$/PPh$_3$ catalyst activated with KF and performed under microwave irradiation, followed by end-capping, to give corresponding copolymer **84** with M_n = 6200 in 62% yield.[82] Another example of polymerization with **76** utilizes monomer **85**, phase transfer catalyst Aliquat, and K$_3$PO$_4$ as the base followed by end-capping to give copolymer **86** with M_n ~ 33 000 (or M_n ~ 43 000 depending on the solvent) in very high yield.[83] In general, the use of Aliquat as a phase transfer catalyst along with a Pd(0) source are very active conditions for Suzuki polymerization, as discussed below for the examples in Scheme 7.12.

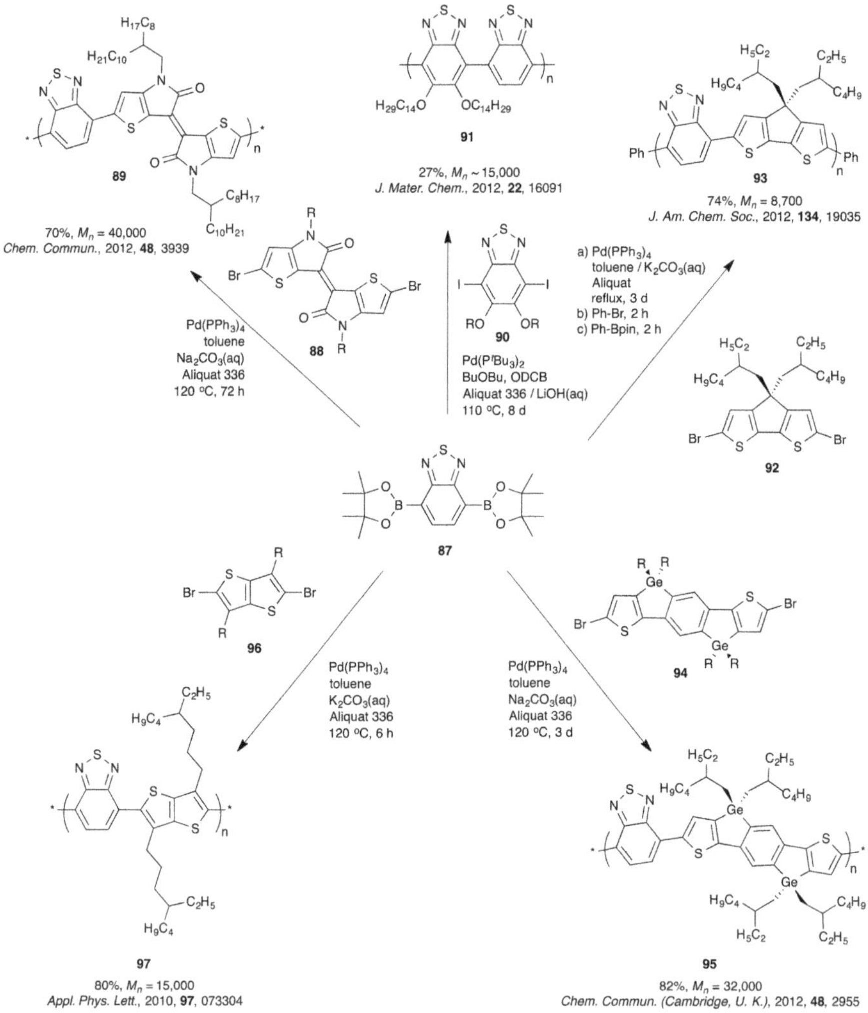

Scheme 7.12 Suzuki polymerization from an electron acceptor benzo-2,1,3-thiadiazole diboronate.[84–86,107,108]

One of the advantages of Suzuki coupling reactions in comparison to Stille coupling reactions for polymerization is that electron acceptor Suzuki reagents are generally more stable and more reactive than the corresponding Stille reagents. Scheme 7.12 illustrates some Suzuki coupling polymerizations with relatively electron-accepting benzothiadiazole (BT) monomer **87**. The first example utilizes an electron-accepting dithienylisoindigo dibromide monomer **88** with $Pd(PPh_3)_4$ and Aliquat 336 to give the corresponding copolymer **89** with $M_n = 40\,000$ in 70% yield.[84] It may be comparatively difficult to synthesize polymer **89** *via* Stille coupling since the Stille reagent would need to be an electron acceptor. Diiodo BT monomer **90** was polymerized with **87** using a very active $Pd(P^tBu_3)_2$ catalyst, Aliquat, and LiOH(aq) in a dibutylether and *o*-dichlorobenzene (ODCB) solvent mixture to give polymer **91** with $M_n \sim 15\,000$ in 27% yield, where less active $Pd(PPh_3)_4$ did not give satisfactory polymerization.[85] The dibutylether was utilized for higher reactivity of the catalyst and *o*-dichlorobenzene for solubility of the polymer. The low reactivity of **90** may be due to the electron-releasing, deactivating alkoxy substituents *ortho* to the iodides and steric hindrance from the bulky C-14 alkyl groups. However, the other examples in Scheme 7.12 were effective with $Pd(PPh_3)_4$ as the catalyst. Polymerization of electron-donor dibromo monomer **92** with **87** with $Pd(PPh_3)_4$ and Aliquat with end-capping gave corresponding copolymer **93** with an $M_n = 8700$ in 74% yield,[86] and electron donor monomers **94** and **96** also gave polymers **95** ($M_n = 32\,000$) and **97** ($M_n = 15\,000$), respectively, in high yields under similar conditions. Although the use of a biphasic system with Suzuki coupling has disadvantages, Suzuki polymerization may often be preferred over Stille polymerization since: (1) a large number of conditions can be modified/optimized—including organic co-solvents, catalysts, bases, and phase transfer catalysts; (2) both electron-donating and electron-accepting organometals potentially can be utilized; and (3) the by-products are less toxic than the Sn-containing by-products of the Stille coupling; however, dithienosilole (DTS) monomers may be challenging to incorporate into Suzuki polymerization since the silicon in DTS compounds is often a source of base sensitivity.

7.4.3 Kumada Polymerization

Kumada coupling was introduced in Section 5.4.4. One of the main uses of Kumada coupling in organic electronics and photonics is the synthesis of π-conjugated polymers *via* Grignard metathesis (GRIM) polymerization.[29] An example of the mechanism underlying GRIM polymerization is outlined in Scheme 7.13 and begins when a Ni(II)

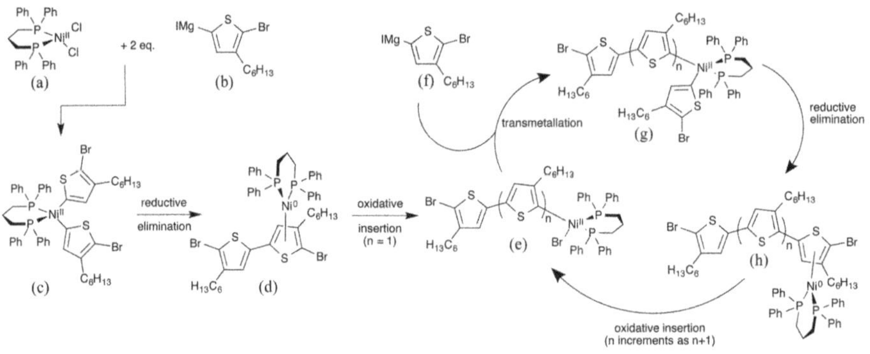

Scheme 7.13 General mechanism for Grignard metathesis (GRIM) living-like polymerization of thiophenes.

complex (S7.13a) is transmetallated with two equivalents of a Grignard monomer (S7.13b, which may be formed by Grignard metathesis) to give a Ni(II) complex (S7.13c) that reductively eliminates to give a dibromo bithiophene with a close, probably bonding, association with the Ni(0) catalyst (S7.13d). The dibromo bithiophene is the initiator of the polymerization that begins when the associated Ni(0) catalyst inserts into the adjacent carbon–bromine bond to give a Ni(II) complex (S7.13e) that can enter a catalytic cycle by transmetallating with another equivalent of Grignard monomer (S7.13f) to give a complex S7.13g that undergoes reductive elimination to give the Ni(0) associated complex on the growing chain (S7.13h). Oxidative insertion of the Ni(0) complex into the adjacent carbon–bromine bond produces a complex ready to enter into another catalytic cycle. Since the Ni catalyst remains associated with the end of the growing polymer chain throughout the reaction, and since the monomer only reacts with the end of the growing chain (S7.13f), GRIM polymerization has living polymerization-like characteristics, and low polydispersity indices (PDI) and block copolymers are achievable.[87] GRIM polymerization is one of the few examples of a controlled chain polymerization in the synthesis of π-conjugated polymers, and in general controlled polymerizations are an active area of research as an alternative to the Stille and Suzuki step polymerizations.[30]

Examples of Kumada polymerization are shown in Scheme 7.14. The first is a prototypical GRIM polymerization where 2-bromo-3-hexyl-5-iodothiophene (**98**) undergoes a Grignard metathesis with isopropylmagnesium chloride to give Grignard monomer **99** that polymerizes to homopolymer **100** on addition of the Ni(II) catalyst Ni(dppp)Cl$_2$ to give polymer **101** on quenching with HCl.[88] The M_n

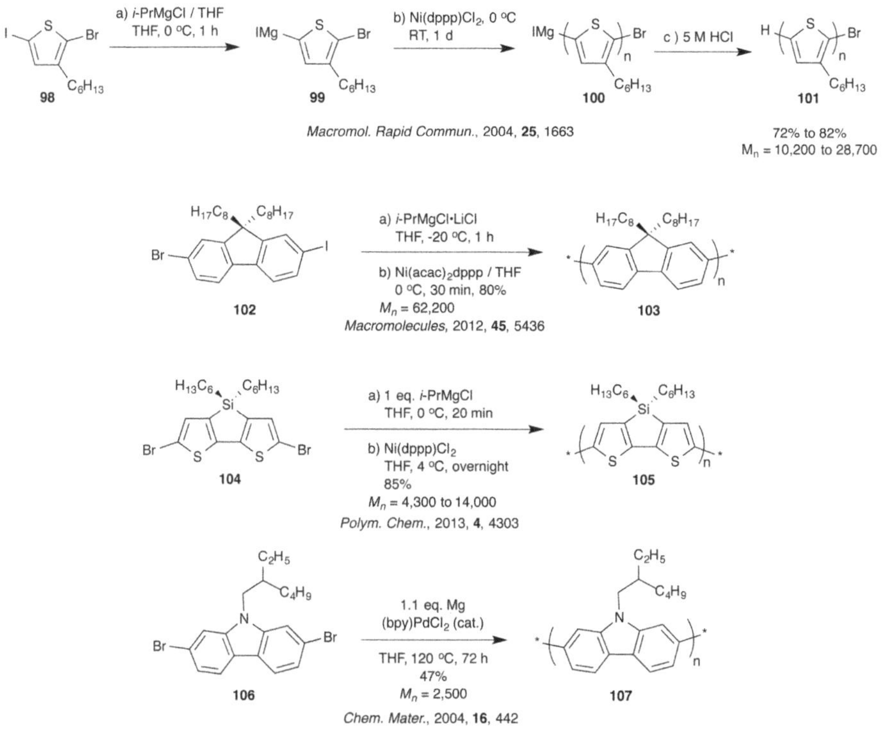

Macromol. Rapid Commun., 2004, **25**, 1663

72% to 82%
M_n = 10,200 to 28,700

Macromolecules, 2012, **45**, 5436

Polym. Chem., 2013, **4**, 4303

Chem. Mater., 2004, **16**, 442

Scheme 7.14 GRIM polymerization of electron donor substrates by various conditions.[88,90–92]

achievable ranged from ~10 000 to ~28 000, and was linearly dependent on conversion of the monomer. Further mechanistic studies by the groups of McCullough[89] and Yokozawa[27] provided support for the GRIM polymerization being essentially a living polymerization. Another example of a GRIM polymerization that exhibited controlled chain growth is the GRIM polymerization of 2-bromo-7-iodo fluorene **102** to give the corresponding polyfluorene **103** with M_n ~ 62 000 in 80% yield.[90] Another GRIM polymerization includes the polymerization of dibromo monomer **104** with one equivalent of the metathesis agent isopropylmagnesium chloride to give homopolymer **105** with M_n ranging from 4300 to 14 000 in high yields.[91] Finally, another example generates the Grignard from approximately one equivalent of Mg metal and dibromo carbazole **106** and utilizes a Pd(II) catalyst to give polycarbazole **107** with M_n = 2500 in 47% yield.[92] In this case, a slight excess of Mg metal was used to compensate for incomplete conversion of the solid metal in the reaction, which may be due to some oxidation of the Mg metal. It should be noted that other systems showing chain-growth mechanisms but not living-like polymerization have been reported for Pd-catalyzed coupling of Zn reagents of naphthalene diimide (NDI, Section 6.5.5) dithiophene derivatives.[93,94]

7.4.4 Direct Arylation Polymerization

Direct arylation by C−H activation was presented in the context of small molecule π-conjugated materials synthesis in Section 5.5. The advantages of reduced synthetic steps and avoidance of hazardous reagents has led to active development of C−H activation as a polymerization method.[95,96] Since C−H activation in the synthesis of materials is in a relatively early stage, recently there have been a number of different catalysts and optimization conditions utilized in the synthesis of polymers; in this respect, recent C−H activated polymerizations may provide good case studies for the effort required to increase the yield of a promising reaction (*e.g.*, 85% isolated yield of a small molecule product) to high *conversion efficiency* in a step polymerization to enable the synthesis of high molecular weight polymers. As such, in the examples for C−H activated polymerization given in this section, a summary of optimization parameters have been given when reported by the authors; however, a lack of reported optimization parameters *should not* be taken to mean optimization was either not necessary or not carried out in any particular study.

Scheme 7.15 illustrates some examples of C−H activated polymerizations using several different dihalide and diaryl C−H substrates. Dibromo benzothiadiazole **108** can be polymerized with donor **109** using Pd(II) catalysis and K_2CO_3 in DMAC to give the corresponding D–A copolymer **110** with an $M_n \sim 40\,000$ in 70% yield.[97] Dibromide **108** was also polymerized with donor **111** using the more active Pd(0) catalyst $Pd_2dba_3/P(o\text{-MeOPh})_3$ with Cs_2CO_3 in THF to give copolymer **112** with an $M_n = 16\,000$ in high yield.[98] In the case of copolymer **112**, optimization reportedly included the use of three different catalysts, eight different solvents, four different ligands, and twelve(!) different bases. Similarly, polymerization of fluorene dibromide **113** with the difluoro-benzothiadiazole acceptor **114** utilized 21 separate conditions to arrive at $Pd(OAc)_2/P(t\text{-Bu})_2Me$ catalyzed polymerization to give copolymer **115** with an $M_n = 41\,000$ in 86% yield.[99] In the case of polymerization of donor **116** with fluorene dibromide **117**, a much higher $M_n = 147\,000$ was achieved under microwave irradiation compared to the same conditions under conventional heating ($M_n = 47\,500$) or compared to Suzuki coupling (from the EDOT diBpin).[99] In general, the examples in Scheme 7.15 indicate not only the potential utility of C−H

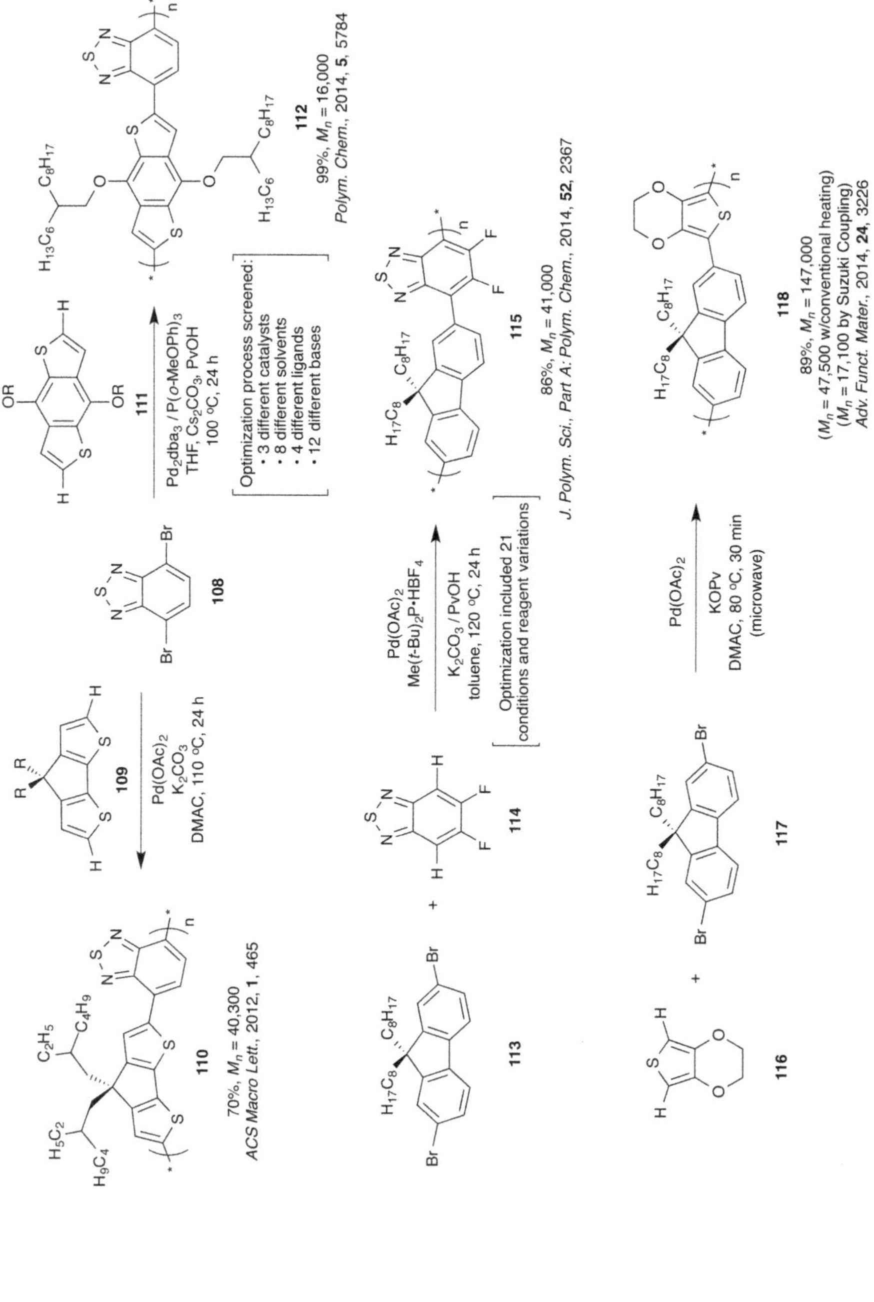

Scheme 7.15 C—H activated direct arylation polymerization of various substrates with optimization summaries.[97–99,109]

activated polymerization but also the amount of optimization (and attention to detail) that is often necessary to increase monomer conversion to a level that is sufficiently high to achieve high molecular weight polymers.

Scheme 7.16 presents some examples of polymerization with thieno[3,4-*c*]pyrrole-4,6-dione (TPD) acceptor monomers using several different catalysts. The first example is the polymerization of TPD monomer **119** with D–A–D dibromo monomer **120** using the Hermann–Beller catalyst (Scheme 7.16 inset) and the ligand P(*o*-MeOPh)$_3$ in toluene at 120 °C to give copolymer **121** with an M_n = 21 000 in 75% yield (with 15 different conditions screened).[100] The same base TPD monomer **119** was used in polymerization with the strong donor **122** screening six different Pd catalysts to give copolymer **123** with an M_n = 68 200 when 2% of the ligand was used.[46] Finally, polymerization of TPD dibromide **124** with donor monomer **125** using a Pd(OAc)$_2$/PCy$_3$ catalyst gave the corresponding polymer **126** with an M_n ~ 27 000 in 63% yield.[101] The generally high yields and relatively high molecular weights of the C−H activated polymerizations illustrated in Schemes 7.15 and 7.16 demonstrate the utility of C−H activated polymerization to the synthesis of π-conjugated polymers; however, the generally high reaction temperatures (>100 °C), longer reaction times (24 h or greater), and the basic conditions may

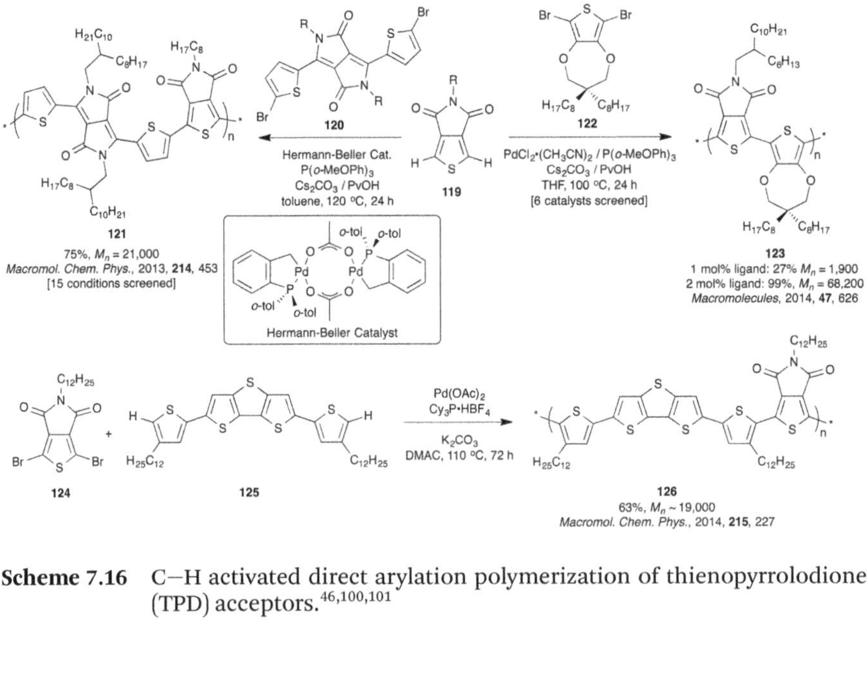

Scheme 7.16   C−H activated direct arylation polymerization of thienopyrrolodione (TPD) acceptors.[46,100,101]

limit the potential scope (*e.g.*, with dithienosilole donors that tend to cleave at silicon in basic and high temperature conditions). Additionally, solvents should be rigorously dried since protodehalogenation to give another C–H coupling site may cause unwanted coupling defects and may imbalance the stoichiometry of the functional groups, which could limit the degree of polymerization.

7.5 CONCLUSION

Most π-conjugated polymers, particularly copolymers, are currently synthesized by step polymerization from two (or more) reactive monomers. In general, to achieve a high degree of polymerization, the conversion of the functional groups to the desired product must be high (>99% for a degree of polymerization of 100) and the monomer stoichiometry must be closely balanced. Practically, this often means that monomers should be highly pure and the reaction conditions should be sufficiently optimized. Alternatively, when a monomer cannot be purified adequately, an accurate percentage of the monomer in the mixture may be used to achieve the correct stoichiometry if the impurities do not negatively affect the reaction and limit monomer conversion. Many reactions that are utilized to create conjugation in small molecule materials can also be used to synthesize π-conjugated polymers. In particular, Stille and Suzuki coupling reactions have been utilized; but the use of C–H activated coupling chemistry has increased in recent years and will likely continue to be developed.

REFERENCES

1. A. C. Arias, J. D. MacKenzie, I. McCulloch, J. Rivnay and A. Salleo, *Chem. Rev.*, 2010, **110**, 3.
2. Y. D. Zhang, C. Zuniga, S. J. Kim, D. K. Cai, S. Barlow, S. Salman, V. Coropceanu, J. L. Bredas, B. Kippelen and S. Marder, *Chem. Mater.*, 2011, **23**, 4002.
3. S.-J. Kim, Y. Zhang, C. Zuniga, S. Barlow, S. R. Marder and B. Kippelen, *Org. Electron.*, 2011, **12**, 492.
4. Z. Xing, J. Zhang, X. Li, W. Zhang, L. Wang, N. Zhou and X. Zhu, *J. Polym. Sci., Part A: Polym. Chem.*, 2013, **51**, 4021.
5. S. Dailey, W. J. Feast, R. J. Peace, I. C. Sage, S. Till and E. L. Wood, *J. Mater. Chem.*, 2001, **11**, 2238.
6. J.-W. Kang, T.-D. Kim, J. Luo, M. Haller and A. K. Y. Jen, *Appl. Phys. Lett.*, 2005, **87**, 071109.

7. X. Piao, X. Zhang, Y. Mori, M. Koishi, A. Nakaya, S. Inoue, I. Aoki, A. Otomo and S. Yokoyama, *J. Polym. Sci., Part A: Polym. Chem.*, 2011, **49**, 47.

8. *Conjugated Polymers: A Practical Guide to Synthesis*, ed. K. Mullen, J. Reynolds and T. Masuda, The Royal Society of Chemistry, Cambridge, UK, 2014.

9. *Design and Synthesis of Conjugated Polymers*, ed. M. LeClerc and J. Morin, WILEY-VCH Verlag GmbH & Co. KGaA, Weinheim, DE, 2010.

10. R. Boyd and P. Phillips, *The Science of Polymer Molecules*, Cambridge University Press, Cambridge, UK, 1996.

11. G. Odian, *Principles of Polymerization*, John Wiley & Sons, Inc., New York, USA, 4th edn, 2004.

12. R. Boyd and P. Phillips, in *The Science of Polymer Molecules*, Cambridge University Press, Cambridge, UK, 1996, ch. 2, pp. 18–59.

13. F. P. V. Koch, J. Rivnay, S. Foster, C. Müller, J. M. Downing, E. Buchaca-Domingo, P. Westacott, L. Yu, M. Yuan, M. Baklar, Z. Fei, C. Luscombe, M. A. McLachlan, M. Heeney, G. Rumbles, C. Silva, A. Salleo, J. Nelson, P. Smith and N. Stingelin, *Prog. Polym. Sci.*, 2013, **38**, 1978.

14. R. J. Kline, M. D. McGehee, E. N. Kadnikova, J. Liu and J. M. J. Fréchet, *Adv. Mater.*, 2003, **15**, 1519.

15. A. M. Ballantyne, L. Chen, J. Dane, T. Hammant, F. M. Braun, M. Heeney, W. Duffy, I. McCulloch, D. D. C. Bradley and J. Nelson, *Adv. Funct. Mater.*, 2008, **18**, 2373.

16. J. K. Park, J. Jo, J. H. Seo, J. S. Moon, Y. D. Park, K. Lee, A. J. Heeger and G. C. Bazan, *Adv. Mater.*, 2011, **23**, 2430.

17. T. Miteva, A. Meisel, W. Knoll, H. G. Nothofer, U. Scherf, D. C. Müller, K. Meerholz, A. Yasuda and D. Neher, *Adv. Mater.*, 2001, **13**, 565.

18. S. Xiao, M. Nguyen, X. Gong, Y. Cao, H. Wu, D. Moses and A. J. Heeger, *Adv. Funct. Mater.*, 2003, **13**, 25.

19. X. Gong, W. Ma, J. C. Ostrowski, K. Bechgaard, G. C. Bazan, A. J. Heeger, S. Xiao and D. Moses, *Adv. Funct. Mater.*, 2004, **14**, 393.

20. M. J. Robb, D. Montarnal, N. D. Eisenmenger, S.-Y. Ku, M. L. Chabinyc and C. J. Hawker, *Macromolecules*, 2013, **46**, 6431.

21. L. Wang, X. Zhang, J. Zhang, H. Tian, Y. Lu, Y. Geng and F. Wang, *J. Mater. Chem. C*, 2014, **2**, 9978.

22. S. Kim, J. K. Park and Y. D. Park, *RSC Adv.*, 2014, **4**, 39268.

23. G. L. Gibson, D. Gao, A. A. Jahnke, J. Sun, A. J. Tilley and D. S. Seferos, *J. Mater. Chem. A*, 2014, **2**, 14468.

24. W. H. Carothers, *Trans. Faraday Soc.*, 1936, **32**, 39.
25. R. C. Coffin, J. Peet, J. Rogers and G. C. Bazan, *Nat. Chem.*, 2009, **1**, 657.
26. G. Odian, *Principles of Polymerization*, John Wiley & Sons, Inc., New York, USA, 4th edn, 2004, ch. 2, pp. 39–100.
27. R. Miyakoshi, A. Yokoyama and T. Yokozawa, *J. Am. Chem. Soc.*, 2005, **127**, 17542.
28. M. C. Iovu, E. E. Sheina, R. R. Gil and R. D. McCullough, *Macromolecules*, 2005, **38**, 8649.
29. I. Osaka and R. D. McCullough, *Acc. Chem. Res.*, 2008, **41**, 1202.
30. K. Okamoto and C. K. Luscombe, *Polym. Chem.*, 2011, **2**, 2424.
31. M. Kubo, C. Takimoto, Y. Minami, T. Uno, T. Itoh and M. Shoyama, *Macromolecules*, 2005, **38**, 7314.
32. T. Ahn, S.-Y. Song and H.-K. Shim, *Macromolecules*, 2000, **33**, 6764.
33. T. Schwalm, J. Wiesecke, S. Immel and M. Rehahn, *Macromol. Rapid Commun.*, 2009, **30**, 1295.
34. A. C. Grimsdale, K. L. Chan, R. E. Martin, P. G. Jokisz and A. B. Holmes, *Chem. Rev.*, 2009, **109**, 897.
35. J. C. Chen, C. J. Chiang, J. C. Chiu and J. J. Ju, *Chem. Commun.*, 2012, **48**, 7756.
36. C. J. Neef and J. P. Ferraris, *Macromolecules*, 2000, **33**, 2311.
37. Y. Xu, Y. Jin, J. Peng, B. Wang and D. Cao, *J. Macromol. Sci., Part A: Pure Appl. Chem.*, 2010, **47**, 1059.
38. Z. Qiao, Y. Xu, S. Lin, J. Peng and D. Cao, *Synth. Met.*, 2010, **160**, 1544.
39. S. Pfeiffer and H. Horhold, *Synth. Met.*, 1999, **101**, 109.
40. S. Wen, J. Pei, P. Li, Y. Zhou, W. Cheng, Q. Dong, Z. Li and W. Tian, *J. Polym. Sci., Part A: Polym. Chem.*, 2011, **49**, 2715.
41. M. Shahid, R. S. Ashraf, E. Klemm and S. Sensfuss, *Macromolecules*, 2006, **39**, 7844.
42. A. V. Patil, H. Park, E. W. Lee and S.-H. Lee, *Synth. Met.*, 2010, **160**, 2128.
43. B. C. Thompson, Y. G. Kim, T. D. McCarley and J. R. Reynolds, *J. Am. Chem. Soc.*, 2006, **128**, 12714.
44. J.-F. Morin, N. Drolet, Y. Tao and M. Leclerc, *Chem. Mater.*, 2004, **16**, 4619.
45. S. Song, Y. Jin, S. H. Kim, J. Y. Shim, S. Son, I. Kim, K. Lee and H. Suh, *J. Polym. Sci., Part A: Polym. Chem.*, 2009, **47**, 6540.
46. M. Wakioka, N. Ichihara, Y. Kitano and F. Ozawa, *Macromolecules*, 2014, **47**, 626.
47. J. Liao and Q. Wang, *Macromolecules*, 2004, **37**, 7061.

48. L. Yu, Y. Lee and Y. Liang, *Synlett*, 2006, **2006**, 2879.
49. J. A. Mikroyannidis, K. M. Gibbons, A. P. Kulkarni and S. A. Jenekhe, *Macromolecules*, 2008, **41**, 663.
50. J. Pei, S. Wen, Y. Zhou, Q. Dong, Z. Liu, J. Zhang and W. Tian, *New J. Chem.*, 2011, **35**, 385.
51. J. Li, M. Yan, Y. Xie and Q. Qiao, *Energy Environ. Sci.*, 2011, **4**, 4276.
52. B. Carsten, F. He, H. J. Son, T. Xu and L. Yu, *Chem. Rev.*, 2011, **111**, 1493.
53. J. K. Stille, *Angew. Chem., Int. Ed. Engl.*, 1986, **25**, 508.
54. S. Zhang, H. Fan, Y. Liu, G. Zhao, Q. Li, Y. Li and X. Zhan, *J. Polym. Sci., Part A: Polym. Chem.*, 2009, **47**, 2843.
55. E. Zhou, J. Cong, Q. Wei, K. Tajima, C. Yang and K. Hashimoto, *Angew. Chem., Int. Ed.*, 2011, **50**, 2799.
56. S. Kola, J. H. Kim, R. Ireland, M.-L. Yeh, K. Smith, W. Guo and H. E. Katz, *ACS Macro Lett.*, 2013, **2**, 664.
57. R. S. Ashraf, J. Gilot and R. A. J. Janssen, *Sol. Energy Mater. Sol. Cells*, 2010, **94**, 1759.
58. X. Zhao, L. Ma, L. Zhang, Y. Wen, J. Chen, Z. Shuai, Y. Liu and X. Zhan, *Macromolecules*, 2013, **46**, 2152.
59. D. H. Lee, J. Shin, M. J. Cho and D. H. Choi, *Chem. Commun.*, 2013, **49**, 3896.
60. X. Guo, M. Baumgarten and K. Müllen, *Prog. Polym. Sci.*, 2013, **38**, 1832.
61. K. H. Hendriks, W. Li, G. H. Heintges, G. W. van Pruissen, M. M. Wienk and R. A. Janssen, *J. Am. Chem. Soc.*, 2014, **136**, 11128.
62. F. C. Krebs, R. B. Nyberg and M. Jørgensen, *Chem. Mater.*, 2004, **16**, 1313.
63. K. T. Nielsen, K. Bechgaard and F. C. Krebs, *Macromolecules*, 2005, **38**, 658.
64. Z. R. Owczarczyk, W. A. Braunecker, A. Garcia, R. Larsen, A. M. Nardes, N. Kopidakis, D. S. Ginley and D. C. Olson, *Macromolecules*, 2013, **46**, 1350.
65. D. G. Patel, K. R. Graham and J. R. Reynolds, *J. Mater. Chem.*, 2012, **22**, 3004.
66. L. Liao, L. Dai, A. Smith, M. Durstock, J. Lu, J. Ding and Y. Tao, *Macromolecules*, 2007, **40**, 9406.
67. M. Seri, M. Bolognesi, Z. Chen, S. Lu, W. Koopman, A. Facchetti and M. Muccini, *Macromolecules*, 2013, **46**, 6419.
68. P. M. Beaujuge, H. N. Tsao, M. R. Hansen, C. M. Amb, C. Risko, J. Subbiah, K. R. Choudhury, A. Mavrinskiy, W. Pisula, J. L. Bredas,

F. So, K. Mullen and J. R. Reynolds, *J. Am. Chem. Soc.*, 2012, **134**, 8944.

69. X. Guo, N. Zhou, S. J. Lou, J. W. Hennek, R. Ponce Ortiz, M. R. Butler, P. L. Boudreault, J. Strzalka, P. O. Morin, M. Leclerc, J. T. Lopez Navarrete, M. A. Ratner, L. X. Chen, R. P. Chang, A. Facchetti and T. J. Marks, *J. Am. Chem. Soc.*, 2012, **134**, 18427.

70. C. M. Amb, S. Chen, K. R. Graham, J. Subbiah, C. E. Small, F. So and J. R. Reynolds, *J. Am. Chem. Soc.*, 2011, **133**, 10062.

71. L. Huo, J. Hou, H.-Y. Chen, S. Zhang, Y. Jiang, T. L. Chen and Y. Yang, *Macromolecules*, 2009, **42**, 6564.

72. Z. Zhu, D. Waller, R. Gaudiana, M. Morana, D. Mühlbacher, M. Scharber and C. Brabec, *Macromolecules*, 2007, **40**, 1981.

73. X. Guo, F. S. Kim, M. J. Seger, S. A. Jenekhe and M. D. Watson, *Chem. Mater.*, 2012, **24**, 1434.

74. W. Zhang, J. Li, L. Zou, B. Zhang, J. Qin, Z. Lu, Y. F. Poon, M. B. Chan-Park and C. M. Li, *Macromolecules*, 2008, **41**, 8953.

75. Y. Lu, H. Chen, X. Hou, X. Hu and S.-C. Ng, *Synth. Met.*, 2010, **160**, 1438.

76. J. W. Jung, F. Liu, T. P. Russell and W. H. Jo, *Energy Environ. Sci.*, 2012, **5**, 6857.

77. J.-Y. Lee, W.-S. Shin, J.-R. Haw and D.-K. Moon, *J. Mater. Chem.*, 2009, **19**, 4938.

78. F. A. Lemasson, T. Strunk, P. Gerstel, F. Hennrich, S. Lebedkin, C. Barner-Kowollik, W. Wenzel, M. M. Kappes and M. Mayor, *J. Am. Chem. Soc.*, 2011, **133**, 652.

79. R. d. B. Aïch, N. Blouin, A. l. Bouchard and M. Leclerc, *Chem. Mater.*, 2009, **21**, 751.

80. S. Wakim, N. Blouin, E. Gingras, Y. Tao and M. Leclerc, *Macromol. Rapid Commun.*, 2007, **28**, 1798.

81. N. Blouin, A. Michaud and M. Leclerc, *Adv. Mater.*, 2007, **19**, 2295.

82. P. Berrouard, F. Grenier, J. R. Pouliot, E. Gagnon, C. Tessier and M. Leclerc, *Org. Lett.*, 2011, **13**, 38.

83. G. W. P. Van Pruissen, F. Gholamrezaie, M. M. Wienk and R. A. J. Janssen, *J. Mater. Chem.*, 2012, **22**, 20387.

84. R. S. Ashraf, A. J. Kronemeijer, D. I. James, H. Sirringhaus and I. McCulloch, *Chem. Commun.*, 2012, **48**, 3939.

85. G. Nagarjuna, A. Kokil, J. Kumar and D. Venkataraman, *J. Mater. Chem.*, 2012, **22**, 16091.

86. A. Saeki, S. Yoshikawa, M. Tsuji, Y. Koizumi, M. Ide, C. Vijayakumar and S. Seki, *J. Am. Chem. Soc.*, 2012, **134**, 19035.

87. M. C. Stefan, M. P. Bhatt, P. Sista and H. D. Magurudeniya, *Polym. Chem.*, 2012, **3**, 1693.

88. R. Miyakoshi, A. Yokoyama and T. Yokozawa, *Macromol. Rapid Commun.*, 2004, **25**, 1663.

89. E. E. Sheina, J. Liu, M. C. Iovu, D. W. Laird and R. D. McCullough, *Macromolecules*, 2004, **37**, 3526.

90. A. Sui, X. Shi, S. Wu, H. Tian, Y. Geng and F. Wang, *Macromolecules*, 2012, **45**, 5436.

91. F. Boon, N. Hergué, G. Deshayes, D. Moerman, S. Desbief, J. De Winter, P. Gerbaux, Y. H. Geerts, R. Lazzaroni and P. Dubois, *Polym. Chem.*, 2013, **4**, 4303.

92. A. Iraqi and I. Wataru, *Chem. Mater.*, 2004, **16**, 442.

93. V. Senkovskyy, R. Tkachov, H. Komber, M. Sommer, M. Heuken, B. Voit, W. T. Huck, V. Kataev, A. Petr and A. Kiriy, *J. Am. Chem. Soc.*, 2011, **133**, 19966.

94. W. Liu, R. Tkachov, H. Komber, V. Senkovskyy, M. Schubert, Z. Wei, A. Facchetti, D. Neher and A. Kiriy, *Polym. Chem.*, 2014, **5**, 3404.

95. L. G. Mercier and M. Leclerc, *Acc. Chem. Res.*, 2013, **46**, 1597.

96. K. Okamoto, J. Zhang, J. B. Housekeeper, S. R. Marder and C. K. Luscombe, *Macromolecules*, 2013, **46**, 8059.

97. S. Kowalski, S. Allard and U. Scherf, *ACS Macro Lett.*, 2012, **1**, 465.

98. X. Wang and M. Wang, *Polym. Chem.*, 2014, **5**, 5784.

99. X. Zhang, Y. Gao, S. Li, X. Shi, Y. Geng and F. Wang, *J. Polym. Sci., Part A: Polym. Chem.*, 2014, **52**, 2367.

100. J.-R. Pouliot, L. G. Mercier, S. Caron and M. Leclerc, *Macromol. Chem. Phys.*, 2013, **214**, 453.

101. E. Zhu, B. Ni, B. Zhao, J. Hai, L. Bian, H. Wu and W. Tang, *Macromol. Chem. Phys.*, 2014, **215**, 227.

102. Z. Liu, Y. Yuan, X. Wen, J. Zhang, G. Lei and P. Zhang, *Polym. Bull. (Heidelberg, Ger.)*, 2012, **70**, 1221.

103. J.-W. Kim, M.-H. Heo, Y.-E. Jin, J.-H. Kim, J.-Y. Shim, S.-H. Song, I. Kim, J.-Y. Kim and H.-S. Suh, *Bull. Korean Chem. Soc.*, 2012, **33**, 629.

104. H.-G. Jeong, B. Lim, S.-I. Na, K.-J. Baeg, J. Kim, J.-M. Yun and D.-Y. Kim, *Macromol. Chem. Phys.*, 2011, **212**, 2308.

105. Y. Zhu, R. D. Champion and S. A. Jenekhe, Macromolecules, 2006, **39**, 8712.

106. X. Zhan, Z. a. Tan, E. Zhou, Y. Li, R. Misra, A. Grant, B. Domercq, X.-H. Zhang, Z. An, X. Zhang, S. Barlow, B. Kippelen and S. R. Marder, *J. Mater. Chem.*, 2009, **19**, 5794.

107. Z. Fei, R. S. Ashraf, Z. Huang, J. Smith, R. J. Kline, P. D'Angelo, T. D. Anthopoulos, J. R. Durrant, I. McCulloch and M. Heeney, *Chem. Commun. (Cambridge, U. K.)*, 2012, **48**, 2955.

108. J. C. Bijleveld, R. A. M. Verstrijden, M. M. Wienk and R. A. J. Janssen, *Appl. Phys. Lett*, 2010, **97**, 073304.

109. J. Kuwabara, T. Yasuda, S. J. Choi, W. Lu, K. Yamazaki, S. Kagaya, L. Han and T. Kanbara, *Adv. Funct. Mater.*, 2014, **24**, 3226.

Appendix

FUNCTIONAL INTERMEDIATES

The following are intermediates or moieties with pendant functional groups such as alcohols, alkyl halides, carboxylic acids, alkynes, and azides that have been used to synthesize "functional" materials, *i.e.*, materials with molecular moieties having some function other than for a particular electrical or optical property (such as crosslinking or sensing). They are roughly divided into donors **1–20**[1–20] (Figure A.1), thiophenes (donors or π-bridges) and cyclohexene π-bridges **21–37**,[21–37] (Figure A.2), and acceptors **38–43**[38–43] (Figure A.3). This compilation is

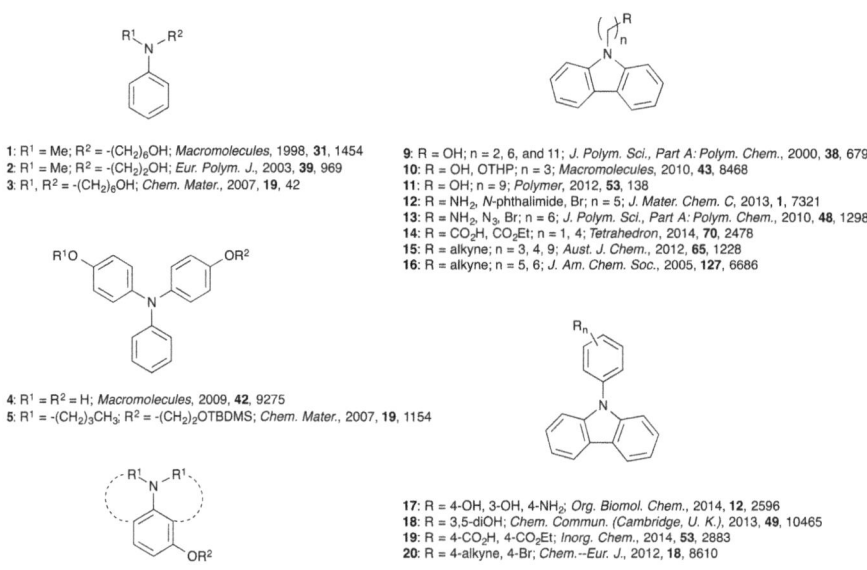

1: R^1 = Me; R^2 = -(CH$_2$)$_6$OH; *Macromolecules*, 1998, **31**, 1454
2: R^1 = Me; R^2 = -(CH$_2$)$_2$OH; *Eur. Polym. J.*, 2003, **39**, 969
3: R^1, R^2 = -(CH$_2$)$_6$OH; *Chem. Mater.*, 2007, **19**, 42

4: R^1 = R^2 = H; *Macromolecules*, 2009, **42**, 9275
5: R^1 = -(CH$_2$)$_3$CH$_3$; R^2 = -(CH$_2$)$_2$OTBDMS; *Chem. Mater.*, 2007, **19**, 1154

6: R^1 = 4-Bu-Ph; R^2 = -(CH$_2$)$_6$OH; *J. Org. Chem.*, 1999, **64**, 4289
7: R^1 = Et; R^2 = -(CH$_2$)$_6$OH; *Macromolecules*, 2004, **37**, 5163
8: R^1 = -CH$_2$CH$_2$- (julolidine); R^2 = -(CH$_2$)$_6$OH; *Chem. Commun. (Cambridge, U. K.)*, 2012, **48**, 9637

9: R = OH; n = 2, 6, and 11; *J. Polym. Sci., Part A: Polym. Chem.*, 2000, **38**, 679
10: R = OH, OTHP; n = 3; *Macromolecules*, 2010, **43**, 8468
11: R = OH; n = 9; *Polymer*, 2012, **53**, 138
12: R = NH$_2$, *N*-phthalimide, Br; n = 5; *J. Mater. Chem. C*, 2013, **1**, 7321
13: R = NH$_2$, N$_3$, Br; n = 6; *J. Polym. Sci., Part A: Polym. Chem.*, 2010, **48**, 1298
14: R = CO$_2$H, CO$_2$Et; n = 1, 4; *Tetrahedron*, 2014, **70**, 2478
15: R = alkyne; n = 3, 4, 9; *Aust. J. Chem.*, 2012, **65**, 1228
16: R = alkyne; n = 5, 6; *J. Am. Chem. Soc.*, 2005, **127**, 6686

17: R = 4-OH, 3-OH, 4-NH$_2$; *Org. Biomol. Chem.*, 2014, **12**, 2596
18: R = 3,5-diOH; *Chem. Commun. (Cambridge, U. K.)*, 2013, **49**, 10465
19: R = 4-CO$_2$H, 4-CO$_2$Et; *Inorg. Chem.*, 2014, **53**, 2883
20: R = 4-alkyne, 4-Br; *Chem.--Eur. J.*, 2012, **18**, 8610

Figure A.1 Functional aryl amines.[1–20]

Synthetic Methods in Organic Electronic and Photonic Materials: A Practical Guide
By Timothy C. Parker and Seth R. Marder
© Timothy C. Parker and Seth R. Marder, 2015
Published by the Royal Society of Chemistry, www.rsc.org

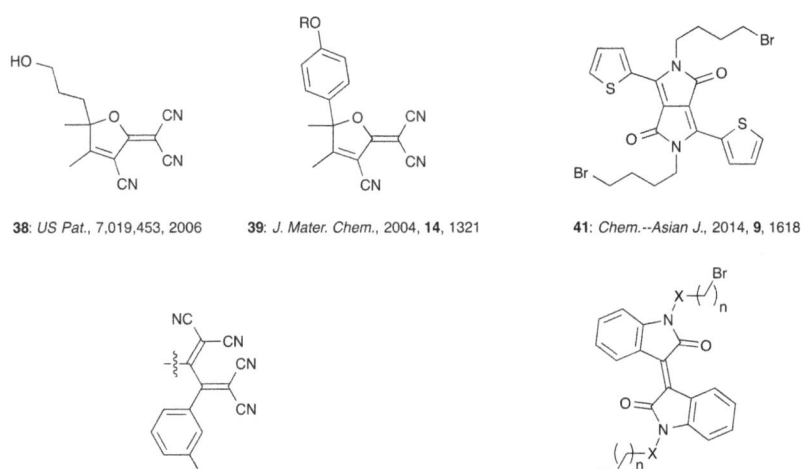

21: R = OH: n = 1; *Chem. Mater.*, 2010, **22**, 5601
22: R = OH; n = 3; *J. Am. Chem. Soc.*, 2010, **132**, 10015
23: R = OH; n = 4; *J. Am. Chem. Soc.*, 2006, **128**, 10930
24: R = OH, Br; n = 6; *ACS Macro Lett.*, 2014, **3**, 738
25: R = OH, OTBDMS; n = 6; *ACS Macro Lett.*, 2012, **1**, 167
26: R = Br; n = 2, 3; *Tetrahedron*, 2006, **62**, 6551
27: R = Br; n = 4; *Macromolecules*, 2013, **46**, 344
28: R = Br; n = 5, 10, 15; *Macromolecules*, 2014, **47**, 1715
29: R = CO$_2$H; n = 2; *Eur. J. Org. Chem.*, 2004, **2004**, 4442
30: R = CO$_2$H; n = 6; *J. Med. Chem.*, 2007, **50**, 3359
31: R = N$_3$, OH; n = 2; *J. Med. Chem.*, 2012, **55**, 9562

33: R = OH, Br; n = 2; *J. Org. Chem.*, 2008, **73**, 8705
34: R = OTHP; n = 2; *J. Am. Chem. Soc.*, 1995, **117**, 9842
35: R = Br; n = 6; *Electrochim. Acta*, 2008, **53**, 3755

36: *Org. Lett.*, 2006, **8**, 1387

32: *US Pat.*, 6,716,995, 2004

37: *J. Mater. Chem.*, 2012, **22**, 951

Figure A.2 Functional thiophenes and π-bridges.[21–37]

38: *US Pat.*, 7,019,453, 2006

39: *J. Mater. Chem.*, 2004, **14**, 1321

41: *Chem.--Asian J.*, 2014, **9**, 1618

40: *Chem. Commun. (Cambridge, U. K.)*, 2002, 888

42: X = CH$_2$; n = 1, 2, 4, 5; *Synthesis*, 2013, **45**, 668
43: X = C=O; n = 3, 5; *Russ. J. Org. Chem.*, 2014, **50**, 906

Figure A.3 Functional acceptors. Taken from ref. 38 (**38**), ref. 39 (**39**), ref. 40 (**40**), ref. 41 (**41**), ref. 42 (**42**) and ref. 43 (**43**).[38–43]

by no means exhaustive. Researchers are encouraged to search the literature for similar compounds if those listed in Figure A.1–A.3 do not suffice for any synthetic or structural plan.

REFERENCES

1. M. Dörr, R. Zentel, R. Dietrich, K. Meerholz, C. Bräuchle, J. Wichern, S. Zippel and P. Boldt, *Macromolecules*, 1998, **31**, 1454.
2. E. Gubbelmans, K. Van den Broeck, T. Verbiest, M. Van Beylen, A. Persoons and C. Samyn, *Eur. Polym. J.*, 2003, **39**, 969.
3. G. W. Kim, M. J. Cho, Y.-J. Yu, Z. H. Kim, J.-I. Jin, D. Y. Kim and D. H. Choi, *Chem. Mater.*, 2007, **19**, 42.
4. K. Aljoumaa, Y. Qi, J. Ding and J. A. Delaire, *Macromolecules*, 2009, **42**, 9275.
5. Y.-J. Cheng, J. Luo, S. Hau, D. H. Bale, T.-D. Kim, Z. Shi, D. B. Lao, N. M. Tucker, Y. Tian, L. R. Dalton, P. J. Reid and A. K. Y. Jen, *Chem. Mater.*, 2007, **19**, 1154.
6. S. Thayumanavan, J. Mendez and S. R. Marder, *J. Org. Chem.*, 1999, **64**, 4289.
7. K. H. Park, R. J. Twieg, R. Ravikiran, L. F. Rhodes, R. A. Shick, D. Yankelevich and A. Knoesen, *Macromolecules*, 2004, **37**, 5163.
8. J. Wu, S. Bo, J. Liu, T. Zhou, H. Xiao, L. Qiu, Z. Zhen and X. Liu, *Chem. Commun.*, 2012, **48**, 9637.
9. F.-S. Du, Z.-C. Li, W. Hong, Q.-Y. Gao and F.-M. Li, *J. Polym. Sci., Part A: Polym. Chem.*, 2000, **38**, 679.
10. H. Jin, Y. Xu, Z. Shen, D. Zou, D. Wang, W. Zhang, X. Fan and Q. Zhou, *Macromolecules*, 2010, **43**, 8468.
11. H. Pei, W. Li, Y. Liu, D. Wang, J. Wang, J. Shi and S. Cao, *Polymer*, 2012, **53**, 138.
12. S. Castelar, J. Barberá, M. Marcos, P. Romero, J.-L. Serrano, A. Golemme and R. Termine, *J. Mater. Chem. C*, 2013, **1**, 7321.
13. M.-C. Yuan, M.-H. Su, M.-Y. Chiu and K.-H. Wei, *J. Polym. Sci., Part A: Polym. Chem.*, 2010, **48**, 1298.
14. T. Zhang, J. Wang, M. Zhou, L. Ma, G. Yin, G. Chen and Q. Li, *Tetrahedron*, 2014, **70**, 2478.
15. J. W. Y. Lam, A. Qin, Y. Dong, J. Liu, C. K. W. Jim, Y. Hong, H. S. Kwok and B. Z. Tang, *Aust. J. Chem.*, 2012, **65**, 1228.
16. A. Krasinski, Z. Radic, R. Manetsch, J. Raushel, P. Taylor, K. B. Sharpless and H. C. Kolb, *J. Am. Chem. Soc.*, 2005, **127**, 6686.
17. T. Q. Hung, N. N. Thang, H. Hoang do, T. T. Dang, A. Villinger and P. Langer, *Org. Biomol. Chem.*, 2014, **12**, 2596.

18. D. Yang, Q. Yang, L. Yang, Q. Luo, Y. Huang, Z. Lu and S. Zhao, *Chem. Commun.*, 2013, **49**, 10465.
19. H. Iden, W. Bi, J. F. Morin and F. G. Fontaine, *Inorg. Chem.*, 2014, **53**, 2883.
20. M. Hayashi, R. Sakamoto and H. Nishihara, *Chem.–Eur. J.*, 2012, **18**, 8610.
21. Z. Shi, W. Liang, J. Luo, S. Huang, B. M. Polishak, X. Li, T. R. Younkin, B. A. Block and A. K. Y. Jen, *Chem. Mater.*, 2010, **22**, 5601.
22. N. T. Jui, E. C. Lee and D. W. MacMillan, *J. Am. Chem. Soc.*, 2010, **132**, 10015.
23. M. Takahashi, K. Masui, H. Sekiguchi, N. Kobayashi, A. Mori, M. Funahashi and N. Tamaoki, *J. Am. Chem. Soc.*, 2006, **128**, 10930.
24. Y. Li, A. Nese, X. Hu, N. V. Lebedeva, T. W. LaJoie, J. Burdyńska, M. C. Stefan, W. You, W. Yang, K. Matyjaszewski and S. S. Sheiko, *ACS Macro Lett.*, 2014, **3**, 738.
25. T. Higashihara, E. Goto and M. Ueda, *ACS Macro Lett.*, 2012, **1**, 167.
26. J. B. Sperry and D. L. Wright, *Tetrahedron*, 2006, **62**, 6551.
27. Z. A. Page, V. V. Duzhko and T. Emrick, *Macromolecules*, 2013, **46**, 344.
28. D. Zeng, I. Tahar-Djebbar, Y. Xiao, F. Kameche, N. Kayunkid, M. Brinkmann, D. Guillon, B. Heinrich, B. Donnio, D. A. Ivanov, E. Lacaze, D. Kreher, F. Mathevet and A.-J. Attias, *Macromolecules*, 2014, **47**, 1715.
29. B. F. Bonini, E. Capito, M. Comes-Franchini, A. Ricci, A. Bottoni, F. Bernardi, G. P. Miscione, L. Giordano and A. R. Cowley, *Eur. J. Org. Chem.*, 2004, **2004**, 4442.
30. C. Hardouin, M. J. Kelso, F. A. Romero, T. J. Rayl, D. Leung, I. Hwang, B. F. Cravatt and D. L. Boger, *J. Med. Chem.*, 2007, **50**, 3359.
31. T. Suzuki, Y. Ota, M. Ri, M. Bando, A. Gotoh, Y. Itoh, H. Tsumoto, P. R. Tatum, T. Mizukami, H. Nakagawa, S. Iida, R. Ueda, K. Shirahige and N. Miyata, *J. Med. Chem.*, 2012, **55**, 9562.
32. D. Huang, T. Londergan, G. K. Todorova and J. Zhu, *US Pat.*, 6,716,995, 2004.
33. A. Martins and M. Lautens, *J. Org. Chem.*, 2008, **73**, 8705.
34. M. J. Marsella, R. J. Newland, P. J. Carroll and T. M. Swager, *J. Am. Chem. Soc.*, 1995, **117**, 9842.
35. Z. Mousavi, T. Alaviuhkola, J. Bobacka, R.-M. Latonen, J. Pursiainen and A. Ivaska, *Electrochim. Acta*, 2008, **53**, 3755.
36. J. Luo, Y. J. Cheng, T. D. Kim, S. Hau, S. H. Jang, Z. Shi, X. H. Zhou and A. K. Jen, *Org. Lett.*, 2006, **8**, 1387.

37. Z. Shi, J. Luo, S. Huang, B. M. Polishak, X.-H. Zhou, S. Liff, T. R. Younkin, B. A. Block and A. K. Y. Jen, *J. Mater. Chem.*, 2012, **22**, 951.
38. D. Huang and B. Chen, *US Pat.*, 7,019,453, 2006.
39. A. J. Kay, A. D. Woolhouse, Y. Zhao and K. Clays, *J. Mater. Chem.*, 2004, **14**, 1321.
40. J. Luo, H. Ma, M. Haller, A. K. Y. Jen and R. R. Barto, *Chem. Commun.*, 2002, 888.
41. K. Tsuto, M. Nakamura, T. Takada and K. Yamana, *Chem.–Asian J.*, 2014, **9**, 1618.
42. A. Bogdanov, G. Yusupova, I. Romanova, S. Latypov, D. Krivolapov, V. Mironov and O. Sinyashin, *Synthesis*, 2013, **45**, 668.
43. A. V. Bogdanov, V. F. Mironov and O. G. Sinyashin, *Russ. J. Org. Chem.*, 2014, **50**, 906.

Subject Index

Page numbers in *italics* refer to tables, schemes or figures.